Tasks for vegetation science 12

Series Editors

HELMUT LIETH

University of Osnabrück, F.R.G.

HAROLD A. MOONEY

Stanford University, Stanford CA, U.S.A.

OTHER BOOKS IN THE TASKS FOR VEGETATION SCIENCE SERIES

1 Box, E.O. *Macroclimate and Plant Forms.* An introduction to predictive modelling in phytogeography.
ISBN 90 6193 941 0
2 Navin Sen, D. & Singh Rajpurohit, K. *Contributions to the Ecology of Halophytes.*
ISBN 90 6193 942 9
3 Ross, J. *The Radiation Regime and Architecture of Plant Stands.*
ISBN 90 6193 607 1
4 Margaris, N.S. & Mooney, H.A. (eds) *Components of Productivity of Mediterranean – Climate, Regions.*
ISBN 90 6193 944 5
5 Müller, M.J. *Selected Climatic Data for a Global Set of Standard Stations for Vegetation Science.*
ISBN 90 6193 945 3
6 Roth, I. *Stratification in Tropical Forests as seen in Leaf Structure.*
ISBN 90 6193 946 1
7 Steubing, L. & Jäger, H.J. *Monitoring of Air Pollutants by Plants: Methods and Problems.*
ISBN 90 6193 947 X
9 Teas, H.J. *Physiology and Management of Mangroves.*
ISBN 90 6193 949 6
10 Feoli, E., Lagonegro, M. & Orlóci, L. *Information analysis of vegetation data.*
ISBN 90 6193 950 X
11 Sesták, Z. (ed) *Photosynthesis during leaf development.*
ISBN 90 6193 951 8

Physiological ecology of plants of the wet tropics

Physiological ecology of plants of the wet tropics

PROCEEDINGS OF AN INTERNATIONAL SYMPOSIUM HELD IN OXATEPEC AND LOS TUXTLAS, MEXICO, JUNE 29 TO JULY 6, 1983

edited by

E. MEDINA, H.A. MOONEY and C. VÁZQUEZ-YÁNES

1984 Springer-Science+Business Media, B.V.

Distributors

for the United States and Canada: Kluwer Boston, Inc., 190 Old Derby Street, Hingham, MA 02043, USA
for all other countries: Kluwer Academic Publishers Group, Distribution Center, P.O.Box 322, 3300 AH Dordrecht, The Netherlands

Library of Congress Catalog Card Number: 83-25567

ISBN 978-94-009-7301-5 ISBN 978-94-009-7299-5 (eBook)
DOI 10.1007/978-94-009-7299-5

Cover design: Max Velthuijs

Softcover reprint of the hardcover 1st edition 1984

Contents

Part Five: Epiphytes and Mycorrhizae

Part Six: Plant-Herbivore Interactions

Part Seven: Species Function and Forest Structure

INTRODUCTION

This book contains the results of a Symposium on the physiological ecology of plants of the lowland wet tropics held in México in June 1983 organized by the Instituto de Biología of the National University of México (U.N.A.M.), and sponsored by UNAM, CONACYT, NSF and UNESCO (CIET). A workshop portion of the Symposium was held at the tropical research station at Los Tuxtlas, Veracruz. This Symposium originated in response to the increasing interest in the physiological ecology of tropical plants, because of the potential of this field to provide a basic understanding of functioning of tropical plant communities.

The study of physiological ecology of tropical plants has been delayed in some cases by the lack of conceptual framework, but also by the absence of appropriate instrumentation and techniques with which to conduct precise measurements under high temperature, high humidity field conditions. Hypotheses and concepts of the physiological ecology of tropical plants have been based mainly on observational data and the analysis of growth forms and leaf anatomy. The early work of A.F.W. Schimper and O. Stocker in Asia, and the extensive surveys made by H. Walter on the osmotic potentials of plants in the tropics and subtropics, constituted, until relatively recently, the only available information on the water and carbon relations of tropical plants. The advent of portable instrumentation which permits the precise measurement of flux density, energy content of incoming radiation within specified wave lengths, the measurement of water potential components, and most importantly, the development of field porometers and gas exchange systems both for H_2O and CO_2 are revolutionizing the extent of our knowledge of the physiological ecology of tropical plants.

This volume reviews the available literature as well as providing new information in a number of areas. Methods for plant physiological ecology in wet tropical areas are discussed, either within the reviews, or in separate chapters.

We hope that the study of the physiological ecology of tropical plants advances from its current rather primitive state to an advanced level in a relatively short time because of the great need for knowledge from this field. We further hope that this book provides, in part, a stimulus for this advancement.

We wish to dedicate this volume to Dr. Peter Raven of the Missouri Botanical Garden, who, through his inspiration and encouragement, has promoted the cooperation of several institutions both outside and within the United States to develop strong biological research projects in the tropics.

E. Medina (Caracas)

H. A. Mooney (Stanford)

C. Vázquez-Yánes (Mexico City)

NUTRIENT REGIME IN THE WET TROPICS: PHYSICAL FACTORS

CARL F. JORDAN
Institute of Ecology, University of Georgia, Athens, Georgia 30602, USA.

ABSTRACT
The wet tropics are characterized by year-round high temperatures and high humidities, resulting in growing seasons which can extend up to 12 months per year. Consequently, primary productivity and nutrient cycling rates are high on an annual basis. However, long periods of high temperature and high humidity also result in high annual rates of soil respiration with consequent high production of carbonic acid in the soil. The acid dissociates and the hydrogen replaces cations exchanged on clay surfaces or bound in clay minerals. Cations thus released are quickly leached from the soil by heavy rains. The low pH of many tropical soils resulting from high concentrations of carbonic acid also results in binding of phosphorus with iron and aluminum. For these reasons, available nutrients are relatively scarce in soils of the wet tropics compared to ecosystems in other regions.

1. INTRODUCTION

Annual rates of net primary productivity are usually higher in the wet tropics than in any other region of the world (Whittaker and Likens, 1975)(Table 1). Rates of litter decomposition too, are higher in the wet tropics (Olson, 1963)(Table 2). The year-round activity of the organisms involved in these processes is both the cause and the effect of a nutrient regime in the wet tropics which is quantitatively different from the nutrient regime of other regions. The quantitative difference in cycling rates is caused by regional differences in the physical factors which control the nutrient cycles. This review discusses the physical factors which control nutrient cycling and how these factors are different in wet tropical forests.

2. CLIMATIC FACTORS IN THE WET TROPICS

2.1. Temperature

Temperature is the most important factor responsible for differences in nutrient cycling between tropical forests and forests at other latitudes. However, it is not extremely high temperature which causes the differences in nutrient cycles in the tropics, because temperatures in tropical regions, including the lowland tropics, are often lower than summertime temperatures in continental temperate regions. It is the distribution of temperatures throughout the year which causes the difference. In the tropics high temperatures occur throughout the year, and the processes resulting from high temperatures go on continuously, if moisture is not limiting.

High temperatures have important effects on rates of ecosystem processes. Within the normal range of temperatures that occur on earth, higher temperatures usually result in higher rates. For example, up to a critical temperature, which is different for each species, photosynthesis of plants increases with increasing temperatures. High year-round rates of photosynthesis result in high rates of annual primary productivity. High primary productivity results in high rates of nutrient uptake by

TABLE 1. Net primary production in ecosystems of the world. (From Whittaker and Likens 1975).

	Net Primary Production (Dry Matter) Normal Range ($g/m^2/year$)	Mean ($g/m^2/year$)
Tropical rain forest	1,000 - 3,500	2,200
Tropical seasonal forest	1,000 - 2,500	1,600
Temperate forest:		
evergreen	600 - 2,500	1,300
deciduous	600 - 2,500	1,200
Boreal forest	400 - 2,000	800
Woodland and shrubland	250 - 1,200	700
Savanna	200 - 2,000	900
Temperate grassland	200 - 1,500	600
Tundra and alpine	10 - 400	140
Desert and semidesert scrub	10 - 250	90
Extreme desert--rock, sand, ice	0 - 10	3

TABLE 2. Decomposition rate factors and turnover times (Olson 1963) for leaf litter from various ecosystems of the world. (From Swift et al. 1979).

Ecosystem	Decomposition rate factor "k", yr^{-1}	3/k (years for 95% decomposition)
Tropical forest	6.0	0.5
Savannah	3.2	1
Temperate grassland	1.5	2
Temperate deciduous forest	0.77	4
Boreal forest	0.21	14
Tundra	0.03	100

plants and abundant available food for herbivores. Secondary productivity also can be high in the tropics, because when temperatures are high, insects are more active. With high herbivory, nutrient movement through the food chain is rapid. High rates of primary productivity also mean more rapid return of nutrients to the soil through leaf and wood litterfall and through death and defecation of herbivores and predators.

Decomposition and other microbiologial processes such as denitrification also are higher at the higher year-round temperatures in the wet tropics. For example, Medina and Zelwer (1972) found a high correlation between soil respiration and temperature in seven different ecosystems in Venezuela. With higher rates of decomposition and nutrient release in the wet tropics, nutrients become available more rapidly in the soil. Because of year-round high temperatures, nutrient cycling processes take place throughout the year in the wet tropics, and, as a result, annual nutrient cycling rates are higher than in regions where cold or drought interrupts these processes.

2.2. Moisture

Although temperatures are high year-round in the tropics, temperatures alone do not cause high annual rates of nutrient cycling. Water also must be available to maintain high primary productivity. Decomposition also is slowed down or stopped by drought conditions, because decomposer organisms are moisture dependent.

Even in most of the "wet" tropics, there are certain months which receive less rain than others. In some areas, such as the northwestern part of the Amazon Basin, the "dry season" may be just a few weeks when average weekly rainfall decreases slightly. In these areas the forest is evergreen, and there is scarcely any noticeable change in monthly rates of leaf fall. Where the length of the dry season is longer and no rain falls at all for several months, deciduous forests and savanna type vegetation occur. During the dry season little growth takes place. Decomposition also slows, but fire can take the place of decomposer organisms in releasing nutrients from litter on the soil surface.

In the temperate regions, there is often seasonality of both temperature and rainfall. In some temperate regions, such as the eastern United States and northern Europe, rains are often frequent during periods of high temperature. However, because high temperatures and high rainfall coincide for only a few months each year, annual rates of production and decomposition are not as high as in the wet tropics. In other regions, such as the Pacific Coast of the United States or the Mediterranean region, the warmest months are the driest months. Under this climate, nutrient cycles have different characteristics. Production and decomposition rates are slow, because temperature and moisture are not optimal at the same time. When temperatures are favorable for growth and decomposition, moisture is not available.

3. INFLUENCE OF CLIMATE ON SOILS

The combined influence of temperature and moisture is important for weathering and leaching processes in the soil. Weathering and leaching of tropical soils strongly influence nutrient cycling in the wet tropics because of the year-round high temperature and water availability.

3.1. Weathering

Water reacts with certain types of minerals in soils, resulting in weathering processes and removal of nutrient elements. For example, the mineral microcline ($KAlSi_3O_8$) undergoes hydrolyses whereby the potassium is removed from the aluminosilicate (aluminum, silica, and oxygen) (Brady, 1974). The potassium then is soluble and can be adsorbed by the soil colloids, used by plants, or removed in the drainage water. Since rates of the reactions are temperature-dependent, this weathering process is more important in lowland tropical areas with year-round high temperatures than at higher latitudes or altitudes. Although the type of clay minerals occurring in soils varies greatly, nevertheless, there are still global patterns which illustrate the point that mineral weathering is generally most intense in the wet tropics. Table 3 shows that, in general, the least weathered minerals are found in cold zones, while the most highly weathered minerals are common in the tropics.

Although dissociation of water is a source of hydrogen for weathering, concentration of hydrogen in soil water from this process in the tropics is relatively low compared to concentrations resulting from the formation of carbonic acid. High concentrations of carbonic acid in soils of tropical wet ecosystems result from year-round root and soil respiration. When the soil is wet, carbon dioxide (CO_2) from root and soil respiration combines with the water to form carbonic acid (H_2CO_3) which then dissociates into bicarbonate (HCO_3^-) and a positively charged hydrogen ion. The hydrogen ion is then available to react with minerals, that is, to "weather" the minerals and to alter them freeing nutrient elements (Johnson et al., 1977, Stumm and Morgan, 1981). If the minerals are in the upper soil horizons, the nutrient elements that are released are readily accessible to roots of plants. However, nutrients in bedrock are not as accessible

Since bedrock is weathered year-round in the wet tropics, the process of nutrient release from bedrock could be expected to contribute to a high soil fertility. Under certain conditions it does, but under others it does not. It does not, for example, in most of the eastern and central Amazon Basin. Lack of uplift of bedrock in the region for tens of millions of years has resulted in the presence of layers of highly leached and partially decomposed bedrock (saprolite) so deep that roots cannot penetrate it. In other regions where there have been recent geological uplifts, such as the Andes, or where lava from recent volcanoes is the parent material, overlying horizons are shallower, and nutrients released from minerals in the bedrock are more easily reached by roots.

TABLE 3. Weathering sequence of clay minerals (Jackson 1965) and zones of frequent occurrence of these minerals (Millot 1979).

Order, beginning with least weathered	Clay mineral	Zone of frequent occurrence
1	Illite	"Cold zone"
2	Vermiculite	Temperate zone
3	Montmorillonite	Mediterranean zone
4	Kaolinite	Humid tropics

3.2. Leaching

The high acidity of soil water in continually wet tropical ecosystems not only results in rapid weathering of minerals, it also causes high leaching of cations such as calcium and potassium which may be exchanged or adsorbed on the surface of clay particles and humus in the upper soil horizons. These cations come from decomposing organic matter on the soil surface and are carried downward into the soil by water during rainstorms. As the positively charged cations come in contact with clay surfaces, they are attracted and held by the negative charges on the clay surfaces. Hydrogen ions from carbonic acid can displace the cations exchanged on the clay surfaces of the soil. The cations combine with bicarbonate resulting from dissociation of carbonic acid and move downward with the soil water during rain storms. Water percolating through the soil carries the cations beyond the range where they can be taken up by roots or mycorrhizal fungi, and, thus, the possibility for uptake and recycling of the nutrients by vegetation is lost. This potential for leaching loss exists year-round in the wet tropics, because soil and root respiration go on year-round. This is the principal reason why nutrients are often critical in tropical rain forest ecosystems.

Other acids also occur in the soil. Dilute concentrations of inorganic acids such as sulfuric and nitric acid occur. Sulfuric acid can be formed in the atmosphere from sulfur dioxide and can be carried into the soil by rain storms, or it may be formed in the soil atmosphere from gaseous sulfur compounds formed by microbiological activity. Nitric acid may result from thunderstorm activity in the atmosphere and from nitrification in the soil. Organic acids also can play a role in soil acidity. Organic compounds from decomposing leaf and wood litter often form organic acids. These acids can be important in nutrient leaching processes when they are carried down from decomposing organic matter on the soil surface into the mineral soil by rain water.

The relative importance of carbonic acid and other acids changes, in general, as a function of latitude or altitude. Johnson et al. (1977) compared the various anions in soil solution at a tropical rain forest site, a temperate site, an alpine site, and an arctic site. They found that at the tropical site, carbonate was by far the most important anion, while at the other sites, nitrate, sulfate, and chloride were relatively more important. This result indicates that carbonic acid is probably the most important leaching agent in tropical wet forests, while it is less important in other regions. Carbonic acid is the most important acid in tropical areas, because the year-round soil respiration results in high production of CO_2. High amounts of carbonic acid result in large amounts of bicarbonate dissolved in soil solutions of tropical ecosystems compared to soils at other latitudes and altitudes. The bicarbonate ion combines with soluble cations and is carried off by drainage water. In contrast, organic acids are often less important in the tropics because high respiration in the tropics often results in breakdown of organic compounds. At higher latitudes and altitudes, production of carbon dioxide is less, but decomposition of organic compounds also is less. Consequently, organic acids can be important in these regions.

Organic acids are not always less important in the tropics. In nutrient poor regions of the tropics, such as on sandy podsol soils of the Rio Negro tributary of the Amazon, rates of soil respiration and organic matter decomposition are relatively low (Herrera, 1979). Consequently, organic acids can be relatively more important in leaching. These organic compounds give the distinctive color to the so-called "blackwater" rivers which are found not only in the nutrient poor regions of the tropics, but also in many other areas such as temperate zone pine forests on sandy soils where low nutrient status inhibits breakdown of soil organic compounds.

4. SOIL FACTORS IN THE WET TROPICS

Soil is an important factor in nutrient cycles, because soil is an important storage compartment for nutrients as well as an important source of nutrients.

4.1. Nutrient storage

4.1.1. Clay. Part of the nutrients in the soil are stored on the surfaces of clays and colloidal organic matter (humus). Clays and organic soil colloids carry negative charges on their surfaces. Nutrients, such as calcium and potassium, have a positive charge when they are ionized, and they are attracted to the negative surface charges of the clay and humus. The ability of a soil to retain cations on the surface of clay and humus is called cation exchange capacity. Soils with a high clay content or a high organic matter content often have a high cation exchange capacity, but the type of clay present also is important. Highly weathered clays such as kaolinite, common in geologically old areas of the tropics like the Amazon Basin, have relatively low cation exchange capacities (Jackson, 1965).

Although potential leaching of cations is great, actual leaching in tropical rain forests is often quite low, as long as the forest is undisturbed, because roots in tropical rain and moist forests are concentrated near or on top of the soil surface. Because they are on or near the surface, the roots and their associated mycorrhizae can intercept and take up nutrients as fast as nutrients are released from decomposing organic matter such as fallen leaves and logs on the soil surface. The theory that nutrients are intercepted and taken up before they percolate down to the mineral soil has been called the theory of direct nutrient cycling (Went and Stark, 1968). It is called direct, because nutrients move directly from decomposing organic matter to roots, by-passing the mineral soil. Although mycorrhizae have been emphasized in discussions of direct nutrient cycling, all the decomposer organisms in close contact with roots serve to prevent leaching of nutrients into mineral soils (Stark and Jordan, 1978).

Direct nutrient cycling prevents major nutrient loss in undisturbed forests. Conversion of forests to agriculture destroys the direct nutrient cycling mechanisms. The root network of crop plants is much smaller than that of the primary forest, and roots are not concentrated near the soil surface. Consequently, cation leaching is frequently a problem when tropical rain forests are converted to agricultural land.

While high acidity of many tropical soils results in high leaching potential for cations, the acidity results in immobilization of phosphorus. At low values of soil pH, soluble phosphorus reacts with iron, aluminum, and manganese to form insoluble compounds (Sanchez, 1976). The lower the soil pH, the greater the proportion of phosphorus in the soil that becomes fixed. Critical levels of pH for plant growth vary depending on the soil type and the species of interest. Most naturally occurring species in tropical rain forests appear to be able to take up sufficient phosphorus despite naturally low pH of the soil. Mutualistic associations that have developed between native forest species and mycorrhizae probably increase the capacity of trees to take up phosphorus under the acid conditions of wet tropical forests (Herrera et al. 1978). High concentrations of soil organic matter such as occur in the undisturbed forest also can increase phosphorus availability for the trees. For example, compounds synthesized by micro-organisms in decomposing organic matter can release phosphorus from insoluble compounds and render the phosphorus available for uptake by plants (Sollins et al. 1981).

In contrast to native vegetation under undisturbed forest conditions, many crop plants under cultivation are not able to take up sufficient phosphorus when soil pH is low. Perhaps the chief reason that slash and burn agriculture is an effective technique in areas of acid tropical soils is that the ash raises soil pH to a level where phosphorus is more readily available to crops (Sanchez, 1976). As the ash leaches away during heavy rainstorms, pH decreases, phosphorus availability declines, and crop production decreases. Secondary successional or "weed" species, which apparently are better adapted to taking up phosphorus at low pH, then increase and out-compete crop vegetation. Consequently, agriculture is abandoned.

Soil acidity, then, common in much of the lowland wet tropics, has opposite effects on cations and phosphorus. High soil acidity results in high leaching of cations and their consequent loss from the ecosystem. High soil acidity, in combination with high concentrations of iron and aluminum common in tropical soils, results in fixation of phosphorus. Both cation leaching and phosphorus fixation are reasons that agriculture in the wet tropics has proven so difficult and unrewarding following removal of the native forest.

Low soil pH also causes aluminum in the soil to become soluble and consequently available for plant uptake. Crop plants will take up aluminum and accumulate it in root systems, where it interferes with the translocation of other nutrient elements (Sanchez, 1976). This is the so-called "aluminum toxicity" common in acid soils of tropical regions.

4.1.2. Soil organic matter. Part of the nutrients in the soil are stored on the surface of clays and colloidal organic matter. Cations are bound by the negative charges on the surfaces of the particles. Phosphorus is bound by iron and aluminum compounds in the soil. Besides being bound on clay surfaces, cations and phosphorus as well as sulfur also are stored in the soil as part of the soil organic matter. These nutrients become available for plant uptake as soil organic matter decomposes.

Soil organic matter is even more important for nitrogen storage than cation and phosphorus storage. In contrast to cations and phosphorus, the greatest proportion of nitrogen in tropical soils is stored as part of the soil organic matter. As soil organic matter decomposes, nitrogen becomes available for plant uptake. At high latitudes or dry regions, slow decomposition often causes nitrogen to be a factor limiting to plant growth. In contrast, in many tropical rain forests, decomposition of soil organic matter is rapid and nitrogen supply often is not limiting. Another reason that nitrogen often is not critical in tropical rain forests is that nitrogen-fixing species such as blue-green algae commonly live on the leaves and in the soil of undisturbed tropical forests, and this nitrogen is believed to become available to the forest trees.

In contrast to most tropical ecosystems, nitrogen may be limiting in unusually poor tropical soils such as tropical podsols or under anaerobic conditions where extremely low nutrient content and high soil acidity inhibit breakdown of soil organic matter (Vitousek, 1982). In overgrazed tropical pastures, where soil organic matter is severely depleted, nitrogen also can be a limiting factor.

Because of the high concentration of nutrients in organic matter, soils with high concentrations of organic matter are often fertile soils and have high agricultural productive capacity. The soils of the mid-western states of the United States are good examples of soils with high productive capacity because of high organic matter content. However, soils high in organic matter are not necessarily highly productive soils, especially in certain wet tropical areas. High accumulation of organic matter could indicate anaerobic conditions are inhibiting the breakdown of organic matter. Soil organic matter must decompose for nutrients to become available and stimulate productivity of plants. For example, soils of cloud forests on tropical mountain tops are often high in organic matter, yet tree growth on these soils is very slow (Tanner, 1980). The reason probably is that the breakdown of the organic matter is very slow due to low temperatures and anaerobic conditions and, consequently, tree growth is severely limited.

4.2. Nutrient supply.
Soils in the tropics are not only relatively poor in their capacity to store nutrients, many tropical soils are also low in their ability to supply nutrients. Most nutrient elements, except nitrogen, occur in primary minerals which comprise various types of rocks. These nutrient elements are released and become available for plant uptake as primary minerals in bedrock are weathered and transformed.

As severity and length of time of weathering increase, the proportion of nutrient elements making up the minerals decreases. Nutrient elements are removed as minerals are transformed through chemical weathering processes. The most highly weathered clay minerals such as kaolinite consist only of silica, aluminum, hydrogen, and oxygen. Continued weathering and leaching of these minerals results in removal and leaching downward of the silica, leaving oxides of iron and aluminum in the upper soil horizons (Jackson et al., 1948, Brady, 1974). This removal of silica is the so-called laterization process which can result in formation of a hard-pan or impermeable crust of "laterite" when the iron and aluminum oxides are exposed to air and become dry (Jenny, 1980).

The lowland, highly weathered, tropical soils capable of forming laterite have been previously classified as "Latasols," and they still are called this in the Brazilian soil taxonomy system. In the FAO (Food and Agriculture Organization) system, the name has been changed to Orthic Ferralsols and Xanthic Ferralsols. The FAO system is convenient for relating soil types to geological soil-forming processes. For example, the central Amazon Basin consists of materials eroded and transported from the Guyana Shield to the north and the Brazilian Shield to the south. The Shields are remnants of Pre-Cambrian mountains. Soils that have formed in place from rocks of these Shields are Orthic Ferralsols, while those transported and deposited in the central part of the Basin are Xanthic Ferralsols. In the U.S. soil taxonomy system, these soils have been reclassified as Oxisols, Ultisols, or Alfisols, depending upon the relative amount of weathering that has taken place (Aubert and Tavernier, 1972).

While these soils are often poor in nutrient supplying power, the iron and aluminum oxides form stable aggregates, that is, granules or crumbs, which give the soils a very good physical structure. The granules can be round or block-shaped, from a millimeter to a centimeter or more in diameter. These aggregates give the soil a porous structure favorable for root growth, aeration, and water drainage. Often the physical properties of highly weathered lateritic soils are well-suited for agriculture, if fertilization is possible, because cultivation does not readily destroy the porous structure.

Allophane is the general term for another type of aluminosilicate (aluminum, silicon, oxygen, and hydrogen) material found where weathered volcanic ash is an important component of the soil. Allophane binds organic matter, and the material can have a relatively high cation exchange capacity. In mid-elevational valleys in tropical mountains, soils of this type can be highly productive of agricultural crops. At lower elevations, higher temperatures cause the organic matter to decompose rapidly, and higher rates of organic acid production in the soil result in higher leaching rates of cations and a lower potential productive capacity. Volcanically derived soils are usually classified as Inceptisols in the U.S. soil taxonomy system (Aubert and Tavernier, 1972).

Since high rates of weathering processes occur more frequently in tropical regions than at higher latitudes, there tends to be a general world pattern of the occurrence of the different types of clay minerals in soils. The least weathered minerals, in general, occur at the highest latitudes, while the highly weathered minerals occur in the tropics (Millot, 1979). There are many exceptions to this pattern, but, in general, the clay minerals in tropical regions are often highly weathered, and, for this reason, soils in the tropics are often relatively low in nutrient supplying ability.

5. CONCLUSION

Nutrient cycling processes in the wet tropics have rates, on an annual basis, which are higher than rates in other regions of the world. The reason is that, in the wet tropics the processes go on year-round, due to continual warm temperatures and abundant moisture. However, these same processes which result in high rates of productivity and decomposition in the wet tropics also result in high potential for weathering of parent material and leaching of nutrients held in the mineral soil. As long as the forest is undisturbed, serious nutrient losses do not occur, because the native forest has evolved nutrient conserving mechanisms which counteract the leaching and weathering processes. However, when the forest is removed for agricultural or other purposes, nutrient losses are more serious than in other regions, because the high rates of decomposition and leaching continue, but recycling of the nutrients does not.

REFERENCES

Aubert G and Tavernier R (1972) Soil survey. In Soils of the humid tropics, pp. 17-44. Washington, DC, National Academy of Science

Brady NC (1974) The nature and properties of soils. 8th ed. New York, Macmillan.

Herrera, RA (1979) Nutrient distribution and cycling in an Amazon caatinga forest on Spodosols in southern Venezuela. Ph.D. dissertation, Dept. of Soil Science, University of Reading.

Herrera R, Merida T, Stark N and Jordan C. (1978) Direct phosphorus transfer from leaf litter to roots. Naturwissenschaften 65, 208-209.

Jackson ML (1965) Chemical composition of soils. In Bear FE, ed. Chemistry of the soil, pp. 71-141 ACS Monograph 160. New York, Reinhold.

Jackson ML, Tyler SA, Willis AL, Bourbeau GA and Pennington RP. (1948) Weathering sequence of clay-sized minerals in soils and sediments. I. Journal of Physical and Colloidal Chemistry 52, 1237-1260.

Jenny H (1980) The soil resource. Ecological studies. 37. New York, Springer-Verlag.

Johnson, DW, Cole DW, Gessel SP, Singer MJ and Minden RV (1977) Carbonic acid leaching in a tropical, temperate, subalpine, and northern forest soil. Arctic and Alpine Research 9, 329-343.

Medina E and Zelwer M (1972) Soil respiration in tropical plant communities. In Golley PM and Golley FB, eds. Tropical ecology with an emphasis on organic productivity, pp. 245-267. Institute of Ecology, University of Georgia, Athens, Georgia. USA.

Millot G (1979) Clay. Scientific American 240, 109-118.

Olson JS (1963) Energy storage and the balance of producers and decomposers in ecological systems. Ecology 44, 322-332.

Sanchez PA (1976) Properties and management of soils in the tropics. New York, Wiley.

Sollins P, Cromak K, Fogel R and Van Li C (1981) Role of low-molecular-weight organic acids in the inorganic nutrition of fungi and higher plants. In Wicklow DT and Carroll GC, eds. The fungal community, its organization and role in the ecosystem, pp. 607-620. New York, Mariel Kekker.

Stark N and Jordan C. (1978) Nutrient retention in the root mat of an Amazonian rain forest. Ecology 59, 434-437.

Stumm W and Morgan JJ (1981) Aquatic chemistry. New York, Wiley.

Swift, MJ, Heal OW, and Anderson, JM. (1979) Decomposition in terrestrial ecosystems. University of California Press, Berkeley.

Tanner EVJ (1980) Studies on the biomass and productivity in a series of montane rain forests in Jamaica. Journal of Ecology 68:573-588.

Vitousek P (1982) Nutrient cycling and nutrient use efficiency. The American Naturalist 119, 553-572.

Went F and Stark N (1968) Mycorrhiza. BioScience 18, 1035-1039.

Whittaker RH and Likens GE (1975) The biosphere and man. In Lieth H and Whittaker RH, eds. Primary productivity of the biosphere, pp. 305-328. Ecological Studies 14. New York, Springer-Verlag.

PHYSICAL ASPECTS OF THE WATER REGIME OF WET TROPICAL VEGETATION

J.J. LANDSBERG
CSIRO Division of Forest Research Canberra, A.C.T. Australia

ABSTRACT

The water regime of wet tropical vegetation is discussed in terms of the standard hydrological equation. The proportion of rainfall which is effective in replenishing soil moisture depends on the amount lost by interception and evaporation from plant canopies; a model is discussed. Redistribution of rainfall by stemflow is considered in some detail; where rainfall is heavy stemflow may cause considerable modifications in the distribution of water to the soil. Transpiration from plant communities is analyzed in terms of canopy energy balance; available data indicate that transpiration rates in the wet tropics are likely to be low. Because of energy absorption by upper layers, low windspeeds and high humidities, transpiration by understory vegetation must be very low. The balance between root surface areas and leaf area, and the consequences of root systems of different types, are discussed briefly in respect to the water relations of under- and overstory vegetation in wet tropical forests.

1. INTRODUCTION

The term 'wet tropics' suggests that water is unlikely to be a limiting factor to plant growth in these regions. However, whether it is seriously limiting or not, the availability and distribution - in time and space - of water must have some effect on the survival, competitive ability and growth patterns of plants in the wet tropics. To analyze these effects we must consider the hydrology of wet tropical ecosystems - the way they accept and dispose of water - and characterize the factors affecting plant water use and status. There is a paucity of empirical information about many of these processes in the wet tropics but the principles of rainfall interception and redistribution and of canopy transpiration rates and microclimate are well established from research elsewhere. They are presented and discussed in this paper together with some relevant experimental information. There is clearly a great need for more research in the tropics, not only because of the inherent ecological interest and importance of tropical systems but also because of the hydrological consequences of the forest clearing proceeding so rapidly in many tropical areas.

2. HYDROLOGICAL CONSIDERATIONS

The 'wet tropics' are not uniformly wet, or wet all the time. There are, in fact, few areas of the world which do not experience marked variations in rainfall or even periods with virtually no rain. To illustrate, Fig. 1 shows the annual pattern of rainfall at three locations in tropical Australia and one in New Guinea. The duration and frequency of dry periods may exert a strong influence on the species composition and growth patterns of the vegetation of a region. To analyze this influence it is necessary to consider the type of rainfall, the way it is intercepted and distributed by the vegetation and the storage capacity of the soil in relation to water loss by transpiration. The standard hydrological equation provides a framework for such analysis.

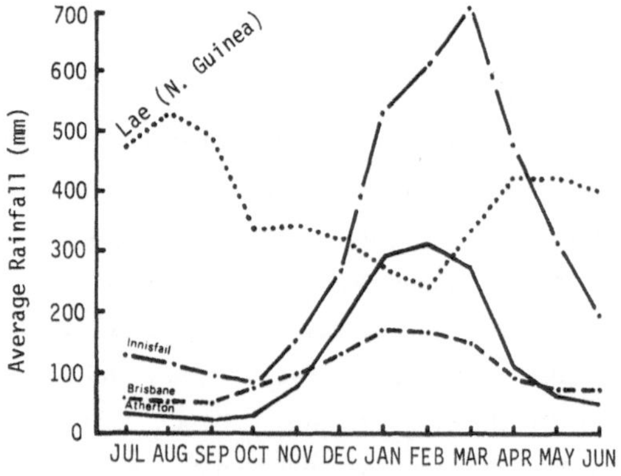

FIGURE 1. Monthly average rainfall (mm) at Atherton, Innisfail and Brisbane in northern Australia, and Lae in New Guinea.

If the amount of water available to the plant community over time interval $t + \Delta t$ is $(\theta(t + \Delta t) - \theta_{min})z$, where θ is the volumetric water content of the soil, θ_{min} is the lower limit of available water, set by soil hydraulic characteristics, and z is the depth of soil exploited by roots, then we may write

$$\theta(t+\Delta t)-\theta_{min})z = (\theta(t) + \theta_{min})z + R' + E' - Dr - RO$$

which re-arranges to

$$\theta(t + \Delta t) = \theta(t) + \frac{R' - E' - Dr - RO}{z} \qquad (1)$$

The symbols R', E', Dr and RO denote amounts of rainfall, evapotranspiration, drainage out of the root zone and run-off, respectively. (When the symbols R and E are used without primes, they denote rates of rainfall and transpiration.)

Drainage and runoff will not be considered in any detail in this paper although they may be of considerable importance in the water balance of a particular site. Run-off from a non-saturated soil occurs when the rate of water input exceeds the soil infiltration capacity; a rough measure of infiltration capacity is given by the saturated hydraulic conductivity of the surface layers. Run-off will obviously also occur when soil is saturated, regardless of the input rate. Loss by drainage includes both vertical and lateral flow in the soil and is only likely to be significant when the root zone is saturated. Drainage is difficult to measure.

Eq.(1) contains the implicit assumption that all rainfall is effective in replacing soil water, which is not the case for any vegetative surface because of interception and evaporation of intercepted water. Furthermore, rainfall is re-distributed by vegetation so that, although the equation may describe the average situation over a relatively homogeneous area, it will be subject to considerable variation and hence inaccuracy, on a microscale. It is important that ecologists appreciate this, and the mechanisms causing variation, because these may affect both the location and the growth patterns of plants.

2.1. Interception losses

Effective precipitation (R'_{eff}) is total precipitation less interception loss (I);

$$R'_{eff} = R' - I \qquad (2)$$

Doley (1981) lists the results of a great many empirical studies of interception loss in relation to rainfall amount and discusses the considerable differences observed in various forest and plantation types. Interception loss has often been described by linear empirical relationships (Gash, 1979) with storm size (total rain per storm). Such studies may provide useful estimates for rough 'broad-scale' calculations, but they are of limited value for detailed hydrological work or studies on plant growth patterns and distribution. For detailed work it is necessary to analyze the process of interception in terms of mechanistic models. The best available is that originally derived by Rutter et al. (1971, 1975) and recently simplified by Gash (1979). E.M. O'Loughlin (personal communication) has refined this to a readily computerized form utilizing the solution to the differential equation describing the rate of change of water stored on the canopy (C, mm depth);

$$\frac{dC}{dt} = (1-p)R(t) - E_c(C) \qquad (3)$$

where p is the proportion of rain which passes through the canopy without touching it and E_c is the instantaneous rate of evaporation of water from the canopy surfaces. There is little information on the magnitude of this latter term but it is clearly significant. O'Loughlin's version of the Gash's model allows it to be estimated.

If S is the canopy storage capacity when saturated (so $C \leqslant S$) eq. (3) can be rewritten

$$\frac{dC}{dt} = (1-p)R(t) - C(t)\,\frac{\bar{E}}{S} \qquad (4)$$

where $\bar{E}$ is the average (assumed constant) rate of evaporation from the canopy during a rain event. If the canopy parameters S and p are known (they can be obtained from conventional in-canopy measurements of throughfall, stem flow and above-canopy rain) eq. (4) can be solved analytically; the solution provides the algorithm for a computer program for which rainfall is the driving variable and canopy parameters and evaporation rate are user-definable. O'Loughlin estimates values of $\bar{E}$ by comparing observed values of R_{eff} with calculated values obtained using a range of values of $\bar{E}$.

For very dense canopies p may be effectively zero, so that $R_{eff} = 0$ until C approaches S. In most cases some water will reach the ground - after the start of a rain event - before canopy saturation because of drip and (possibly) stemflow. The amount will depend on the nature of the canopy and of the precipitation. The direct throughfall component of light, fine rain is likely to be smaller - perhaps much smaller - than that of heavy rain where most of the water is contained in large drops. Furthermore, light rain will wet canopy surfaces more evenly, and in low windspeeds there will be less mechanical displacement of drops and hence - at least initially -less drip than in heavy rain. When raindrop sizes are large rainsplash on rigid surfaces (e.g. branches or leathery leaves) contributes to immediate throughfall.

Interception losses will be much greater if rainfall is intermittent than if it falls continuously for relatively long periods. This can be illustrated by considering the hypothetical situation where each of a series of rainfall events is of order S mm, while the interval between events is long enough for all the water stored on the canopy to evaporate. In this case R_{eff} could be near zero.
The rate of evaporation of water from wet canopies depends mainly on the atmospheric vapor pressure deficit (D), and on windspeed. The effectiveness of wind depends on the aerodynamic properties of the canopy - its height and roughness, which is characterized by the separation of the dominant trees and the density of the foliage. Evaporation rates from wet canopies are up to 4 times transpiration rates (Stewart, 1977). Clearly evaporation of intercepted water (on a land area basis) will be lower from multi-layered canopies than from single-layered ones because of the high humidity and low windspeeds in the understory. (See

'Canopy Microclimate').

The above considerations indicate why water losses by canopy interception vary greatly in different situations and also indicate that, in plant ecological work, detailed study will be necessary to characterize particular habitats in terms of their water balance.

2.2. Redistribution of Rainfall

The distribution of the unintercepted component of rainfall obviously depends on canopy homogeneity. Water intercepted by the canopy and not evaporated either coalesces into large droplets which drip off a multitude of low points (leaf tips, twigs etc.) in the canopy, or reaches the ground as stemflow. Drip may or may not be evenly distributed, depending on canopy type; stemflow will inevitably cause gross distortion of the soil wetting patterns which would have occurred from uninterrupted rainfall.

Stemflow does not commence until the streaks of water collecting on the trunks of trees from the lower side of branches are established as continuous flow lines. Doley (1981) gives a table of data characterizing the partitioning of precipitation of tropical forests and woodlands which indicates that stemflow has been observed to vary between zero and 39% of the total rain in a storm. As in the case of interception losses, such data are of limited value as a guide to what might be observed in any particular situation because stemflow amounts depend so strongly on the physical characteristics of canopies and on rainfall characteristics. It is probably true to say that stemflow only becomes an important component of the water balance when rainfall is heavy. Herwitz (1982), who studied the redistribution of rainfall by trees in a tropical rainforest in Australia, noted that the sheltered undersides of branches represented detention storage that usually requires a high intensity rainfall event (> 100mm/day^{-1}) or several consecutive days of substantial rain before a thoroughly wetted condition is achieved and substantial stemflow commences.

Herwitz (1982) considered stemflow volumes in relation to the basal area of trunks, not the crown area. He argued that the volume of water expected at the base of a tree is that which would have been collected by a rain gauge occupying the same area as the trunk, therefore the volume of water actually delivered by stemflow should be expressed as a ratio of the expected volume. This is the 'funnelling ratio'. Some of Herwitz's data are presented in Table 1, which includes an assessment of the

TABLE I. REDISTRIBUTION OF RAINFALL BY TROPICAL TREES (Data from Herwitz, 1982)

Species	Trunk basal area (m^2)	Stemflow vol. (m^3 s-1 x 10^4)	Min. infiltration area(m^2)*	Funnelling ratios**
Balanops australiana	0.061	157	2.17	156
Ceratopotalum virchowii	0.049	84	1.15	112
Cardwellia sublimis	0.127	24	0.33	12
Elaeocarpus sp.	0.159	20	0.27	10

* Stemflow volume/mean saturated hydraulic conductivity to 0.2 m

** Volume of rainwater which would have been collected in a rain gauge occupying the same area as the trunk minimum infiltration area - the minimum area over which stemflow has to spread in order to infiltrate the soil. This was calculated by dividing the rate of water input from stemflow by the mean saturated hydraulic conductivity of the surface 20 cm. The area is minimum because no allowance was made for throughfall.

The distribution of water resulting from stemflow may have important effects on the water balance of trees, particularly as they enter a dry period. The amount of rain intercepted by the dominant trees of a canopy is likely to be proportional to their canopy area. If the rainfall in the preceding period was heavy enough to cause significant stemflow the soil at the base of the dominant trees of a canopy may be considerably wetter than that around the base of subdominants. The water available in the root zone of the sub-dominants, and understory vegetation, then depends on their location and root distribution relative to the dominants. A compensating factor for these trees may, however, lie in the likelihood that they will experience reduced evaporative demand because of shading by dominant trees, wind speed reduction and high humidities below canopies.

Between rain events, if the drainage term in eq. (1) can be taken to be zero, the water balance of a site is determined by the rate of loss by evaporation from the soil surface and transpiration through plants. Evaporation from soil is negligible if the soil surface is dry, and is (presumably) negligible under a full canopy even if the soil is more or less continuously wet. This component of evapotranspiration will not be treated further in this paper. Water loss from plant leaves - transpiration - extracts water from the soil in a pattern depending on soil moisture content and root distribution so that eq.(1) is, in this respect as in others, over-simplified for some purposes. The rate of transpiration at any time (E(t)) is an important determinant of plant water status because it determines flow rates through plants across the resistances in the flow pathways. The following section provides a more detailed treatment of canopy climate and transpiration from canopies.

3. CANOPY CLIMATE AND TRANSPIRATION

3.1. Canopy energy balance

Transpiration from forests can be considered in terms of transpiration (per unit ground area) from the canopy as a whole or in terms of water loss from individual plants. Both are important. Estimates of water loss by transpiration from canopies, in conjunction with rainfall data, allow estimates of trends and patterns in the water balance of plant communities, and hence (given data on soil depth and plant root zones) evaluation of the probability of drought and its likely severity. (As noted in the section on 'Interception losses', rough estimates of R'_{eff} for 'broad scale' studies can be obtained by assuming I to be a linear function of rainfall amount.)

Canopy transpiration rates for all vegetation types can be estimated from the canopy energy balance, described by the well-known simple equations

$$\phi_n = \phi_s(1-\alpha) \pm \phi_L \tag{5}$$

and

$$\phi_n = \lambda E + H \tag{6}$$

where ϕ_n, ϕ_s and ϕ_L denote net radiation, short-wave incoming radiation and long-wave radiation respectively, is the albedo or short-wave reflectivity of the canopy, is the latent heat of vaporization of water and H is sensible heat. Eq. (5) defines the amount of energy retained by a surface (plant canopy) and eq.(6) states that the energy is conserved as latent and sensible heat. This ignores heat storage in (and released from) biomass and the ground, energy used in photosynthesis and advected energy. However, temperatures in the wet tropics fluctuate very little (hence the amount of heat stored is constant) advected energy is only likely to be important where there are marked discontinuities between wet and dry areas with different surface temperatures, and photosynthesis only accounts for only about 1% of ϕ_s. The errors involved in using eq.(6) as it stands are therefore negligible.

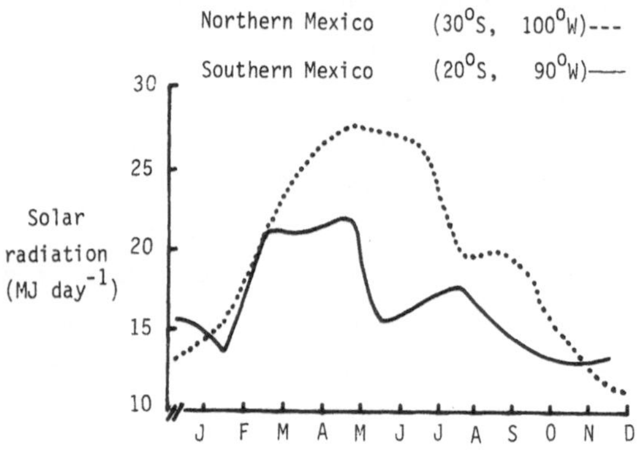

FIGURE 2. Average daily short-wave radiation (MJ m^{-2}) for Sonora and Yucatan, Mexico. (Derived from Black, 1956).

Short-wave radiant energy income in the tropics is obviously more nearly constant through the year than at higher latitudes, but it is also somewhat lower than might be expected because of the large amount of cloud. The contrast between ϕ_s in wet and dry areas is illustrated in Fig. 2, which shows the annual course of ϕ_s (estimated from monthly maps published by Black, 1956) for Mexico's Yucatan Peninsular (20° South, 90° West) and for the Sonora region (30° South, 100° West). Obviously, since radiant energy is the most important environmental factor affecting plant growth, continuous measurements of ϕ_s should be made at study areas of particular interest.

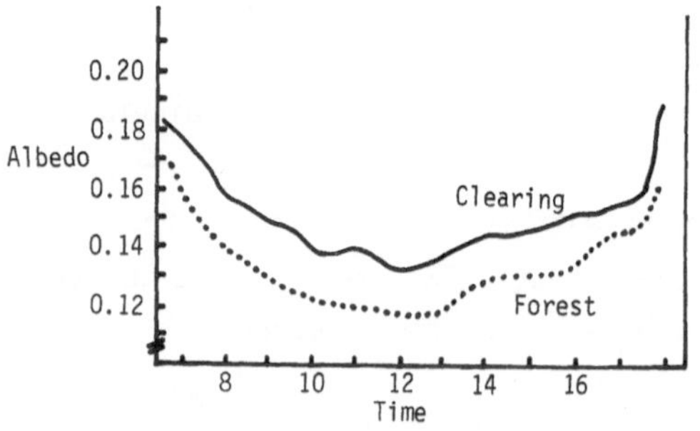

FIGURE 3. Diurnal variation in the albedo of an evergreen tropical forest, and a clearing, in Thailand. (Re-drawn from Pinker, 1982).

Fig. 3 shows the diurnal variation in albedo for a tropical evergreen forest and a clearing in Thailand (from Pinker, 1982). Pinker found the average mid-day albedo for the forest to be 0.12 and 0.13 for the summer and winter monsoons respectively; the corresponding figures for the clearing were 0.11 and 0.15. The amplitude of the diurnal variation was substantially reduced by cloudy conditions.

Pinker et al.(1981) studied the energy balance of the Thai forests and derived linear regressions of ϕ_n on ϕ_s. The constant and coefficient of eq. (7) are averages of the equations they derived from a total of 50 days measurements:

$$\phi_n = 0.88\phi_s - 33.7 \text{ (W m}^{-2}\text{)} \tag{7}$$

(Similar equations have been obtained by others for many surfaces; see Jarvis et al. (1976) for a discussion).

The Bowen ratio, $\beta = H/\lambda E$, can be used to partition ϕ_n. Substituting in eq.(6) gives

$$\lambda E = \frac{\phi_n}{1+\beta} \tag{8}$$

There are very few values of β for wet tropical vegetation in the literature. Based on studies over temperate forests we would expect low values in wet periods and higher values in dry, when water shortage causes stomatal closure and higher canopy temperatures, hence higher heat flux (H). Where stomata are influenced by atmospheric humidity this will also affect the Bowen ratio (See Jarvis et al. 1976).

Pinker et al. (1981) give estimates of β for the Thai forest, based on two days measurements, which they suggest should be regarded as qualitative rather than quantitative. The values are $\beta = 0.45$ in June (the middle of the wet season) and $\beta = 6.4$ in January - after no rain in December and very little during January. These are at least consistent with expectation and despite the uncertainty about values we can use the Bowen ratio to calculate the transpiration likely at given sites, and hence site water balance. As an example, using the ϕ_s data for Yucatan (Fig. 2, and Bryson and Hare, 1974) I have assumed that the average monthly value of β is 0.5 in the months where rainfall exceeds 100 mm, increasing linearly to $\beta = 5$ at $R' = 0$. Eq. (7) was then used to calculate monthly ϕ_n (assuming a uniform daylength of 12 hours to convert the intercept to MJ m^{-2} day^{-1}, i.e. 37.7 x 12 x 3600 = 1.6 MJ) then $E = (\phi_n$ x number of days in the month)/ β+1))/λ.

The values of β used are given on Fig. 4.

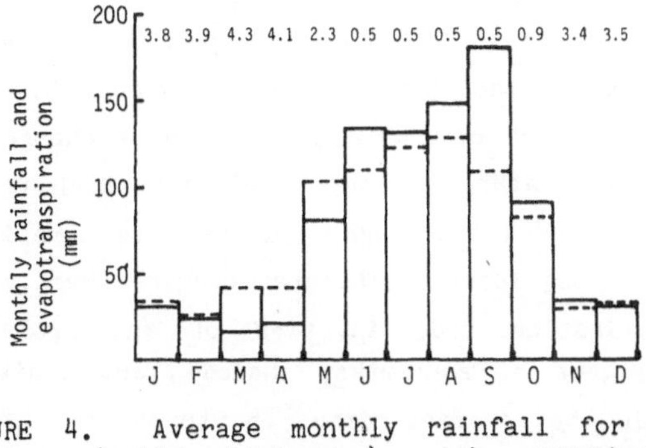

FIGURE 4. Average monthly rainfall for the Yucatan (solid histogram) with evapotranspiration (dotted) estimated from energy balance, assuming the Bowen ratio to be inversely related to rainfall. Estimated ratio values are along the top of the histogram.

The calculations done to produce Fig. 4 are of course only illustrative, but if we had more information about β and how it varies in the tropics such calculations could be usefully made for particular areas where data on ϕ_s and/or ϕ_n and rainfall are available. Measurements of soil water balances would provide independent estimates of such calculations. The direct evaluation of β by measuring air temperature and vapour pressure gradients above canopies is technologically demanding and expensive and requires relatively large uniform reasonably level sites. For these reasons such measurements are often unsuitable for ecological studies.

Where the necessary weather data are available more complex equations can be used to calculate transpiration; the best known and most widely used is probably the so-called Penman-Monteith version of the mass transfer-energy balance combination equation. This is usually written

$$E = \frac{s\phi_n + \rho c_p D g_a}{s + \gamma(1+g_a/g_c)} \qquad (9)$$

where ρ and c_p denote air density and specific heat respectively, s is the slope of the saturation vapour pressure/temperature curve, γ is the psychrometric constant, D is air vapor pressure deficit and g_a and g_c are the aerodynamic (boundary layer) and canopy conductances. The canopy conductance depends on the leaf area index and stomatal conductances. The derivation of the parameters and the use of eq. 9 as a diagnostic and predictive tool have been discussed in detail many times (see Monteith, 1973; Thom 1976; Jarvis, 1981) and will not be discussed further here. To my knowledge it has never been tested for wet tropical vegetation but there is no reason why it should not prove as useful there as elsewhere.

The equation suffers from the fact that it does not contain a term for vapor flux from the soil, which may be important even in full-canopy situations because of the coupling of in-canopy microclimate and fluxes from leaves and soil. However, this - and the detailed micrometeorology necessary to assess its importance - is probably an unnecessary refinement in ecological studies.

3.2. Canopy microclimate and understory transpiration

The discussion in the previous section considered transpiration from plant communities on a unit land area basis. However, much wet tropical vegetation is forest, which often consists of several layers of vegetation. Understory vegetation exists in an environment which is quite unlike that of the overstory in respect to energy, air movement and humidity, and wind speed. Because of these differences in physical environment, understory plants must transpire at rates different from those which are exposed to above-canopy conditions, which in turn has implications for the strategies they might be expected to adopt in response to water shortage. To evaluate the situation we must consider canopy microclimate.

The first and most obvious difference between in- and above-canopy conditions are energy levels. Table 2 is drawn from data of Pinker et al. (1980) and shows that average radiation at the floor of a tropical evergreen forest is about 5 to 6% of that at the canopy top.

TABLE 2. DAILY TOTAL RADIATION (MJ m^2) AT CANOPY TOP (FT) AND FOREST FLOOR (FF) IN A TROPICAL EVERGREEN FOREST (from Pinker et al. 1981)

	Feb.	Mar.	June	Aug.	Sept.	Average
FT	11.6	9.6	11.8	10.8	11.7	11.1
FF	-1.0	0.8	1.1	0.6	0.8	0.6

Pinker et al. found average ϕ_s at the forest floor to be about 8% of that incident on the canopy. (It is worth noting in passing that the average energy income over this forest is quite low compared to that in less cloudy regions.) These data are, again, consistent with results obtained in other vegetation types, although clearly the values in any particular situation depend on canopy structure.

An important facet of the energy environment of forest floors is sunflecks. The forest floor energy values given by Pinker et al. are averages. Björkman et al. (1972), studying the significance of sunflecks for photosynthesis, found that radiation intensity in a sunfleck and in the shade were different by orders of magnitude and Robichaux and Pearcy (1980) found that although the average energy flux density in a forest site at Hawaii was only 1 - 2% of that in the open, sunflecks, on a clear day, may contribute a significant proportion of the total energy received at a location. They confirmed that energy flux densities in sunflecks may be 10 to 50 times those in the shade.

Sunflecks are important for photosynthesis but whether they influence plant water relations greatly is still a matter for speculation. Much depends on the duration of the sunfleck, the response time of stomata, the rate of change of leaf temperature and the speed with which water vapor moves through plants to rapidly transpiring leaves. The area requires much more investigation and, in view of the probable importance of sunflecks, shaded plant habitats should be characterized in terms of their observed number and frequency of occurrence.

The environmental conditions inside a plant canopy are a product of external weather conditions and the plants themselves - there is close coupling between microclimate, leaves and soils. Transpiring leaves or water evaporating from soil increases the water vapor content of the air and hence reduces the vapor pressure deficit, causing a feed-back effect which tends to reduce the rate of water loss. Turbulent transfer - the generally highly effective process by which water vapor and CO_2 are exchanged between leaves and air - is likely to be very ineffective because wind speeds are low in dense canopies. The force of the above-canopy wind is dissipated by momentum absorption by elements of the canopy, therefore the extent of wind reduction will depend on the density and momentum absorption properties of those elements. Landsberg and James (1971) gave an equation which provides an estimate of wind speeds at any level inside plant canopies;

$$u(z) = u(h)\left[1 + m\left(1- \frac{z}{h}\right)\right]^{-2} \qquad (10)$$

Wind speed at the top of the canopy (u(h)) must be known or estimated; the parameter m is empirically determined and has a value of about 3 for forests. Measurements by Thompson and Pinker (1975) in a tropical evergreen forest show that the wind speed profiles there conform to the expected shape, with very low wind speeds (< 1 m s^{-1}) at levels below the main foliage masses.

Thompson and Pinker (1975) also provide temperature profiles for the forest. Because of the low energy levels below the canopy heat fluxes are small and, as would be expected, temperatures show little variation with height. Average air temperatures will usually provide an acceptable estimate of in-canopy temperatures in wet regions.

Air vapor pressures in wet canopies would be expected to be high and vapor pressure deficits small.

Leaf transpiration rates in any environment can be calculated from the leaf energy balance equation

$$\phi_{n.\ell} = \frac{\rho c_p}{\gamma}\,\frac{(e_s(T_\ell)-e_a)}{r_s + r_a} + \rho c_p\,\frac{(T_\ell - T_a)}{r_h} \qquad (11)$$

where ϕ_{nl} is leaf net radiation, $e_s(T_l)$ and e_a denote saturated vapor pressure at leaf temperature (T_l) and actual vapor pressure respectively, T_a is air temperature and r_s, r_a and r_h are stomatal resistance and the boundary layer resistances to water vapor and heat transfer. (Conductance, the reciprocal of resistance, is used for many purposes.) Strictly, $r_a \neq r_h$, but for most purposes they can be taken as equal. The two terms on the right-hand-side of the equation correspond to the two terms of the right-hand-side of eq. 6. Making a simple approximation which allows elimination of the restrictive requirement of eq. 11 - the need to know leaf temperature - eq. 9 can be derived from eq. 11 in a relatively straightforward manner. However, the physical principles involved in transpiration can be seen more easily in eq. 11 and it is a useful tool for exploring the consequences of particular assumptions or conditions.

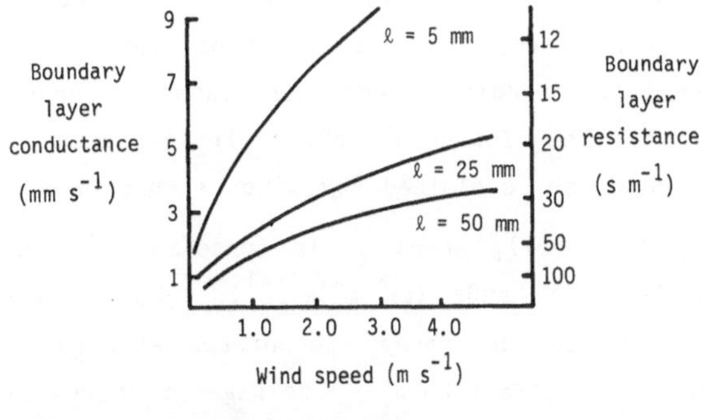

FIGURE 5. Leaf boundary layer conductances as a function of linear dimensions and wind speed. (Re-drawn from Grace, 1981).

The boundary layer resistance of a leaf depends on leaf size (linear dimension) and wind speed. Fig. 5 (adapted from Grace, 1981) shows the way that g_a ($= 1/r_a$) varies with wind speed for various leaf sizes. The curves were derived using iterative solutions of eq. 11 and relationships involving leaf dimensions, tested and refined by Grace. Grace et al. (1980) also studied the boundary layer conductance of large-leaved tropical species (Tectona grandis (teak), Gmelina arborea and Triplociton scleroxylon; linear dimensions $\geq$ 100 mm). They found that with these leaves g_a increased almost linearly - but remarkably slowly - with wind speed. The equation

$$g_a = 2 + 5\,u \qquad (12)$$

where g_a has the units mm s^{-1}, adequately describes their results.

Stomatal conductance ($g_s = 1/r_s$) has been very widely studied in many plant species in recent years. There are models describing stomatal behavior at the cellular, leaf and canopy levels and multitudinous papers on all aspects of it. Inevitably, there is less information on wet tropical species than others, but a paper by Whitehead et al. (1981), also dealing with Tectona grandis and Gmelina arborea, studied in Nigeria, provides useful information although these are not understory species. Whitehead et al. analysed the fluctuations in g_s, measured by porometer in the field, in terms of plant water potential, irradiance and air vapor pressure deficit. They found that 80% of the variance in g_s could be explained by the simple model ($g_s = f(\phi_p, D)$), where ϕ_p is irradiance in the visible wave bands (µE m^{-2} s^{-1}). Leaf water potential in the range encountered exerted no significant effect on g_s. The highest values of g_s observed in Gmelina arborea were about 12 mm s^{-1} and in teak about 35 mm s^{-1}. The relationship with irradiance was non-linear in both cases. At low values of D, g_s in Tectona grandis did not reach its maximum value until $\phi_p \simeq$ 1000 µE m^{-2} s^{-1}; in Gmelina arborea g_s reached its maximum at about 400 µE m^{-2} s^{-1}.

Pearcy and Robichaux (1980) present diurnal trends in g_s for Euphorbia species in shaded situations in Hawaii. There was little diurnal variation and g_s remained at 1-2 mm s^{-1} all day, consistent with the results of Whitehead et al. at $\phi_p \simeq$ 50 µE m^{-2} s^{-1}.

To evaluate the importance of water balance for understory species in wet tropical forests, I have used some reasonable figures in eq. 11 to calculate transpiration rates. Leaf-air temperature difference was assumed negligible (i.e. $\phi_{nl} \simeq \lambda E$) and I set $T_l \simeq T_a = 25^{\circ}C$, RH = 85% - hence ambient vapor pressure (e_a) $\simeq$ 2.7 kPa, and $e_s(T_l) - e_a \simeq 0.5$ kPa. If g_s = 2 mm s^{-1}, r_s = 0.5 mms^{-1} = 500 s m^{-1}, and if u $\simeq$ 1 m s^{-1} then (from eq. 12) $g_a \simeq$ 7 mm s^{-1} and $r_a \simeq$ 140 s m^{-1}. The result is $\phi_{nl} \simeq$ 14 W m^{-2}. If this was the average over a 10 h day it is equivalent to about 0.5 MJ m^{-2} (cf Table 2) and 0.2 mm transpiration.

The calculation provides quantitative support for a result which could have been guessed; water loss from understory vegetation in shaded environments is extremely low and, despite the relatively low canopy transpiration rate in the wet tropics (see Fig. 4), the contribution of understory vegetation to total water flux from the community is negligible. Obviously different parameter values could be used in eq. 11 but as long as these are realistic they would not change the essential result much.

Because of the low transpiration rates we would not expect leaf water potential of understory species to fall very low, and indeed Robichaux and Pearcy's data (1980) show that in shaded Euphorbia it fell to only about -0.2 MPa in winter, although it reached -1 MPa in the drier summer.

It appears, therefore, that as long as the overstory vegetation does not lose enough foliage in dry periods to greatly increase irradiance on the forest floor, water loss from understory vegetation will remain very low even in dry periods. It is likely that the effects of such periods on plant performance will depend strongly on the soil volume exploited by their root systems - and perhaps on the form of their root systems. The following brief section outlines some of the considerations pertinent to this point.

4. ROOT AND LEAF SYSTEMS

The way a plant responds to its physical environment depends to a large extent on its own structure, particularly on the leaf area it has deployed (and the arrangement of those leaves) in relation to the size and form of the root system.

Reductions in leaf water potential are the inevitable result of water loss from leaves. To replace the loss water must move from soil to roots and through the plant across the resistances in the flow pathways. On a daily basis the leaf water potential of tall overstory trees will usually fall lower than that of understory vegetation, both because of the greater transpiration rate and the greater resistance in the flow pathways. The resistance is likely to result from the type of root system and the way roots exploit the soil as much as the length of the pathways.

Tall trees must have large support roots and although they may produce many fine (absorbing) roots the absorbing surface per unit root mass may be smaller than of shrub or herbaceous species. Large deep root systems tap a large reservoir of water and, if that reservoir is not replenished, the decrease in water potential is relatively slow (Fowkes and Landsberg, (1981); see Fig. 6). There is evidence from temperate and cold area forests that, under adverse conditions, a greater proportion of the assimilate available for growth is diverted to the maintenance of root systems - especially fine roots (Persson, 1980; Keyes and Grier, 1981). We may speculate - although there is no evidence for this - that as water potentials begin to fall in dry periods, the first response of tropical trees is to divert assimilate to increase root proliferation in those soil regions where conditions remain suitable, and hence to increase the water absorbing area relative to leaf area.

The short-term response of trees to stress is, in many cases, stomatal closure, which reduces transpiration rates; the longer term response is leaf shedding. When enough leaves have been shed to restore the balance between uptake and loss rates, trees may recommence growth, even if no rain has fallen (Borchert, 1980; Reich and Borchert, 1982). Again speculating, it seems likely that shrubby or herbaceous understory species will have finer, more branched, root systems with a relatively greater number of small roots than overstory trees. This will enable them to exploit microenvironments such as wet areas resulting from drip, or caused by stemflow at the base of large trees, more thoroughly. Fine root systems have the potential disadvantage that, if water in the soil volume they exploit is reduced to ϕ_{min}, there will be a much more rapid increase in stress than will occur with deep coarse root systems (see Fig. 6). In general, because of their low water use rates, it will be rare for understory species to reach this stage.

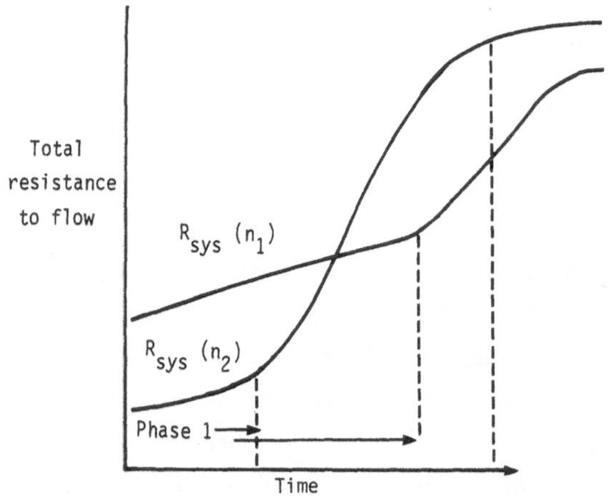

FIGURE 6. Time course of change in root-soil resistance for different type root systems. R sys (n_1) has few large roots, penetrating deeply. R sys (n2) is fibrous, thoroughly exploiting a relatively small soil volume. (Redrawn from Fowkes and Landsberg, 1981).

In summary, root systems must be as important in the adaptation of species to their ecological niches as the structure and arrangement of their aerial parts. The plants edaphic environment is characterized by soil water holding capacity and hydraulic conductivity but these properties do not in themselves determine the influence of soil water on plant water status and consequent growth patterns. Plant water potential at any time is determined by the rate of loss of water from the whole plant (transpiration rate per unit leaf area x leaf area) and the surface area of roots in contact with wet soil, which determines the root-soil resistance. Plants may adapt to changes in the balance between loss and uptake by producing more roots or shedding leaves. The type of root system will affect the speed with which imbalances may occur.

5. CONCLUDING REMARKS

Most of the physical relationships considered here are well documented for crops and, in some cases, for various temperate or arid zone ecosystems. There is much less information on wet tropical systems but, in most cases, there seems no reason to expect surprises.

There is little temperature variation and it seems likely that, ironically, vegetation in the wet tropics is most strongly influenced by dry periods. This is quite well illustrated by the studies of Borchert (1980) and Reich and Borchert (1982), who demonstrated that the flowering patterns of some tropical trees are, apparently, almost completely determined by water shortages.

In general, transpiration rates in the wet tropics are not high but rates of water use by tropical vegetation are not well known and we need much more information on the energy balance of canopies, stomatal behavior of both over-and understory species, and in-canopy microclimates. There seems to be considerable scope for studies on stem flow and the microenvironments in forests in terms of the distribution of water and the (related?) distribution of understory vegetation. Many questions about root system types and their relation to the leaf area they support arise; these must be considered in terms of rates of water loss of vegetation exploiting different niches in the aerial environment.

REFERENCES

Björkman O, Ludlow MM and Morrow PA (1972) Photosynthetic performance of two rainforest species in their native habitat and analysis of their gas exchange. Carnegie Inst Washington Yearb 71, 94-102.

Black JN (1956) The distribution of solar radiation over the earth's surface. Arch. Met. Geoph. Biokl. B.7.165 - 189.

Borchert R (1980) Phenology and ecophysiology of tropical trees : Erythrina poeppigiana O.F. Cook. Ecology 61, 1065 - 1074.

Bryson RA and Hare FK (1974) World survey of climatology. Vol. 11 Climates of North America. Elsevier, Amsterdam.

Doley D (1981) Tropical and sub-tropical forests and woodlands. In Kozlowski TT ed. Water deficits and plant growth V1, pp 209-307. Acad. Press Inc.

Fowkes ND and Landsberg JJ (1981) Optimal root systems in terms of water uptake and movement. In Rose DA and Charles-Edwards DA, ed. Mathematics and plant physiology, pp 109-125. Acad. Press.

Gash JHC (1979) An analytical model of rainfall interception by forests. Quart. J.R. Met. Soc. 105, 43-55.

Grace J (1981) Some effects of wind on plants. In Grace J, Ford ED and Jarvis PG, ed. Plants and their atmospheric environment, pp 31-56. Blackwell Sci. Pubs.

Grace J. Fasehun FE and Dixon M (1980) Boundary layer conductance of the leaves of some tropical timber trees. Plant, Cell and Environment 3, 443-450.

Herwitz SR (1982) The redistribution of rainfall by tropical rainforest canopy tree species. In O'Loughlin EM and Bren LJ, ed. First National Symposium on Forest Hydrology, pp 26-29. Institution of Engineers, Australia.

Jarvis, PG (1981) Stomatal conductance, gaseous exchange and transpiration. In Grace J, Ford ED and Jarvis PG, ed. Plants and their atmospheric environment, pp 175-204. Blackwell Sci. Pubs.

Jarvis, PG, James GB and Landsberg JJ (1976) Coniferous forest. In Monteith JL ed. Vegetation and the atmosphere. Vol. 2 Case studies, pp 171-264. Acad. Press Inc.

Keyes MR and Grier CC (1981) Above- and below-ground net production in 40-year-old Douglas fir stands on low and high productivity sites. J. For. Res. 11, 599-605.

Landsberg JJ and James GB (1971) Wind profiles in plant canopies. J. appl. Ecol. 8, 729-741.

Monteith JL (1973) Principles of environmental physics. Edward Arnold

Persson H (1980) Fine-root dynamics in a Scots pine stand with and without near-optimum nutrient and water regimes. Acta Phytogeogr. Suec. 68, 101-110.

Pinker RT (1982) The diurnal asymmetry in the albedo of tropical forest vegetation. Forest Sci. 28, 297-304.

Pinker RT, Thompson OE and Eck TF (1980) The energy balance of a tropical evergreen forest. J. appl. Meteorol. 19, 1341-1349.

Reich PB and Borchert R (1982) Phenology and ecophysiology of the tropical tree, Tabebuia neo chrysantha (Bignoniaceae). Ecology 63, 294-299.

Robichaux RH and Pearcy RW (1980) Environmental characteristics, field water relations and photosynthetic responses of C_4 Hawaiian Euphorbia species from contrasting habitats. Oecologia (Berl.) 47, 99-105.

Rutter AJ, Kershaw KA, Robins PC and Morton AJ (1971) A predictive model of rainfall interception in forests I. Derivation of the model from observations in a plantation of Corsican pine. Agric. Meteorol.9, 367-383.

Rutter AJ, Morton AJ and Robins PC (1975) A predictive model of rainfall interception in forests II. Generalization of the model and comparison with observations in some coniferous and hardwood stands. J. appl. Ecol. 12, 367,380.

Stewart JB (1977) Evaporation from the wet canopy of a pine forest. Water Resources Res. 13, 915-921.

Thom AS (1975) Momentum, mass and heat exchange of plant communities. In Monteith JL, ed. Vegetation and the atmosphere. Vol.1 Principles, pp 57-109. Acad. Press Inc.

Thompson OE and Pinker RT (1975) Wind and temperature profile characteristics in a tropical evergreen forest in Thailand. Tellus 27, 562-573.

Whitehead D, Okali DUU and Fasehun FE (1981) Stomatal response to environmental variables in two tropical forest species during the dry season in Nigeria. J. appl. Ecol. 18, 571-587.

LIGHT ENVIRONMENTS OF TROPICAL FORESTS

R. L. CHAZDON (Cornell University; Ithaca, NY; USA)
N. FETCHER (Duke University; Durham, NC; USA)

"All light measurement involves some compromise between accuracy and possibility"

-- M.C. Anderson (1964)

ABSTRACT

Measurements of photosynthetically active radiation (PAR) in tropical forests are reviewed and discussed. Many studies of the light environment in tropical forests have emphasized the importance of sunflecks to the total light energy reaching the forest floor. Much of the spatial variation in total daily PAR in understory environments has been attributed to localized sunfleck activity, but levels of diffuse radiation were also found to vary spatially. Values of percent transmission differs greatly among forests throughout the world, ranging from 0.4-3.8, depending on forest structure and weather conditions. Total daily photosynthetic photon flux density (PPFD) in understory habitats is often below 1.0 mol m^{-2} d^{-1}, and measurements as low as 0.15 mol m^{-2} d^{-1} have been recorded.

Light environments in four rain forest habitats were compared over a 13-day period during wet and dry seasons in a lowland rainforest in Costa Rica. Daily total PPFD in the understory, 200 m^2 gap, and 400 m^2 gap were 1-2%, 9%, and 20-35%, respectively, of PPFD in a 5000 m^2 clearing. Daily total PPFD in the 400 m^2 gap was 2-3 times greater than in the 200 m^2 gap and 15-35% greater than in the understory. Seasonal differences in light availability occurred only in the clearing habitat, where decreased cloudiness during the dry season resulted in higher PPFDs.

In the understory, more than 70% of the daily 10-min averages of PPFD were below 10 μmol m^{-2} s^{-1}; on clear days, sunflecks contributed 55-77% of the total quantum flux. The 200 m^2 gap center received little direct radiation; total daily PPFD ranged from 1.52-3.07 mol m^{-2} d^{-1}. Total daily PPFDs in the 400 m^2 gap center ranged from 3.86-13.6 mol m^{-2} d^{-1} and 11% of the 10-min averages were above 1000 μmol m^{-2} s^{-1} on the day with maximum PPFD. In the clearing, daily total PPFDs ranged from 13.7-33.9 mol m^{-2} d^{-1}, and over 40% of the 10-min averages were above 1000 μmol m^{-2} s^{-1} on the maximum day.

Measurements of the spectral distribution of radiation from 300-1100 nm were made in 2 gaps, a clearing, during sunflecks, and in deep shade. The median red:far-red ratio in gaps and sunflecks was 0.86 and 0.99, respectively, as compared to 0.42 in the shade and 1.23 in the clearing. The median % of total energy in the photosynthetic waveband was 53% in the clearing, 39% in sunflecks and gaps, and 18% in the shade.

1. INTRODUCTION

Of the above-ground environmental factors affecting the life of tropical rain forest plants, light is undoubtedly the most variable, most complex, and least readily quantified. It is therefore not surprising that the study of light relations of rain forest species has offered a great challenge to plant ecologists for almost a century. Because light affects plant growth in a multitude of ways, it is not possible to consider simultaneously all aspects of the light environment in one investigation In this review, measurements of solar radiation in the photosynthetically active waveband, 400-700 nm (PAR), and the spectral distribution of light energy from 300-1100 nm are discussed.

The major factors affecting the light regime at any particular location within the forest are 1) position of the sun (solar angle, distance); 2) atmospheric conditions (cloudiness, particulates); and 3) vegetation structure (canopy closure, stratification, height of peripheral vegetation). These three sets of factors interact in complex ways, producing a heterogeneous pattern of light microclimates within the forest. Forest light regimes are as dynamic as the vegetation that they support.

In describing the light environment under dense forest cover one cannot escape from the problem of sampling a continuously variable phenomenon. Forest light conditions are variable on every scale, spatially and temporally. In moist tropical forests, where water generally does not limit plant growth, variability in the quality and quantity of available light strongly affects vegetation patterns. The relevance of light measurements to ecophysiological studies depends not only on adequate sampling of microsite variability, but also on an understanding of plant responses to variability on several scales. Descriptions of light environment are only the first step towards understanding the light relations of tropical plants.

2. REVIEW: LIGHT MEASUREMENTS IN TROPICAL FORESTS

The first quantitative measurements of rain forest light conditions were made using photoelectric cells with color filters, connected to a galvanometer. Due to changes in the spectral composition of direct and diffuse light, only relative intensities could be measured. Another limitation of this equipment was the low sensitivity of the photocells in spectral regions relevant to photosynthesis (Evans 1939).

During the 20's and 30's, the importance of sunflecks to understory vegetation was a subject of considerable controversy (Evans 1939, 1956; Whitmore, Wong 1959). Because of the difficulty involved in measurement of sunfleck conditions, many early workers chose to ignore this component of the light environment (Evans 1956). Questions regarding the distribution and relative intensity of sunflecks motivated the first studies of light conditions in tropical forests (Ashton 1958; Evans 1939, 1956, Evans et al. 1960; Whitmore, Wong 1959). Despite the lack of absolute units of measurement, these studies provided much quantitative information on the contribution of sunflecks to the total light energy reaching understory vegetation.

In a lowland rain forest in Southern Nigeria, Evans (1939) found that 5% of the total number of observations were twice the mean intensity of shade, and about 2% were over five times shade intensity. Observed sunflecks were of short duration and lower intensity than direct sunlight in the open. In a later study in another Nigerian forest he determined that sunflecks occupied 20-25% of the area on the forest floor during midday (Evans 1956). Sunflecks contributed approximately 80% of the total spectral energy when the sun was shining at midday. Allowing for cloudiness and non-sunfleck periods, sunflecks contributed about 70% of the total light energy on the forest floor during the 3-month measurement period. Whitmore and Wong (1959), using the area survey apparatus of Evans (1956) in a rainforest near Singapore, estimated that over an entire year

50% of the total incident energy was derived from sunflecks. During midday under sunny conditions, sunflecks contributed 65% of the total light energy. In a lowland rainforest in Ecuador, Grubb and Whitmore (1967) found that 70% of the light energy from 1000-1400 hrs in sunny conditions was contributed by sunflecks. The frequency of sunflecks was found to increase with height in a Brazilian rainforest from ground level to 17 m (Ashton 1958).

Diffuse radiation reaches the forest floor as direct sky light penetrating through small holes in the canopy and as transmitted light, which is altered in spectral composition as it filters through foliage layers. Evans (1956) found that only 17% of the light in the forest was unfiltered sky light, a value similar to the 13% value determined by Whitmore and Wong (1959). Evans, Whitmore, and Wong (1960) were unable to find any correlation between diffuse light readings and sunny or cloudy conditions, but Whitmore and Wong (1959) found that under sunny conditions the daylight factor (the ratio of diffuse light of forest interior to exterior) was twice the value measured under cloudy conditions. The relevance of the daylight factor (an instantaneous measure) to studies of forest light environments was questioned by Evans (1956) due to the inconstancy of atmospheric conditions and the dominant role of sunflecks.

Anderson (1964) introduced the term "diffuse site factor," the average percent of diffuse light (of specified wavelengths) transmitted through a vegetation canopy over a specified time period. Diffuse site factors are also highly variable spatially and temporally. Under cloudy conditions a reproducible pattern of light readings on the forest floor was obtained by Evans et al. (1960) and Grubb and Whitmore (1967). Calculation of diffuse site factors was later simplified through the use of hemispherical photographs of the canopy above a specified point on the forest floor (Anderson 1964, 1971).

Absolute measurements of solar radiation in tropical forests followed technological advances in radiation measurement. Studies in a Puerto Rican montane rainforest showed that less than 1% of solar energy in the visible spectrum penetrated to the forest floor (Odum et al. 1970). Periods of direct radiation over 0.36 cal cm^{-2} min^{-1} ranged from 0-16 min per day, with an average of 8 min over a 16-day period. The percent of solar energy (400-700 nm) reaching ground level in a Malaysian lowland rainforest ranged from 0.12-0.99%, with an average of 0.41% over a 12-day period (Yoda 1974). Daily integrated PAR on the forest floor averaged 1.05 cal cm^{-2} d^{-1}.

The advent of the quantum sensor brought about a new era in light measurement (Biggs et al. 1971). This sensor directly measures quantum flux in the waveband 400-700, has a rapid response time, and is small enough to reduce problems of spatial averaging. The first measurements of quantum flux in a rainforest setting were made over an 11-day period in Queensland, Australia (Björkman, Ludlow 1972). Measurements were made in a deeply-shaded microsite and therefore reflect the most extreme shade conditions that support plant life. On a clear day, the quantum flux on the forest floor was 0.15% of that above the canopy. Sunflecks contributed 62% of the total daily quantum flux, but only 10% of the total daily energy. Individual sunflecks were generally less than 2 minutes long, and momentarily reached intensities up to 150 times greater than diffuse shade light. Values of total daily quantum flux and canopy transmission are summarized in Table 1. The fraction of above-canopy PAR reaching

Table 1. A summary of understory light conditions in subtropical and tropical forests.

Location (latitude)	Mean percent transmission (range)	Mean total daily PPFD (range) (μmol m^{-2} s^{-1})	Max. % total PPFD due to sunflecks	Reference
Queensland, Australia (28 15'S)	0.5 (0.4-1.1)	0.21 (0.15-0.24)	62	Björkman, Ludlow (1972)
Oahu, Hawaii (21 30'N)	2.4 (1.5-3.8)	0.86 (0.55-1.38)	80	Pearcy (1983)
La Selva, Costa Rica (10 26'N)	ca. 1.0-2.0	0.32 (0.18-0.55)	55-77	Chazdon, Fetcher (1983)

the forest floor was greatest on overcast days, although absolute measurements were highest on overcast days (Björkman, Ludlow 1972). Overcast skies have greater zenith brightness than clear skies, allowing more diffuse light to penetrate through small holes at the top of the canopy (Anderson 1964).

Pearcy (1983) made extensive measurements of photosynthetic photon flux density (PPFD) in a Hawaiian evergreen subtropical forest understory using 18 quantum sensors and 3 microloggers for data acquisition. He also used hemispherical photographs (Anderson 1971) to characterize direct and diffuse site factors. Diffuse PPFD was between 10-30 μmol m^{-2} s^{-1}. Daily total quantum flux and % transmission values are summarized in Table 1.

Sunfleck dynamics were measured using a portable strip-chart recorder (Pearcy 1983). Sunflecks were very short-lived; two-thirds were less than 0.5 min in duration. Maximum flux density during sunfleck periods reached an average of 410 μmol m^{-2} s^{-1} and 90% of all sunflecks had peak PPFDs higher than 150 μmol m^{-2} s^{-1}. On relatively clear days, sunflecks contributed as much as 80% of total daily PAR; over the 5 week measurement period, an estimated 40% of the total quantum flux was contributed by sunflecks (Pearcy 1983).

The frequency of sunflecks (total minutes of sunflecks received) was highly variable from day-to-day at any given location and from site-to-site during any given day. Cloudiness appeared to be the greatest cause of day-to-day variation whereas canopy structure (as interpreted through hemispherical photographs) was throught to be the major determinant of site-to-site variation in sunfleck frequency (Pearcy 1983).

There are few measurements of the spectral distribution of light in rainforest habitats. Stoutjesdijk (1972) measured transmission spectra from 400-900 nm in a montane rainforest in Indonesia. Transmission values under sunny conditions in the waveband 400-700 were about 0.7%, and increased to 5% above 700 nm. Under overcast conditions, transmission values were higher at all wavelengths, which was attributed to higher relative intensities of light. Transmission from 400-700 nm during a sunfleck was about 1% (Stoutjesdijk 1972).

3. A COMPARATIVE STUDY OF LIGHT MICROENVIRONMENTS IN A COSTA RICAN LOWLAND RAINFOREST

Although some studies of rainforest light environments have included measurements in canopy gaps and large clearings (Schulz 1960; Odum et al 1970), most have focused on the understory environment (Bazzaz, Pickett 1980). Consequently, very little is known about how light regimes vary in different forest microhabitats. The study discussed here was undertaken to compare diurnal and seasonal patterns of photosynthetically active radiation in a range of forest habitats within a single rainforest location (Chazdon, Fetcher 1983).

3.1 Methods

The study was conducted at Finca La Selva, Costa Rica, a tropical premontane wet forest (Holdridge et al. 1971) at 37-100 m elevation. Four sites were selected for detailed measurements of photosynthetic photon flux density (PPFD): heavily shaded understory, a 200 m^2 gap, a 400 m^2 gap, and a 5000 m^2 clearing. Further details of study sites are described by Chazdon and Fetcher (1983).

Continuous measurements of PPFD were made using quantum sensors (Biggs et al. 1971) and a Campbell Scientific CR21 Micrologger. Data were reduced to 10-minute PPFD averages to facilitate between-site comparisons. Rapid changes in PPFD, as discussed by Gross and Chabot (1979) and Pearcy and Calkin (1983) were obscured by averaging. Measurements were made for 6 days during the 1981 wet season and for 7 days during the 1982 dry season, except for the 200 m^2 gap, which was measured only during the dry season. Two sensors were used at each site; sensors were supported horizontally within 3 m of each other, at a height of 0.7-2 m. Sensors were placed in the center of the gaps.

Spectral measurements in a range of microhabitats were made during the 1983 dry season using a LiCor 1800 spectral radiometer. Instantaneous measurements of quantum flux and total energy from 300-1100 nm were recorded at ground level between the hours of 0900-1500 during sunny and overcast conditions. Two parameters were used to compare microhabitats: 1) red:far-red ratios (PPFD between 668-672 nm divided by PPFD between 728-732 nm) and 2) the proportion of total solar energy (W m^{-2}) in the waveband 400-700 nm.

3.2. Results

3.2.1 The understory light environment

Daily average PPFD in the understory ranged from between 4.12 to 12.7 μmol m^{-2} s^{-1}. Instantaneous readings rarely exceeded 500 μmol m^{-2} s^{-1}, even during sunfleck periods. Diffuse PPFD during midday in shaded sites ranged from 5-20 μmol m^{-2} s^{-1} (Fig. 1; Chazdon, unpublished). Daily total PPFD and % transmission values are summarized in Table 1.

In the understory, 74-100% of the 10-min averages were below 10 μmol m^{-2} s^{-1} (Fig. 2). On clear days, sunflecks contributed 55-77% of the daily total PPFD, despite their low frequency during the day (Table 1; Chazdon, Fetcher 1983). Total daily PPFD was not significantly different between seasons, but intensities of diffuse radiation were higher during the wet season (Fig. 2d). The lack of seasonal differences in understory light levels may be due to the interaction of increased cloudiness with higher solar angles during the wet season measurement period (Chazdon, Fetcher 1982). Light conditions in the understory were not a simple function of incident PPFD. No correlation was found between total daily PPFD

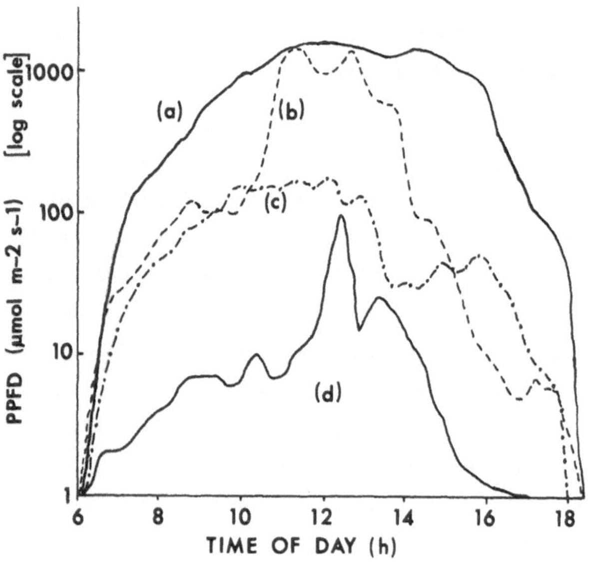

Fig. 1. Log plot of the daily pattern of photosynthetic photon flux density (PPFD; μmol m^{-2} s^{-1}) in tropical rain forest in Costa Rica for one clear, sunny day in a (a) 5000 m^2 clearing, (b) 400 m^2 gap, (c) 200 m^2 gap, and (d) understory. From Chazdon and Fetcher (1983).

measured in the center of a 400 m^2 gap and in the adjacent understory on the same days. This may be due to greater transmission of radiation through the forest on overcast days, but may also reflect the small number of sensors used. During both seasons, the understory had a higher coefficient of variation in total daily PPFD than the clearing. The understory environment also exhibited the greatest spatial variability; instantaneous readings from sensors less than 3 m apart occasionally differed by a factor of two or more (Chazdon, Fetcher 1983). This microscale variability was primarily due to localized sunfleck activity, rather than to diffuse light factors.

3.1.2. Gap and clearing light environments

During the dry season, daily total PPFD in the 400 m^2 gap center was, on average, 25 times

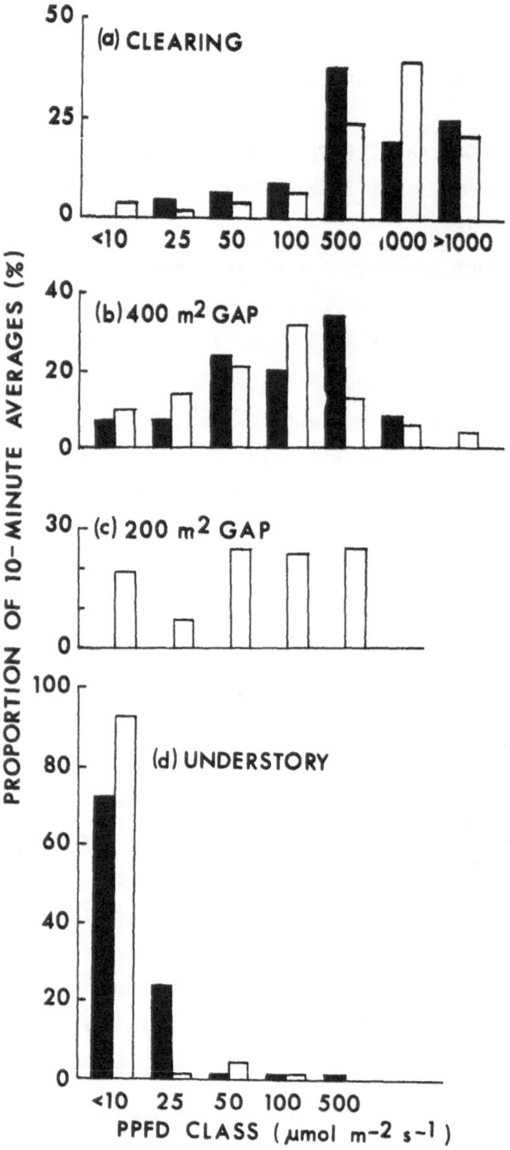

Fig. 2 The proportion of 10-min average readings in different photosynthetic photon flux density classes on the median day measured in tropical rain forest in Costa Rica during (■) wet and (□) dry seasons. Sites are (a) 5000 m^2 clearing, (9b) 400 m^2 gap, (c) 200 m^2 gap, and (d) understory. From Chazdon and Fetcher (1983).

greater than in the understory, and 2-3 times greater than in the 200 m^2 gap. During the wet season, values of daily total PPFD in the 400 m^2

gap center were, on average, 20 times greater than in the adjacent understory. Total daily PPFD in the 200 m^2 gap was 9 times that in the understory. The 400 m^2 and 200 m^2 gaps received 20-35% and 9% of the PPFD in the clearing, respectively (Chazdon, Fetcher 1983). Total daily PPFD in the 400 m^2 gap ranged from 3.86-13.6 mol m^{-2} d^{-1} with no significant difference between dry and wet seasons. This may be due to the relatively high solar angles during the wet season measurements in this gap. During the dry season, total daily PPFD in the 200 m^2 gap ranged from 1.52-3.07 mol m^{-2} d^{-1}. In this gap, only 32% of the 10-min averages were above 500 μmol m^{-2} s^{-1} on the maximum day measured, and no averages were above 1000 μmol m^{-2} s^{-1}. The 400 m^2 gap center received more direct light (Figs. 1,2b-c); 11% of the 10-min averages were above 1000 μmol m^{-2} s^{-1} on the day with maximum PPFD (Chazdon, Fetcher 1983). Gap edges and small gaps (less than 100 m^2) generally have higher intensities of diffuse light and greater frequency of sunflecks than heavily shaded understory habitats. Total daily PPFD in small gaps often exceeds 1.0 mol m^{-2} d^{-1} (Chazdon, unpublished). Spatial variability in PPFD within and between gap habitats is due to both direct and diffuse light factors.

Daily total PPFD in the clearing ranged from 13.7-33.9 mol m^{-2} s^{-1}. A significant difference was found between wet and dry seasons with 24% higher daily total PPFD during the dry season. Higher PPFDs during the dry season were due to longer periods of direct radiation, rather than to higher solar angles (Chazdon, Fetcher 1983). On a clear day, PPFD exceeded 1000 μmol m^{-2} s^{-1} for a 5-hour period (Fig. 1). Over 50% of the 10-min averages were above 1000 μmol m^{-2} s^{-1} on the maximum days measured, and few readings were below 100 μmol m^{-2} s^{-1}.

3.2.3. Microhabitat variation in light quality

Data on light quality are summarized in Table 2 (Lee, unpublished). Shade light is severely depleted in photosynthetically active wavelengths as evidenced by the low red:far-red ratios and the low proportion of solar energy in the 400-700 nm waveband. Measurements made in gap habitats and during sunflecks in understory habitats are very similar in both red:far-red ratios and % PAR.

Correlations between PPFD and spectral characteristics were highly significant ($r2 = 0.89$; $p < 0.01$). Both % PAR and red:far-red ratios can be predicted based on instantaneous PPFD measurements using regression equations.

The spectral distribution of quantum flux at 3 locations is illustrated in Figure 3. In the clearing, the instantaneous PPFD was 1737 μmol m^{-2} s^{-1}; this represented 45.4% of the total flux density under (μmol m^{-2} s^{-1}) under full sun quantum flux (Fig. 3a). The sunfleck had a PPFD of 743.2 μmol m^{-2} s^{-1}, which was 38.2% of total quantum flux (Fig. 3b). Only 5.6% of the total quantum flux was in the photosynthetic waveband in the deep shade location, where PPFD was as low as 5.1 μmol m^{-2} s^{-1} (Fig. 3c).

4. DISCUSSION AND SYNTHESIS

Rainforest light environments vary widely between forests, between habitats within a forest, and within a single habitat. Nevertheless, some generalizations can be made that are relevant to studies of tropical vegetation and the dynamics of tropical forests. Light conditions in the understory at La Selva are intermediate between those described from Queensland, Australia and Oahu, Hawaii (Table 1; Björkman, Ludlow 1972; Pearcy 1983). The low % transmission and PPFD values of Björkman and Ludlow (1972) are representative of the most extreme shade microsites in that forest, and

Table 2. Summary of spectral characteristics measured in a lowland tropical rainforest in Costa Rica*

	Clearing	Gaps	Understory sunflecks	Understory shade
# readings	6	26	11	31
PPFD (μmol m^{-2} s^{-1})				
range	422.2-2045	24.1-1304	24.2-743	1.5-34.5
median	837.5	70.8	155.5	7.33
Red:far-red ratio				
range	1.17-1.28	0.59-1.14	0.37-1.25	0.17-0.69
median	1.23	0.86	0.99	0.42
% total W m^{-2} between 400-700 nm				
range	49.3-54.8	26.7-52.6	22.4-59.8	6.4-31.7
median	52.9	39.5	38.9	17.6

*data courtesy of David Lee

should not be considered "typical" of understory conditions. The higher values of PPFD and % transmission presented by Pearcy (1983) reflect the relatively open canopy conditions in the 10-20 m high forest at Pahole Gulch. Extensive measurements of PPFD in many understory locations at La Selva indicate that the site described by Chazdon and Fetcher (1983) is quite typical of understory environments in this forest (Chazdon, unpublished). Measurements of PPFD in understory habitats on Barro Colorado Island, Panama, indicate that the range and modal values of daily total PPFD there are very similar to those measured at La Selva (Chazdon, unpublished). Without further study, it is difficult to attribute differences in understory light regimes to any one factor, although differences in forest structure presumably play a major role.

Studies dating back to Evans (1939) repeatedly illustrate the significant contribution of sunflecks to the total light available in understory habitats. Recent studies have established that understory species are able to respond rapidly to short-term changes in light level (Gross, Chabot 1979; Pearcy, Calkin 1983; Chazdon, unpublished). Patterns of sunflecks are not random and can be related to the growth of understory species through the use of hemispherical photographs (Pearcy 1983). Direct measurements of photosynthesis under sunfleck conditions have shown that sunflecks make an important contribution to daily photosynthesis (Björkman et al, 1972; Pearcy, Calkin 1983; Chazdon, unpublished). Using computer simulations of photosynthesis, it is possible to estimate the percentage of total daily carbon gain due to sunflecks, based on light measurements made in the field (Chazdon, unpublished).

Seasonal changes in light availability in the tropics can result from changes in cloud cover

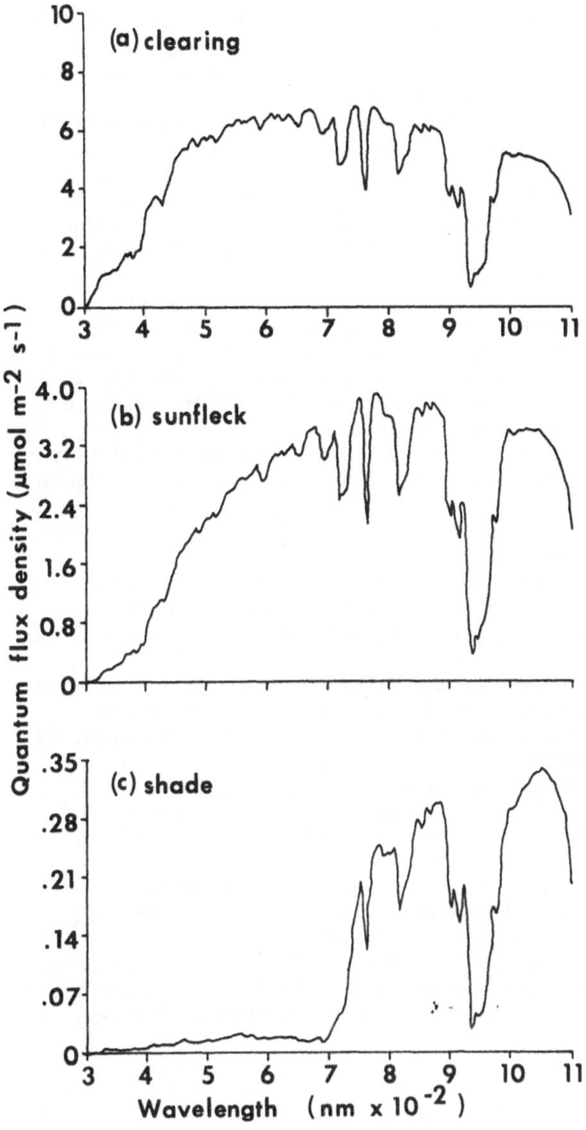

Fig. 3. The spectral distribution of quantum conditions in a (a) clearing, (b) sunfleck, and (c) shade of a tropical rain forest in Costa Rica.

and/or solar angle. Few generalizations can be made about season patterns in PPFD at La Selva, due to the high day-to-day variability within seasons and the high year-to-year variability in seasonality (Holdridge et al., 1972). In Hawaii, where annual changes in solar angle are greater than in more equatorial regions, seasonal changes in light availability are more pronounced (Pearcy 1983).

Light environment of canopy gaps are highly variable. Much of the variability in light conditions between gaps can be attributed to gap size (Barton, Redhead 1982; Chazdon, Fetcher 1983; Denslow 1980). Gap orientation, the height of neighboring vegetation, and solar angle are other important factors influencing gap light regimes (C. Canham, personal communication). The increased light availability in gaps is extremely important in the regeneration of most tropical tree species (Hartshorn 1980). Successional pioneer species in Surinam required light levels 10-20 times greater than in the understory to regenerate (Schulz 1960). Differences in red:far-red ratios between gaps and shaded understory (Table 2) may also affect germination (see Vasquez-Yanes & Segovia, this volume) and growth of gap-requiring species. In terms of availability of light and other resources, conditions are sufficiently heterogeneous within individual gaps that descriptions of "gap environments" have limited utility. For instance, average total daily PPFD in the center of a 150 m^2 gap was comparable to values measured at the edge of a large 600 m^2 gap during the same period (Barton, Redhead 1982).

The task of measuring variability in the light environment between and within forest habitats is certainly easier than relating this variability to the plants that dwell therein. The window of plant responses determines which aspects of environmental variability are relevant to a particular individual or species. For example, in heavily shaded understory habitats, where PPFD is often below 10 $\mu mol\ m^{-2}\ s^{-1}$, variations in PPFD of a magnitude of 10 $\mu mol\ m^{-2}\ s^{-1}$ may have a large effect on carbon gain due to the fact that many understory species show a linear photosynthetic response in this PPFD range (Chazdon, unpublished). Small changes of this nature in a clearing are not as likely to affect carbon gain in early successional species. Ultimately, studies of light relations of tropical species must be considered in a broader context, incorporating other important biotic and abiotic environmental factors.

REFERENCES

Anderson MC (1964) Light relations of terrestrial plant communities, Biol. Rev. 39, 425-486.

Anderson MC (1971) Radiation and crop structure. In Sestak Z, Catsky J, and Jarvis PG, eds. Plant photosynthetic production, manual of methods, pp. 412-466. Junk, The Hague.

Ashton PS (1958) Light intensity measurements in rain forest near Santarem, Brazil, Journal of Ecology 46, 65-70.

Barton A and Redhead S (1982) The relationship between treefall gap size and light regime. OTS Coursebook 82-1: tropical biology, an ecological approach, pp. 450-452, OTS, Durham, NC.

Bazzaz FA and Pickett STA (1980) Physiological ecology of tropical succession: a comparative review, Ann. Rev. Ecol. Syst. 11, 287-310.

Biggs WW, Edison AR, Easton JD, Brown KW, Maranville JW, and Clegg MC (1971) Photosynthesis light sensor and meter, Ecology 52, 125-131.

Björkman O and Ludlow MM (1972) Characterization of the light climate on the floor of a Queensland rainforest, Carnegie Inst. Wash. Yearbook 71, 85-94.

Björkman O, Ludlow MM, and Morrow PA (1972) Photosynthetic performance of two rainforest species in their native habitat and analysis of their gas exchange, Carnegie Inst. Wash. Yearbook 71, 94-102.

Chazdon RL and Fetcher N (1983) Photosynthetic light environments in a lowland tropical rain forest in Costa Rica, Journal of Ecology, in press.

Denslow JS (1980) Gap partitioning among tropical rainforest trees, Biotropica, 12 (supplement), 47-55.

Evans GC (1939) Ecological studies on the rain forest of southern Nigeria. II. The atmospheric environmental conditions, Journal of Ecology, 27, 436-482.

Evans GC (1956) An area survey method of investigating the distribution of light intensity in woodlands, with particular reference to sunflecks, Journal of Ecology, 44, 391-428.

Evans GC, Whitmore TC, and Wong, TK (1960) The distribution of light reaching the ground vegetation in a tropical rainforest, Journal of Ecology, 48, 193-204.

Gross LF and Chabot BF (1979) Time course of photosynthetic response to changes in incident light energy, Plant physiology 63, 84-93.

Grubb PJ and Whitmore TC (1967) A comparison of montane and lowland rainforest in Ecuador. III. The light reaching the ground vegetation, Journal of Ecology 55, 33-57.

Hartshorn GS (1980) Neotropical forest dynamics, Biotropica, 12 (supplement), 23-30.

Holdridge LR, Grenke WC, Hatheway WH, Liang T, and Tosi JA Jr (1971) Forest environments in tropical life zones; a pilot study. Pergamon Press, San Francisco.

Odum HT, Drewry G, and Kline JR (1970) Climate at El Verde, 1963-1966. In Odum HT and Pigeon RF, eds. A tropical rain forest, pp. B347-418. U.S. Atomic Energy Commission, Oak Ridge, Tennessee.

Pearcy RW (1983) The light environment and growth of C3 and C4 tree species in the understory of a Hawaiian forest, Oecologia, 58, 19-25.

Pearcy RW and Calkin H (1983) Carbon dioxide exchange of C3 and C4 tree species in the understory of a Hawaiian forest, Oecologia, 58, 26-32.

Schulz JP (1960) Ecological studies on rainforest in northern Surinam, Verh. K. Ned. Akad. Wet. Afd. natuurk'd. Tweede Reeks 53, 1-367.

Whitmore TC and Wong TK (1959) Patterns of sunfleck and shade in tropical rain forest, Malayan Forester 22, 50-62.

Yoda K (1974) Three dimensional distribution of light intensity in a tropical rain forest of West Malaysia, Japanese Journal of Ecology, 24, 247-254.

ECOPHYSIOLOGY OF SEED GERMINATION IN THE TROPICAL HUMID FORESTS OF THE WORLD: A REVIEW.

C. VAZQUEZ-YANES AND A. OROZCO SEGOVIA
(Departmento de Botánica, Instituto de Biología, UNAM, Apartado Postal 70-233, 04510 México, D.F.)

ABSTRACT

Most forest tree seeds from the humid tropics germinate soon after dispersal forming carpets of semi-dormant seedlings. Delayed germination and enforced dormancy is frequent among light-gap adapted species, some emergent trees and species with hard-coated seeds. Dormancy is more important as an adaptative trait in marked seasonal forests.

The seed bank of the forest soils studied are composed mainly of fast growing pioneer tree seeds and some weed seeds. The dormancy of pioneer trees and other light-gap colonizers is often enforced by light and temperature conditions of forest soil and finishes when a gap is formed.

Much research on seed ecophysiology is needed in order to understand the seed germination regulating mechanisms in tropical forests.

1. INTRODUCTION

Most of the available information on the germination of seeds from tropical rain forest plants consist of data on the time and percentage of germination and the duration of seed longevity mainly from plants of economical value. Very few of the published papers relate the physiology of germination processes with the natural environment where it takes place. The impressive work done by Malaysian scientists on tree seeds and information from other sites in Asia, Africa, and America are utilized here to establish the general patterns of seed germination processes of tropical forest plants. A previous review by Bazzaz (1980) also provided information on this topic.

2. GERMINATION OF THE SEEDS OF FOREST TREES

The basic works on the tropical rain forests (Richard, 1952; Whitmore, 1975) provide a description of the seeds of the mature forests; these indicate that the seeds are big, with a high humidity content and rapid germination, and with short or nonexistent dormancy. The seeds are difficult to store and longevity is, in general, short. But there are also some long lasting seeds with hard coats. Mature forest seeds are not sensitive to light and have a thermal optimum of germination between 20° to 30°.

A detailed study of the available information shows that although a high percentage of the species of the trees from the humid forests have the characteristics described above, there are also other very different patterns of germination and longevity, depending, in great measure, on the degree of seasonality of the forest, and on the peculiar establishment behavior of the tree species. The massive and simultaneous germination of the seeds to produce a continuous and homogeneous carpet of seedlings is frequent among the canopy species, but understory species, emergents and light-gap colonizers might not conform to this pattern.

The seeds from different species of wet forest trees may be morphologically and physiologically very variable. An example of this is shown by Defresne (1982) who compared in great detail the seeds from Cedrela odorata and Simphonia globulifera; the first has a low humidity content, low stored reserves but their longevity may be maintained for relatively long periods in appropriate conditions of storage. Germination takes place quickly when humidity is provided. Simphonia, on the other hand, produces seeds with a high humidity content, with large stored reserves and they are very difficult to keep alive in artificial conditions. They show very low germination percentages unless located on the floor of the forest under very deep shade and high environmental humidity. The ensemble of characteristics of Cedrela are linked with germination in light gaps after dispersal and Simphonia shows the typical behavior of a seedling carpet-forming species. Another study showing examples of these differences is the one by Maury-Lechon, et al. (1981), who worked on two species of the Dipterocarpaceae from Malaysia. Ng (1973, 1978, 1980) studied the germination pattern of more than 330 tree species from Malaysian forests. He found that 65% of the woody flora shows rapid germination, with the seeds completing the process during the first twelve weeks after dispersal. No viable seeds remain after that time. Twenty-eight percent of the flora studied had germination during the first twelve weeks after dispersal but it was not completed during that time. Those species with delayed germination included 7% of the woody flora, and in this case germination started some time after twelve weeks from dispersal. Seeds of this type can be stored longer more successfully than the fast and intermediate germinating seeds (Fig. 1).

Ng's interpretation of his results is that rapid germination gives plants the advantage of escaping seed predation to produce seedlings which have greater chance to survive herbivory. Intermediate and retarded germination may favor distance dispersal and the arrival, in time or space, to microenvironments which are favorable for germination and establishment of the plant species that show this kind of seed behavior. The selective pressures which originated the three types of seed behavior are different. Among the Dipterocarpaceae germination is always fast, but other families of trees may contain species with different behaviors, for example Leguminoseae have species with hard-coated seeds and retarded germination, as well as species with soft-coated seeds and therefore fast germination.

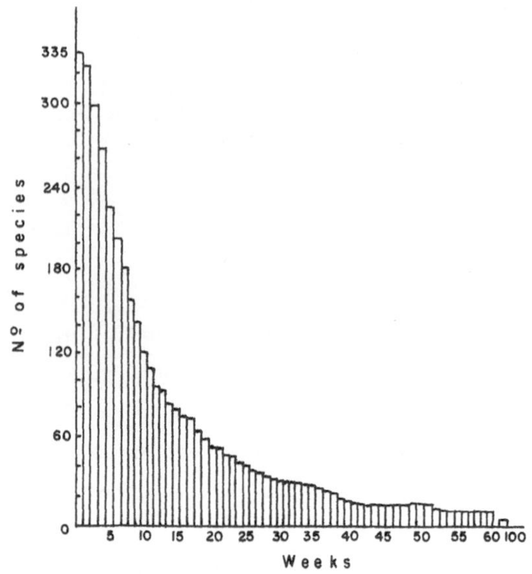

FIGURE 1. The number of species (y) with viable seeds after (x) weeks under conditions normally suitable for germination (Ng, 1980)

Trees producing rapid or retarded germinating seeds may have a very complex origin as shown by Gupta and Pattanath (1975) in their work on teak seeds. They found that seeds coming from different plants and populations growing in different areas and climates show different degrees of germination velocity depending on the environment around the parent trees, the presence of an inhibitor in the mesocarp of the fruit, after-ripening conditions, and the seed positions on the soil or under its surface.

Rapid germination as described by Ng is the most characteristic trait of tropical rain forest trees. Many species show this behavior not only in Asia but also in America and Africa. In Brazil, Macedo (1977) studied 37 woody species from the Amazonian campinna. She found that germination took place in not more than two weeks under artificial conditions. In the field, germination was retarded a little in comparison with laboratory results. In Africa, Alexandre (1980) working in the Ivory Coast found that 79% of 61 species showed rapid germination, and the remaining 21% had retarded germination. Turreanthus africana (Alexandre, 1977) is a typical example of a species with fast germinating seeds, but only if they lie on the forest litter. If they do not, germination is difficult and the seed soon dies.

Gilbert (1952) worked with 10 tree species of the Belgian Congo (now Zaire) among which nine showed rapid germination and only one retarded germination. Other studies describing rapid germinating seeds are those from Alencar and Magalhaes (1979), Johnson and Morales (1972), Tang (1971), Yap (1980), Alencar (1981), Chai (1973).

Maury-Lechon, et. al. (1980) described in detail the germination of Simphonia globulifera, which is a model type species with fast germinating seeds. The optimum germination temperature is between 25° and 30°C. There is no light sensitivity. The humidity content surpasses 40% but there are wide differences among the seeds coming from different trees. Temperatures near 40°C are lethal. Dehydration takes place very often if environmental humidity drops under 90%, and the only way to increase the longevity of the seed is to keep them in a humid atmosphere at 15°C.

An extreme case of fast germination is vivipary where germination takes place in the fruits while they are still attached to the branches of the trees. Vivipary is a typical trait of mangrove species, but it is also frequent among rain forest plants and has been described for Pithecolobium racemosum (Leite, Rankin, 1981), Pouteria ramiflora and Magonia pubescens (Laboriau, et. al., 1964) and Dryobalanops aromatica (Sasaki, et. al., 1979). In the case of Pithecolobium 80% of the fruits show germinating seeds on the parent tree, and the remaining ones germinate as soon as they reach the soil. In this case, vivipary is interpreted as a predation escape mechanism. Fruits and seeds are eaten by birds when they are still on the tree. When the seeds that fall have pre-germinated and start to grow immediately, they escape a second wave of predation. The phenomena of vivipary is also frequent in the varzea (or seasonally flooded forest) of Brazil (Leite, Rankin, 1981) and in this case it is linked with efficient water dispersal, as in the case of mangroves.

In every forest there are a certain number of species that have retarded germinating seeds and this trait is often linked to the presence of hard seed coats. Hard-coated seeds have low humidity contents and long longevity; they are also resistant to high temperatures and to the attack of fungi and insects. Many can withstand a passage through the digestive tract of warm-blooded animals. Several very typical rain forest genera like Dialium and Ceiba (Ponce de

Leon, 1982), Koompasia, Intsia and Sindora (Sasaki, 1980a; Ng, 1977) and Podocarpus (Chamshama, Downs, 1982) have hard-coated seeds. The seeds from these plants are not eaten by animals and thus scarified. There is no information on the nature of the mechanisms that scarify seeds in the soil. Probably the coat is slowly destroyed by microorganisms until water can come into the seed, triggering germination. Other triggering agents are in some cases the thermal fluctuation which occurs in bare soils in gaps (Váquez-Yánes, Orozco-Segovia, 1982). This mechanism applies mainly to pioneer trees like species from such genera as Ochroma and Heliocarpus which will be discussed later.

It seems that retarded germination is frequent among many pioneer emergent trees but we do not have enough data to support this hypothesis.

Some seeds showing retarded germination do not have hard coats, as the case Euterpe globosa (Bannister, 1970) which is an understory palm tree. Its seeds germinate from three to six months after seeding, in spite of having a high humidity content (72%). In this case delayed germination might be possible through the presence of toxic secondary compounds or other kinds of defenses in the seeds which prevent their predation. Almost nothing is known on the relationship between toxicity of the seeds and retarded germination in the forest. Many species have differential dormancy. Ng (1978, 1980) found that most of the species studied by him, even those with rapid germination, show a certain degree of differential dormancy. That is, the seeds start to germinate at different times during the germination period of the particular seed population. The delay between seed germinating within a cohort can be from several hours to several months. Part of this difference in time might be due to genetic differences between the seeds or to environmental differences produced by a certain degree of heterogeneity in the area where germination is taking place.

The delay between the arrival of the seeds to the soil and germination is also closely related to the degree of seasonality of the forest. In forests where there is an alternation of well defined dry and wet seasons, retarded germination is often linked to the survival of the seeds through the dry season.

3. SEASONALITY AND DORMANCY

The seed germination study at the community level in the Barro Colorado seasonal forest by Garwood (1982, 1983) showed several patterns of germination which are closely related with the interaction between fruiting periods and the seasonality and distribution of the species in relation to the gap phase dynamics of the community. Garwood indicates that of the 185 species of germinating seeds identified in the forest, 75% germinated during the first two months of the rainy season. That group included the majority of the species which set fruit at the end of the previous rainy season and during the dry season. All those species show some degree of dormancy in the soil. Forty percent have rapid germination because they are dispersed and germinated during the same rainy season. In this case germination occurs in periods from less than two weeks to 16 weeks from dispersal. Very fast germination takes place in the light gaps at the beginning of the rainy season where there is strong competition for the colonization of open spaces. A carpet of seedlings is formed in the rest of the community during all the rainy season. In this particular forest the dormancy of the seeds may last from 2 to 370 days and half of the species have a dormancy longer than four weeks.

In seasonal flooded forests, the few data available indicate the existence of very peculiar germination processes like the vivipary previously described. Couhtihno and Struffaldi (1971) describe the case of Parkia auriculata from the "varzea." These trees produce hard-coated seeds with retarded germination but as soon as the coat is broken, germination takes place very fast during the dry season. The emerging seedling are able to survive submerged in the water for as long as 7 months.

4. GERMINATION ACROSS ECOTONES

Another area of seed research in tropical forests is in the ecotone between the forest and the savannas, caatingas and cerrados. The main objectives of some of these studies have been to verify if germination of the plants from each of the ecotonal communities takes place in the other one, in order to understand the processes that maintain the well defined limits between the communities. The work of Ponce de Leon (1982) in the Ivory Coast indicates that many of the seeds and seedlings of the forest trees do not survive when they are exposed to the high temperature fluctuations, lower humidity, and occasional fires that are typical of the nearby savanna environment. These prevent the establishment of the trees in the savanna, but when the fire is excluded from the herbaceous community, some of the resistant seed species can become established, therefore initiating the colonization of the savanna by the forest. When the microclimatic conditions are changed by the presence of the woody colonizers the remaining forest plants can germinate under their shade. The destructive effect of the heat from fires on most of the seeds from the forest plants has been also demonstrated experimentally by Brinkman and Vieira (1971) in Costa Rica.

In the American savannas, caatingas and cerrados the effect of the grassland microclimatic conditions on forest seed and seedlings has been described (Macedo, Prance, 1978; Medina, 1982).

5. SEED LONGEVITY

As a consequence of the particular characteristics of tropical rain forest seeds, their longevity is normally very short and they are very difficult to store for reforestation purposes. Few species can survive dehydration and cold storage and therefore seed banks in the tropics limit their stocks to a few species and in many cases to exotic plants that can grow in tropical rain forest areas. Moreno-Casasola (1976) compared the available data on artificially-determined longevity of seeds from temperature and tropical forests (Fig. 2).

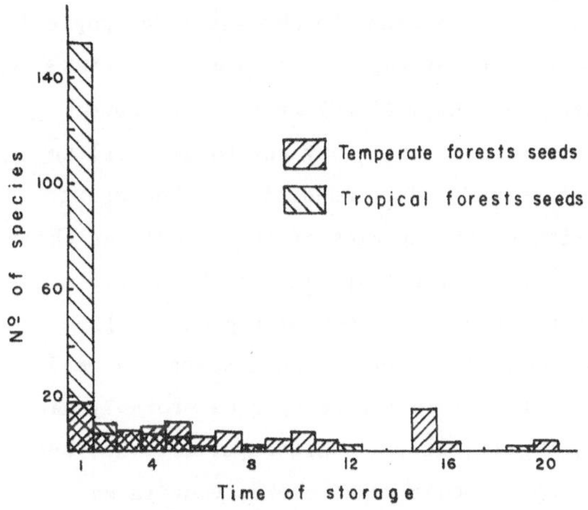

FIGURE 2. Mean longevity of seeds from tropical and temperate forest trees under storage conditions (Moreno-Casasola, 1976).

Tropical forest seeds show a mean longevity of 10.7 months compared to 88.8 months for temperate seed species.

There have been several large studies on storability of tropical tree seeds like that of Marrero (1949) in Puerto Rico, Maury-Lechon "et al." (1981), Sasaki (1980a, 1980b), Tang and Tamari (1973), Tang (1971), Yap (1981) and others in Malaysia and finally Vega, et al. (1981), in Mexico. All of these show that a high proportion of the seed species studied cannot be stored successfully unless they have a hard coat. Ng (1974) suggests a strategy for sustaining supplies of seeds for reforestation in Malaysia based on the maintenance of an intact forest as germplasm areas where a careful study of the phenology of the trees would permit the recollection of fresh seeds for nurseries when they are needed. This would work as a living seed bank.

6. COMPOSITION OF THE SEED BANKS IN THE FOREST SOIL

Several experiments have been performed to determine the composition of the seed bank (or dormant seeds present in the soil) in tropical humid forests of the world. Symington (1933) in Malaysia and Keay (1960) in Nigeria found similar results when they put forest soil to germinate under open conditions. The great majority of the species of tree seedlings which germinated were not species of the mature forest but nomadic pioneer species typical of light gaps and early secondary successions as well as a few mature forest tree species probably having delayed germination. This experiment has been repeated by other researchers; Guevara and Gomez-Pompa (1972), in Veracruz, found a great number of weed seeds in the forest soil probably due to the high degree of disturbance of the area near the forest where it was done.

In Asia the predominant species in the seed bank according to Liew (1973) in Malaysia and Cheke "et al." (1979) in Thailand are pioneer trees from the genera Macaranga, Mallotus, Trema, Anthocephalus and others. The remaining species constitute a very small proportion of the total number of seedlings germinated from the soil samples.

In Africa (Ghana) Hall and Swaine (1980) also found a large proportion of seeds from pioneer trees like Musanga and Trema but also weed seeds and a few seeds from forest trees, like Albizia, Celtis, Ceiba, Terminalia. Some of these have hard-coated seeds.

In America the predominant genus found in all the studies on seed banks is Cecropia (Blum, 1968; Bell, 1970; Kellman, 1974; Prevost, 1981). Other important genera in American soil seed banks are Laetia, Palicourea, Trema, Piper, Didymopanax and several Melastomataceae.

It is possible that the sampling technique employed in all the studies has favored the detection of secondary species as opposed to primary ones, nevertheless the results strongly suggest that there are fundamental differences in the characteristics of the dormancy and longevity in the soil between forest species and many of the pioneer light-gap colonizers.

The seed bank studies show very low amounts of seeds with delayed germination. This might be due to the fact that the conditions of germination provided are not appropriate for those species or to a real low number of these seeds in the soil. Ng (1978) mentioned that it is always possible to find viable seeds from several hard-coated legumes like Anisophylea, Barringtonia, Intsia, and Sindora on the soil. The small amount of mature forest tree seeds in the soil bank may be due to the rapid germination, the strong predation and fast decomposition of the big seeds and finally to the fact that germination conditions in an exposed environment might kill many seeds instead of promoting germination.

7. GERMINATION AND LONGEVITY OF THE SEEDS FROM LIGHT GAP COLONIZERS

The presence of abundant seeds of pioneer trees

in the soil bank demonstrate that there are fundamental differences in dormancy and germination triggering factors between the pioneer trees defined in a previous paper (Vázquez-Yánes, 1980a) and the rest of the woody flora of the forest. The seeds from the pioneer short-lived trees are small, with a low humidity content. They are produced in great numbers and often continuously during the year and are dispersed widely by several agents. Longevity in the soil of these seeds seems to be longer than in the other woody plants and germination is retarded by an imposed or enforced dormancy caused by inappropriate conditions for establishment (Budowski, 1965). Germination-triggering factors are related with light-gap microclimatic conditions such as light quality and temperature fluctuations of the soil. The viability in the soil of the seeds of some of the species has been determined. Holhiutzjen and Boerboom (1982) kept viable seeds from two Cecropia species buried in the soil for more than 5 years. Some seeds of Ochroma lagopus from herbarium specimens have been germinated after 49 years (Moreno-Casasola, 1976).

There are descriptions of essentially two kinds of enforced dormancy types among pioneer tree seeds, photoblastic dormancy and temperature-regulated dormancy. A well studied example of photoblastic dormancy exists for Cecropia obtusifolia (Vázquez-Yánes, 1980b; Vázquez-Yánes, Smith, 1982) and Cecropia glaziovi (Valio, Joly, 1979). In the first case the phytochrome of the seeds acts as an environmental sensor which detects the R/FR ratio of the light. When this ratio increases, due to the destruction of the canopy, germination of the seeds is triggered. The long periods of radiation with a high R/FR ratio required for germination has been related to germination only in large light gaps. The enforcement of dormancy occurs even under the possible triggering effects of the normal sunflecks of the forest (Fig. 3).

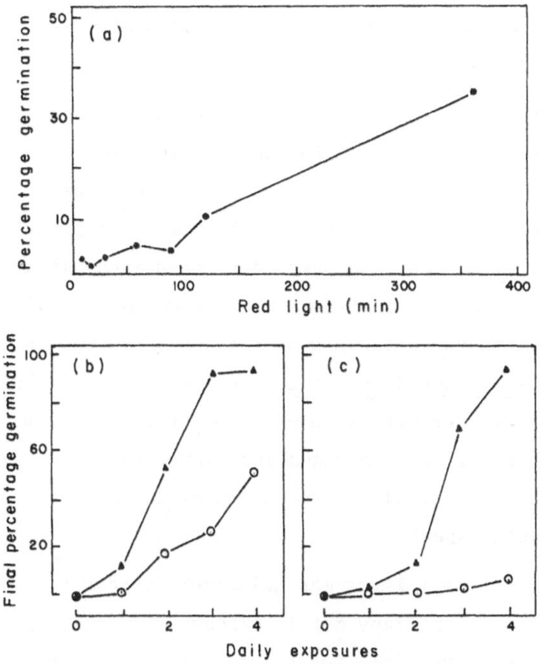

FIGURE 3. Germination in Cecropia obtusifolia seeds. (a) is the relationship between the final percentage germination and period of a single exposure to red light. (b-c) final percentage (after 1 month) of seed samples given up to four repeated exposures to red light (either 10 or 120 min duration) at 24 h intervals (▲) 120 min; (0) 10 min. (Vázquez-Yánes and Smith, 1982).

Photoregulated germination is frequent in many pioneer plants from the genus Macaranga (Ng, 1980), Musanga (Aubreville, 1947), Trema (Vázquez-Yánes, 1977, Alexandre, 1978a), several species of the genus Piper (Vázquez-Yánes, 1976a) and Chlorophora excelsa (Fandisi, Olofinbora, 1975).

Germination favored by temperature fluctuations is found among several light-gap colonizers like Didymopanax (Aubreville, Leroy, 1970), Ochroma

Vázquez-Yánes, 1976b and Heliocarpus (Vázquez-Yánes, Orozco-Segovia, 1982). This type of dormancy is often related to the presence of a waterproof hard coat that becomes permeable under the effect of the heat produced on the soil surface (Vázquez-Yánes, Pérez-García, 1976).

Both mechanisms seem to be efficient for the persistence of seeds in the soil at least between the fruiting seasons of the species, which is an important survival trait for light-gap colonizers. In some cases seed accumulation in the soil is indicated when large amounts of seeds in the soil have been found as in the case of Cecropia and Macaranga. Cheke, et al. (1979) compared the seed rain with the stock of seeds in the soil and they concluded that in fact there is accumulation of seeds of the most abundant species.

Seeds of many light-gap colonizers appear to arrive to the gaps mainly through dispersal of the seeds when the formation of the gap takes place and, because of this, some authors are skeptical about the importance of dormant seeds in the soil and of dormancy in general as a factor determining the colonization of open spaces in the forest. However, the sophistication of certain dormancy mechanisms, as in Cecropia, and the abundance of pioneer tree seeds in the soil are convincing evidence of the adaptive value of dormancy for certain if not all the species which germinate in light gaps. Certain species like Cordia alliodora (Johnson, Morales, 1972) and Pithecolobium racemosum (Laite, Rankin, 1981) germinate and become established in light gaps but do not have any specialized dormancy. These might represent cases of colonization through dispersal.

More detailed studies on seed rain, seed banks, relations between enforced dormancy and microsite environment and seed longevity in the soil are needed in order to understand the relative importance of dispersal and dormancy in light-gap colonization from the point of view of the whole set of species adapted to establish in that particular environment.

The detailed physiological study of the enforced dormancy mechanisms in pioneer species is a promising area of research with tropical rain forest seeds because of the very sophisticated triggering systems already discovered among some of these plants.

8. SEED GERMINATION IN OTHER KINDS OF FOREST PLANTS

The available information on the seeds and germination of forest weeds, shrubs, vines, hemi-epiphytes, epiphytes and parasitic plants is very scarce with the exception of the orchids (see Arditti, 1967). The herbaceous plants of the understory have not been the object of any known seed physiological research. Seeds of shrubs of the understory like Hybanthus prunifolius (Augspurger, 1979) and Palicourea riparia (Bell, 1970; Lebron, 1979) seem to germinate freely and quickly in the forest soil when humidity is available after their arrival. An enforced dormancy occurs during dry periods.

Most understory shrubs must also have rapid germination after dispersal because of the possibilities of seed predation and parasitism. In some cases the understory plants take advantage of the light gaps as does Piper hispidum (Vázquez-Yánes, Orozco-Segovia, 1982). In this case the seeds are photoblastic and establishment occurs in light gaps but the adult plants can survive in the deep shade of the forest. Germination can occur at low R/FR ratios, suggesting that it may take place in small gaps or under sparse canopies, but in the darkness the seeds remain dormant for a long time. The available data on the germination of the Piper species from Los Tuxtlas, Veracruz, show how each species has a different degree of adaptation to light gap establishment which is reflected in the germination properties of its seeds (Vázquez-Yánes, 1976b).

Among climbers and vines there are also light gap adapted species as well as those which germinate in the shade of the forest. There is very little information on the germination of the first type but it appears that their seeds are frequent in the seed bank (Guevara, Gómez-Pompa, 1972). In the case of Smilax germination is triggered by light on bare soil (Quarterman, 1970). The best studied vines are the rattans from Asia. Manokaran (1978) worked with 33 species from 6 genera and found that 10 species had rapid germination, 8 intermediate germination and 11 delayed germination. In four cases there were wide differences in germinability between seeds from different individuals of the same species. In Calamus mannan, from 15 parent plants, 10 produced seeds with intermediate germination and 5 with fast germination. The germination percentage varied between 0.2 and 83 percent. This indicates wide genetic or physiological variability among seeds from different parent plants. Germination only takes place in the shade and high humidity conditions of the forest because the seeds must maintain their high humidity content to remain viable. Mori, et. al. (1980), working with Calamus mannan showed that the removal of the seed sarcotesta greatly enhanced germination.

The strangler figs of the genus Ficus have adaptations to favor germination on other trees or rocks. The ripe achenes have an outer layer of cells that form a mucilage with water. This substance helps to stick the seeds on the surfaces but it has to be digested by microorganisms before germination can take place. When the seeds are sown on sterile media they do not germinate (Ramirez, 1976).

The seeds of Ficus obtusifolia (Orozco-Segovia, et. al., unpublished) collected directly from fruits do not germinate, but those from the droppings of the bats (Artibeus jamaicensis) have a rapid and abundant germination on an agar surface. The seeds keep their sticky surface even after passing through the digestive tract of the bats.

Among the epiphytes, the most specialized germination mechanisms have been described for the orchids. In some of these plants the invasion of mycorrhizal fungi is required to start germination. There is an extensive review of the subject by Arditti (1967).

Parasites like Psittacanthus, Phthirusa and Loranthus germinate on woody surfaces. Their germination is different from other angiosperms in relation to the morphology of the emerging embryo which promotes the rapid invasion of the host tissues. Bird-dispersed seeds of many species also have a sticky surface and low humidity requirements for germination (Wellman, 1964).

9. BIOTIC INTERACTIONS AND SEED GERMINATION

There are many reports of the germination triggering effects of transit of seeds through the digestive tract of animals. The seeds that show this behavior normally have a hard coat or an external germination inhibitor on the seed surface which is eliminated when subjected to the digestion process. Many forest animals act as dispersal agents and germination enhancers. Some of the more dramatic examples have been described by Alexandre (1978b). He found that the elephant droppings in the Ivory Coast serve as a nursery for many forest seedlings. The seeds from certain species pass through the elephants. The seeds of some fruits remained the same in germination capacity and others were destroyed. The very hard coated seeds of Enterolobium cyclocarpon require drastic scarification to enhance germination. The main factors promoting the breaking of the seed coat seems to be the action of soil microorganisms and rodent gnawing, more than the effect of digestion by large animals (Janzen, 1982).

An interesting case of germination enhancement by microorganisms was reported by Edmisten (1970) in Puerto Rico. The seeds from Ormosia krugii have enhanced germination when they are contaminated with bacteria of the genus Rhizobium which in a later stage of seedling development forms nodules in the roots.

Ants also play a role in germination as was demonstrated by Horvitz (1981) for Calathea microcephala in Mexico. Germination of the seeds of this plant improves when the aril is removed by the ants thus breaking the dormancy of the seeds. A similar case has been reported for the Brazilian caatinga plant Protium heptaphyllum (Macedo, 1977).

Several Malaysian authors (for example, Sasaki, 1980b) note that seeds collected directly from mature fruits still attached to the parent plant, normally have a much higher germination potential than those seeds of the same species collected from the forest floor where the seeds undergo attack by parasites and predators which reduce their viability. Most of the biotic interactions taking place on the soil have negative effects on the seeds. The viability data obtained with seeds collected on the trees and on the soil are very different, suggesting that the biotic interactions are the selective environmental factors which promote rapid germination as the typical trait of most humid forest seeds.

10. METHODOLOGICAL DIFFICULTIES WHEN WORKING WITH TROPICAL FOREST SEEDS

Most studies on tropical humid forest seed germination were not designed with the purpose of establishing a relationship between the characteristics of the natural environment, the appropriate conditions for seedling growth, and the physiology of the regulation of the germination process. Few experiments accurately reproduce the natural conditions where these processes take place and many of them lack an adequate control of the experimental materials and the conditions of the experiments. As an example of this many experiments lack control of environmental moisture which appears to be an extremely important factor as mentioned by Tang (1971), Tang and Tamari (1973), Mori, et. al. (1980), Sasaki (1980b) and Yap (1981). The survival of Dipterocarpaceous seeds require an environmental moisture near 95% which is difficult to provide unless the seeds are lying on the forest litter or are carefully protected from the sun and the wind by mean of covers which simultaneously permit ventilation. Light is also important because many pioneer seed species are photoblastic and extremely sensitive to the R/FR ratio of the light (Vázquez-Yánes, Smith 1982). Therefore the light microclimate where germination experiments are conducted should be controlled carefully. With respect to these problems it should be mentioned that pioneer tree seeds may have rapid germination in open nurseries but this characteristic does not mean that there is rapid germination in natural conditions because enforced dormancy may exist under forest conditions. There must be an adequate balance between experiments conducted under natural conditions and experiments under controlled conditions to understand the mechanisms controlling seed germination.

Careful observations of the conditions where germination is taking place under natural conditions must precede the design of experiments because there are many factors that must be taken into account like the position of the seeds on or in the soil (Sasaki, Ng, 1981), the distance between them, how deep they are buried (Campbell, 1979), and the effect of the dispersal agents on the seeds. Voluntary or involuntary pretreatments given to the seed must also be well known. There are cases where seeds are subjected to excessive heat or dryness before experiments, which could alter the results. Transportation and storage conditions can also be important (Maury-Lechon, et. al., 1981).

The great variability among seeds which come from different individuals and populations of parent plants (Gupta, Pattanath, 1975; Manokaran, 1978) is also important to consider and should be the subject for detailed research.

Many of the experiments reported in the literature lack an adequate control of the maturity state and storage conditions of the seeds. The sample size is often too small. The experimental conditions are frequently poorly documented, such as temperature, humidity and light. Many of these studies lack any statistical testing. There is a general need of more research on the forest microenvironment, to reproduce more closely the environment of the forest in germination experiments.

11. CONCLUSIONS

Based in the conceptual framework given by Angevine and Chabot (1979) we may conclude that in a forest community composed of polycarpic species where there is only slight seasonal temperature fluctuations and where the amount of rainfall defines the different growing seasons, the strongest selective pressures determining dormancy and longevity will be the effects of parasites and predators on the seeds and the ecological position of the species in the dynamics of the forest. There are two main groups of seeds according to their germination behavior. The first group is formed by those species with seeds which germinate in the microclimatically stable conditions of the forest floor and the second group includes species with seeds which germinate in discontinuous and sometimes unstable microhabitats of the forest such as light gaps. Among the species which germinate well under stable forest-floor conditions, the general trend of germination is the rapid formation of partially or totally dormant seedlings that are less susceptible to the damaging effects of predation. Seeds with delayed germination will have a hard coat or some mechanism to escape predators such as Janzen has described for many species (1969, 1971, 1974, 1977, 1978, 1981).

Plants adapted to germinate in gaps or special microhabitats will have specialized dormancy mechanisms that detect the best conditions for establishment, such as the opening of a light gap, or the presence of the appropriate symbient, in the case of an orchid.

In seasonally humid forests the seeds produced during less favorable periods for germination will have a partially endogenous or enforced dormancy until the beginning of the following rainy season. Information on these types of dormancy, apart from hard seed coats, is practically nonexistent.

REFERENCES

Alencar JC (1981) Estudos silviculturais de uma populacão natural de Copaifera multijuga. Hayne Leguminosae, na Amazônia Central. 1 - Germinação, Acta Amazonica 11, 3-11.

Alencar JC and Magalhaes LMS (1979) Poder germinativo de sementes de doze espécies florestàis da regiào de Manaus I, Acta Amazonica 9, 411-418.

Alexandre DY (1977) Régéneration naturalle d'un arbre caracteristique de la fôret èquatoriale de Côte d'Ivoire, Oecol. Plant. 12, 24-262.

Alexandre DY (1978a) (Observations sur l'écologie de Trema guineènsis en basse Côte d'Ivoire, Cah. O.R.S.T.O.M. ser. Biol. 13, 261-266.

Alexandre DY (1978b) Le role disseminateur des elephants en forèt de Tai, Côte d'Ivoire, Terre et Vie 32, 47-72.

Alexandre DY (1980) Caractere saisonnier de la fructification dans une foret hygrophile de Côte d'Ivoire, Rev. Ecol. (Terre et Vie) 34, 335-350.

Angevine MW and Chabot BF (1979) Seed germination syndromes in higher plants. In Solbrig, OT, Jain S, Johnson GB and Raven PH, eds. Topics in plant population biology, pp. 188-206. N.Y., Columbia University Press.

Arditti J (1967) Factors affecting the germination of orchid seeds, Bot. Rev. 33, 1-98.

Aubreville A (1947) Les brousses secondaires en afrique équatoriale, Côte d'Ivoire, Cameroun-A.E.F., Bois et Forets Trop. 2, 24-49.

Aubreville A and Leroy JF (1970) Contribution a L'étude biologique d'une Araliaceae d'America tropicale: Didymopanax morototoni Adansonia Ser. 2, 10, 383-407.

Augspurger CK (1979) Irregular rain cues and the germination and seedling survival of Panamanian shrub (Hybanthus prunifolius) Oecologia Berl. 44, 53-59.

Bannister BA (1970) Ecological life cycle of Euterpe globosa Gaertn. In Odum HT, and Pigeon RF, eds. A tropical rain forest, pp. B-299-314. Oak Ridge Tenn. U.S., Atomic Energy Commission.

Bazzaz FA and Pckett STA (1980) Physiological ecology of tropical succession: a comparative review, Ann. Rev. Ecol. Syst. 11, 287-310.

Bell CR (1970) Seed distribution and germination experiment. In Odum HT and Pigeon RF, eds. A tropical rain forest, pp. D-177-182. Oak Ridge Tenn. U.S., Atomic Energy Commission.

Blum KE (1968) Contributions toward an understanding of vegetational development in the Pacific lowlands of Panama, Ph.D. Thesis Fla. State University Tallahassee Fla. 119 p.

Brinkmann WLF and Vieira AN (1971) The effect of burning on germination of seeds at different soil depths of various tropical tree species, Turrialba 21, 77-82.

Budowski G (1965) Distribution of tropical American rain forest species in the light of successional process, Turrialba 15, 40-42.

Campbell U de Araujo (1979) Sobre a germinação do mongo (aguano) Swietenia macrophylla King, Acta Amazonica 9, 23-29.

Chai DNP (1973) A note on Parashorea tomentella (Urat mata beledu) Seed and its germination, Malay. Forest. 36, 202-204.

Chamshama SAO and Downs RJ (1982) Germination behavior of Chlorophora excelsa (Welw.) Benth & Hook and Podocarpus usambarensis, Pilger, Indian Forest. 108, 397-401.

Cheke AS Nanokorn W and Yankoses C (1979) Dormancy and dispersal of seeds of secondary forest species under the canopy of a primary tropical rain forest in Northern Thailand, Biotropica 11, 88-95.

Coutinho LM and Struffaldi Y (1971) Observacões sobre a germinação das sementes e o crescimiento das plàntulas de una leguminosa da mata amazônica de igapó (Parkia auriculata Spruce Mss.), Phyton 28, 149-159.

Defrense S (1982) Principales caracteristiques de la germination des graines e du devélopment des plantules de deux species tropicales Simphonia globulifera L.F. (Guttiferae) et Cedrela odorata L. (Meliaceae), Memoire. Universite Pierre et Marie Curie.

Edmiston J (1970) Some autecological studies of Ormosia krugii. In Odum HT and Pigeon RF, eds. A tropical rain forest, pp. B-291-298. Oak Ridge Tenn. U.S., Atomic Energy Commission.

Fasidi IC and Olifonbora MO (1975) Effects of light and herbicides on seed germination of Chlorophora excelsa (Welw Benth), Nigerian J. Forest 5, 9-19.

Ferreira SDN (1982) Observacao da germinanção de sementes de Aracà-Pèra (Psidium acutangulum D.C.), Acta Amazonica 13, 503-507.

Garwood NC (1982) Seasonal rhythm of seed germination in a semideciduous tropical forest. In Leigh EG Rand AS and Windsor DM, eds. The ecology of a tropical forest, pp. 173-185. Washington, D.C., Smithsonian Institution Press.

Garwood NC (1983) Seed germination in a seasonal tropical forest in Panama: a community study, Ecol. Monog. (in press).

Gilbert G (1952) Contribution à la biologie des essences forestières congolaises, Bull. Soc. Bot. Belg. 84, 289-296.

Guevara S and Gómez-Pompa A (1972) Seeds from surface soils in a tropical region of Veracruz, Mexico, J. Arnold Arbor. 53, 312-335.

Gupta BN and Pattanath PG (1975) Factors affecting germination behavior of teak seeds of eighteen Indian origins, Indian Forest. 101, 583-588.

Hall, JB and Swaine MD (1980) Seed stocks in Ghanian forest soils, Biotropica 12, 256-263.

Holthuijzen AMA and Boerboom JHA (1982) The Cecropia seedbank in the Surinam lowland rain forest, Biotropica 14, 62-68.

Horvitz CC (1981) Analysis of how ant behavior affect germination in a tropical myrmecochore Calathea microcephala (P&E). Koernicke (Marantaceae): microsite selection and aril removal by neotropical ants, Odontomachus, Pachycondyla and Solenpsis (Formicidae), Oecologia Berl. 44, 47-52.

Johnson P and Morales R (1972) A review of Cordia alliodora (Ruiz & Pav.) Oken, Turrialba 22, 210-220.

Jansen DH (1969) Seed eaters versus seed size, number toxicity and dispersal, Evolution 23, 1-27.

Jansen DH (1971) Seed predation by animals, Ann. Rev. Ecol. Syst. 2, 465-492.

Janzen DH (1974) Tropical blackwater rivers, animals and mast fruiting by the Dipterocarpaceae, Biotropica 6, 69-103.

Janzen DH (1977) Why fruits rot, seeds mold and meat spoils. Amer. Nat. 111 (9801), 619-713.

Janzen DH (1978) Seeding patterns of tropical trees. In Tomlinson PH and Zimmermann MH, eds. Tropical trees as living systems, pp. 83-128. Cambridge University Press.

Janzen DH (1981) The defenses of legumes against herbivores. In Rolhill RM and Raven PH, eds. Advances in legumes systematics, pp. 951-977.

Janzen DH (1982) Differential seed survival and passage rates in cows and horses, surrogate pleistocene dispersal agents, Oikos 38, 150-156.

Keay RWJ (1960) Seeds in forest soils, Nigerian Forestry Inf. Bull. (N.S.) 4, 1-4.

Kellman MC (1974) The viable weed seed content of tropical agricultural soils, J. Appl. Ecol. 11, 669-677.

Laboriau LG, Valio IM and Heringer EP (1964) Sôbre o sistema reproductive de plantas dos cerrados, Ann. Acad. Brasil. Cien. 36, 449-464.

Lebron ML (1979) An autecological study of Palicourea riparia Bentham as related to rain forest disturbance in Puerto Rico, Oecologia Berl. 42, 31-46.

Leite AMC and Rankin JM (1981) Ecologîa de sementes de Pithecolobium racemosum Ducke, Acta Amazonica 11, 309-318.

Liew TC (1973) Occurrence of seeds in virgin forest top soil with particular reference to secondary species in Sabah, Malay. Forest. 36, 185-193.

Macedo M (1977) Dispercão de plantas lenhosas de uma campina amazônica, Acta Amazonica 7, Supp 1-69.

Macedo M and Prance GT (1978) Notes on the vegetation of Amazonian II. The dispersal of plants in Amazonian white sands campinas: The campinas as functional islands, Brittonia 30, 203-215.

Manokaran N (1978) Germination of fresh seeds of malaysian rattans, Malay. Forester 41, 319-324.

Marrero J (1949) Tree seed data from Puerto Rico, Caribb. Forest 10, 11-30.

Maury-Lechon G. Corbineau F and Come D (1980) Données preliminaires sur la germination des graines et la conservation des plantules de Symphonia globulifera L. F. (Guttifère), Bois et Forets Tropiq. 193, 35-40.

Maury-Lechon G, Hassan AM and Bravo DR (1981) Seed Storage of Shorea parvifolia and Dipterocarpus humeratus, Malay. Forest. 44, 267-280.

Medina E (1982) Physiological ecology of neotropical savanna plants In Huntley BJ and Walker BH, eds. Ecological studies 42: Ecology of Tropical Savannas, pp. 308-335. Berlin Heidelberg New York, Springer-Verlag.

Moreno-Casasola P (1976) Viabilidad de semillas de árboles tropicales y templados: una revisión bibliográfica. In Gómez-Pompa A, Vázquez-Yánes C, Del Amo S, and Butanda A, eds. Regeneración de selvas, pp. 471-526. Mexico, Editorial Continental.

Mori T. Zollfatah bin H, Add Rahman and Tan CH (1980) Germination and storage of rattan manau (Calamus manan) seeds, Malay. Forest. 43, 14-55.

Ng FSP (1973) Germination of fresh seeds of Malaysian Trees, Malay. Forest. 36, 54-65.

Ng FSP (1977) Germination of fresh seeds of Malaysian Trees III, Malay. Forest. 40, 160-163.

Ng FSP (1978) Strategies of establishment in Malayan forest trees. In Tomlinson PB and Zimmermann MH, eds. Tropical trees as living systems, pp. 129-162. Cambridge University Press.

Ng FSP (1980) Germination ecology of Malaysian woody plants, Malay. Forest. 43, 406-437.

Ponce de Leon L (1982) L'ecophysiolgie de la germination d'especes forestieres et de savanne, en rapport avec la dynamique de la végétation en Cote d'Ivôire, Bull. Liaison Cher. Lamto, N.S.1, 1-44.

Prevost MF (1981) Mise en évidence des graines d'especes pionnières dans le sol de foret primaire en Guyane, Turrialba 31, 121-127.

Quaterman E (1970) Germination of seeds of certain tropical species. In Odum TH and Pigeon RF, eds. A tropical rain forest, pp. D-173-175. Oak Ridge Tenn. U.S., Atomic Energy Commission.

Ramirez WB (1976) Germination of new world uvostigma Ficus, Rev. Biol. Trop. 24, 1-6.

Richards PW (1952) The tropical rain forest, Cambridge at the University Press.

Sasaki S (1980a) Storage and germination of some Malaysian legume seeds, Malay. Forest. 43, 161-165.

Sasaki S (1980b) Storage and germination of Dipterocarp seeds, Malay. Forest. 43, 290-308.

Sasaki S and Ng FSP (1981) Physiological studies on germination and seedling development In Intsia palembanica (Merbau), Malay. Forest. 44, 43-59.

Sasaki S, Tan CH and Zolfatah bin H (1979) Some observations on unusual flowering and fruiting of Dipterocarps, Malay. Forest. 42, 38-45.

Symington CF (1933) The study of secondary growth on rain forest sites in Malaya, Malay. Forest. 2, 107-117.

Tang HT (1971) Preliminary tests on the storage and collection of some Shorea species seeds, Malay. Forest 34, 84-98.

Tang HT and Tamari C (1973) Seed description and storage tests of some dipterocarps, Malay. Forest. 36, 38-53.

Valio IFM and Joly CA (1979) Light sensitivity of the seeds on the distribution of Cecropia glaziovi Snethlage (Moraceae), Z. Pflanzenphysiol. Bol. 915, 371-376.

Vázquez-Yánes C (1974) Studies on the germination of seeds of Ochroma lagopus Swartz, Turrialba 24, 176-179.

Vázquez-Yánes C (1976a) Seed dormancy and germination in secondary vegetation tropical plants: The role of light, Comp. Physiol. Ecol. 1, 30-34.

Vázquez-Yánes C (1976b) Estudios sobre ecofisiología de la germinación en una zona cálido-húmeda de México. In Gómez-Pompa A, Vázquez-Yánes C, Del Amo S, and Butanda A, eds. Regeneración de selvas, pp. 279-387. México, Editorial Continental.

Vázquez-Yánes C (1977) Germination of a pioneer tree (Trema guineensis Ficahlo, from equatorial Africa, Turrialba 27(3), 301-302.

Vázquez-Yánes C (1980a) Notas sobre la autecología de los árboles pioneros de rápido crecimiento de la selva tropical lluviosa, Trop. Ecol. 21, 103-112.

Vázquez-Yánes C (1980b) Light quality and seed germination in Cecropia obtusifolia and Piper auritum from tropical rain forest in Mexico, Phyton 38, 33-35.

Vázquez-Yánes C and Pérez-García B (1976) Notas sobre la morfologia y la anatomía de la testa de las semillas de Ochroma lagopus Sw, Turrialba 26, 310-311.

Vázquez-Yánes C and Orozco Segovia A (1982) Seed germination of a tropical rain forest pioneer tree (Heliocarpus donellsmithii) in response to diurnal fluctuations of temperature, Physiol. Plant. 56, 295-298.

Vázquez-Yánes C and Orozco-Segovia A (1982) Germination of the seeds of a tropical rain forest shrub, Piper hispidum Sw (Piperaceae (under different light qualities, Phyton 42, 143-149.

Vázquez-Yánes C and Smith H (1982) Phytochrome control of seed germination in the tropical rain forest pioneer trees (Cecropia obtusifolia and Piper auritum and its ecological significance, New Phytol. 92, 477-485.

Vega ECF, Patino V and Rodríquez P (1981) Viabilidad de semillas en 72 especies forestales tropicales almacenadas al medio ambiente. pp. 325-345 In Reunion sobre problemas de semillas forestales tropicales, Tomo I. Pub. Inst. Nal. Invest. Forest. México No. 35.

Wellam FL (1964) Parasitism among neotropical phanerogams, Ann. Rev. Phytol. 2, 43-56.

Whitmore TC (1975) Tropical rain forest of the far east, Oxford Claredon Press.

Yap SK (1980) Jelutong, phenology, fruit and seed biology, Malay. Forest. 43, 309-315.

Yap SK (1981) Collection, germination and storage of dipterocarp seeds Malay. Forest. 44, 281-300.

LEAF AND CANOPY ADAPTATIONS IN TROPICAL FORESTS

T. J. GIVNISH (The Biological Laboratories, Harvard University, Cambridge, MA 02138 USA)

ABSTRACT

Ecological patterns in several aspects of leaf form and canopy structure in tropical forest plants are reviewed and analyzed in terms of their potential significance for competitive ability. Cost/benefit models for traits that directly influence gas exchange — such as the size, inclination, and reflectivity of leaves and the profile and aerodynamic roughness of canopies — suggest a basis for the paradoxical duality of morphological adaptations to drought and nutrient poverty. Models based on the balance between photosynthesis and mechanical efficiency predict various patterns in leaf shape, and analyze the functional significance of orthotropy and plagiotropy, asymmetric leaf bases, anisophylly, alternate vs. opposite leaves, and simple vs. compound leaves. Brief comments are made on the potential importance of biotic interactions for trends in plant form.

1. INTRODUCTION

Plants in tropical moist forests show several characteristic features in the form and arrangement of their photosynthetic organs. Many of these traits — such as leaf size and shape (figure 1) — show strong convergence among unrelated species within a community or tree stratum, and well-marked trends along gradients of rainfall, soil fertility, and elevation (Richards 1952, Webb 1968, Sarmiento 1972, Whitmore 1975, Grubb 1977). Other traits — such as growth form or branching pattern — show considerable variation within forests and less recognizable patterns along environmental gradients (Ashton 1978, Hallé et al. 1978).

In this paper I review trends in several aspects of leaf form and canopy structure, and examine briefly how these traits might influence competitive ability and lead to ecological patterns in the morphology of terrestrial plants in tropical moist forests. Throughout I assume that natural selection favors plants whose form and physiology tend to maximize their net rate of carbon gain, since such plants should have the greatest resources with which to reproduce and compete for additional space (Horn 1971, Orians and Solbrig 1977, Givnish 1978a, 1979, Mooney and Gulmon 1979; but see discussion by Cohen 1970, Givnish 1982). In evaluating the net contribution of a trait to whole-plant carbon gain, I focus on three basic kinds of energetic tradeoffs involving the economics of gas exchange, the economics of support and supply, and the economics of biotic interactions.

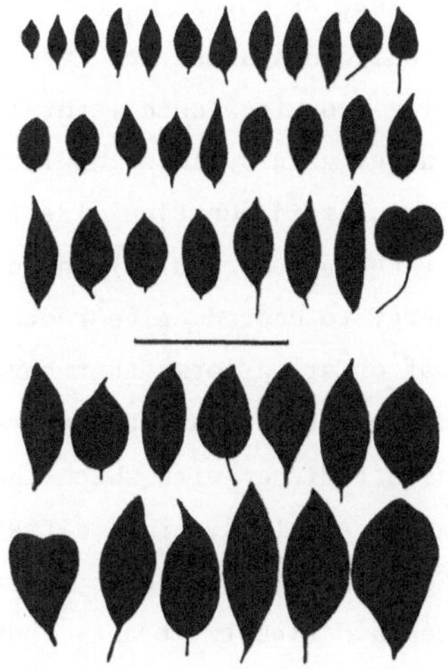

Figure 1. Leaf size and shape in woody species of Macuira cloud forest, Guajira Peninsula, Colombia (Sugden 1982). Silhouettes represent largest leaves or leaflets in herbarium specimens of adult individuals of all 42 woody dicots in 18 families. Bar equals 20 cm.

The first tradeoff results from the ineluctable

association of carbon gain with water loss — any passive structure that permits the passage of large, slow-moving CO_2 molecules will also allow the diffusion of smaller, faster water molecules. Thus, the photosynthetic benefit of any trait that increases the rate at which CO_2 can diffuse into a leaf must be weighed against the transpirational costs resulting from increased water loss, which can reduce energy gain by lowering leaf hydrature and inhibiting plastid function directly, or by forcing (over evolutionary time) an increased allocation of energy to unproductive roots. Second, among the leaf or crown forms that have equivalent effects on photosynthesis and transpiration, many differ in the efficiency with which the leaves can be supported and supplied. Such differences imply a tradeoff between photosynthetic benefits and mechanical costs. Finally, traits that influence a plant's potential rate of growth may also affect its attractiveness to herbivores, implying a tradeoff between photosynthetic benefits and biotic costs. These three sets of tradeoffs are addressed below first at the level of individual leaves, and then at the level of entire canopies.

2. LEAF LEVEL

2.1. Economics of gas exchange.

Several aspects of leaf form, phenology, and biochemistry can influence gas exchange and thus the balance between photosynthetic profits and transpirational costs. Three such traits that show prominent ecological patterns in the tropics — effective leaf size, leaf inclination and/or reflectivity, and seasonal pattern of photosynthetic activity — are considered in the following sections.

2.1.1. Effective leaf size. It has long been known that the average width of leaves or their lobes or leaflets tends to decrease toward dry, sunny, or nutrient-poor habitats (Volkens 1872, Schimper 1898, Raunkiaer 1934, Shields 1950, Webb 1968, Walter 1973, Hall and Swaine 1981). On relatively fertile, well-drained sites in the lowland tropics, average leaf width increases with the logarithm of annual rainfall (figure 2). The index of leaf width used here is:

$$\bar{w} = \sum_i p_i w_i \quad , \qquad (1)$$

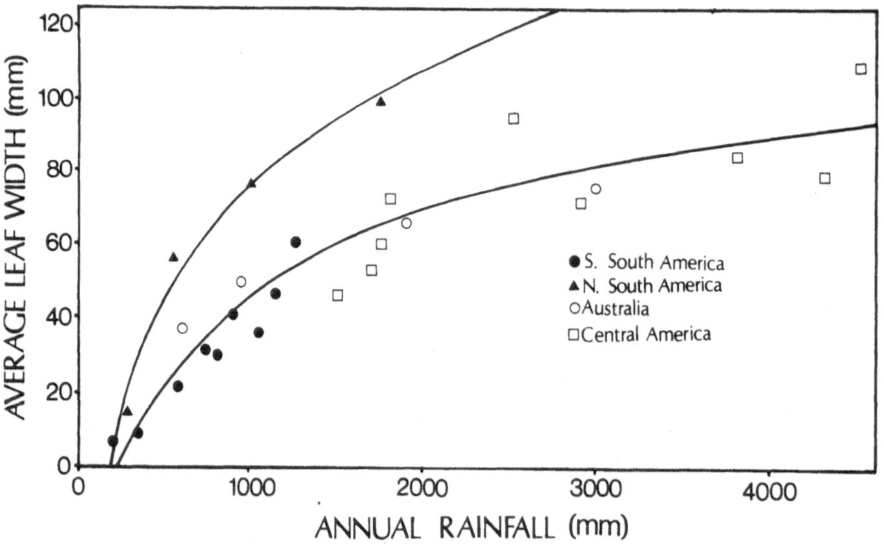

Figure 2. Average leaf width ($\bar{w}$) as a function of annual rainfall in northern and southern South America, Central America, and Australia. Curves represent the relationships y = 32.7 ln (x/244.5) for sites in southern South America, Central America, and Australia ($r^2 = 0.82$, $P < 0.001$ for 20 d.f.), and y = 45.0 ln (x/188.6) for sites in northern South America with a less marked dry season ($r^2 = 0.98$, $P < 0.01$, 2 d.f.). Replicates at a given site and rainfall in Central America have been averaged.

where w_i is the characteristic width of leaves in the ith Raunkiaer-Webb category of leaf area (sensu Webb 1968), and p_i is the proportion of species falling into that category. The characteristic leaf width for each category is taken as the square root of three-quarters of the geometric mean of the upper and lower bounds of leaf area for that category (see table 1). This calculation assumes that leaves are about twice as long as wide, and have an area of roughly two-thirds leaf length times width (cf. Webb 1968). The index $\bar{w}$ is a useful single number with which to summarize the usual tabulation of species into leaf area categories, and could be used as a supplement for such tabulations in future studies.

Average leaf width shows roughly the same response to rainfall in seasonal portions of South America, Central America, and Australia studied by Sarmiento (1972), Dolph and Dilcher (1980a) and Webb (1968). Leaves from these regions achieve a maximum average width of about 90 mm at 5000 mm of annual rainfall (figure 2). Sites in northern South America, experiencing less seasonality in rainfall and thus less exposure to seasonal drought (Sarmiento 1972),

Table 1. Area and characteristic width of leaves in the Raunkiaer-Webb categories of leaf size (see text for definition of w_i).

Raunkiaer-Webb category	Leaf area (mm^2)	w_i (mm)
Leptophyll	0-25	2.5*
Nanophyll	25-225	7.5
Microphyll	225-2025	22.5
Notophyll	2025-4500	47.6
Mesophyll	4500-18225	82.4
Macrophyll	18225-164025	202.5
Megaphyll	164025+	607.5*

*Assumes lower bound equals nine times upper bound where actual lower or upper bound would cause w_i to equal zero or be undefined, paralleling the original definition of the categories by Raunkiaer (1934). The characteristic width for Raunkiaer's mesophylls (including the notophylls) is 67.5 mm.

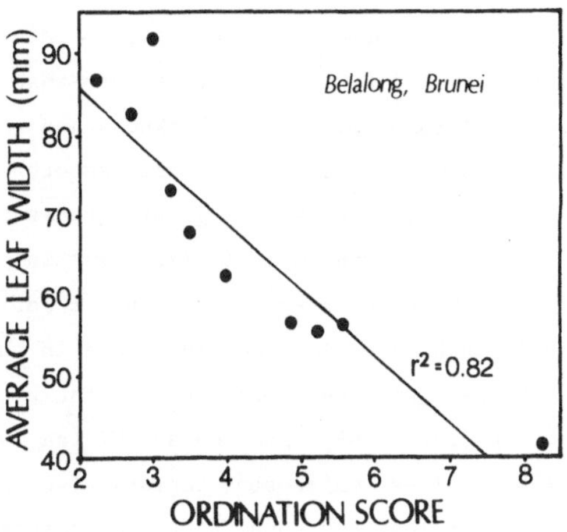

Figure 3. Average leaf width ($\bar{w}$) as a function of ordination score along a riverbank-ridge gradient of soil depth and moisture-holding capacity in lowland dipterocarp forests at 100-600 meters, Brunei (calculated from data of Ashton 1964). Sites with deep soils are at left, and sites with shallow, slightly less fertile soils are at right.

have larger leaves at a given rainfall. Within an area receiving a given amount of rainfall, average leaf width appears to decrease with increasing site exposure and decreasing soil-moisture holding capacity. This is nicely illustrated by Ashton's (1964) study of leaf size in lowland dipterocarp forests along a gradient from deep soils on river banks and moist slopes to shallow, slightly less fertile soils in ridges at 100 to 600 meters in Brunei (figure 3). Gentry (1969) obtained similar results for a slope-ridge gradient in Costa Rican rain forests of the Osa peninsula. Finally, Hall and Swaine (1981) show that leaf size decreases along a gradient of decreasing rainfall and/or site moisture in Ghana.

Leaf width decreases with increasing soil poverty, even on sites receiving heavy rainfall. Webb (1968) found that subtropical Australian rain forests on oligotrophic sites had smaller, more scleromorphic leaves than those on adjacent sites with more fertile substrates. Brünig (1970, 1971, 1976, 1983) showed that leaf size decreases from mesophyll to leptophyll along an edaphic gradient from dipterocarp rain forests on relatively fertile latosols to heath forests and kerangas on infertile podsols in parts of Borneo receiving more than 3000 mm annual rainfall. Leaf thickness and reflectivity, as well as the aerodynamic smoothness of plant canopies, increase along this gradient. Parallel trends in reduced leaf size occur in peat swamps that develop over extremely mineral-deficient soils in Borneo (Anderson 1964) and in the caatinga forests and bana formations of the Rio Negro region (Ferri 1961, Sobrado and Medina 1980, Brünig 1983). These white sand regions also receive a more or less aseasonal supply of rainfall exceeding 3000 mm. The expected average leaf width at this rainfall on relatively fertile sites is about 80 mm (figure 2). This compares with a $\bar{w}$ of 47.5 mm for shrubs of the Rio Negro bana (calculated from data of Sobrado and Medina 1980), and much lower widths for plants of peat swamps and heath forests (cf. Anderson 1964, Brünig 1970).

Similar trends toward small leaves on infertile soils occur outside the tropics. Beadle (1962, 1966) found that shrubs on mesic, phosphate-deficient soils in southeastern Australia had smaller, thicker leaves than those on adjacent richer sites. Cowling and Campbell (1980) showed that shrubs on sterile soils in South African fynbos have smaller leaves than species on richer sites in comparable climatic regions of Chile and California. Finally, the phenomenon of "bog xeromorphism", the tendency for plants in nutrient-poor habitats to display many of the same traits as plants under dry conditions, is known globally (Schimper 1898, Shields 1950, Walter 1973).

Leaf size decreases with elevation in the humid tropics, with the characteristic leaf class shifting from mesophyll in lowland rain forest, to notophyll in lower montane forest, microphyll in upper montane forest, and nanophyll in elfin forest or subalpine forest; similar trends are known for all three major rain forest areas in Southeast Asia and Malesia, Africa, and South America (Brown 1919, Beard 1943, Hedberg 1951, Grubb et al. 1963, Howard 1960, Grubb 1974, 1977). However, leaves are somewhat broader in lower and montane forests on Jamaica than elsewhere (Tanner and Kapos 1982). The general elevational trend may partly parallel the response of leaf size to soil fertility, since the excess of precipitation over evapotranspiration increases with elevation on moist mountains at low and moderate altitudes, thus promoting leaching and soil poverty (Whitmore 1975). In semi-arid areas, leaf size may initially increase with elevation in response to increased orographic rainfall, as seen in the transition from seasonal deciduous forest to premontane rain forest in the Coastal Cordillera of Venezuela (Walter 1973), the Pacific slope of the Central Cordillera in Costa Rica (Dolph and Dilcher 1980a, 1980b), and the lower slopes of the mountains of East Africa (Hedberg 1951). Dolph and Dilcher (1980a, 1980b) claim that the elevational trend in leaf size is discontinuous, and propose recognition of four foliar belts in which there is little or no systematic variation in leaf size. However, their conclusion may be an artifact of using an insensitive index of modal leaf size (% of species with leaves $\geq$ mesophyll size class) and sampling at only a few different elevations. A re-examination of their data using the index proposed in this paper shows no apparent elevational discontinuity in leaf size.

Leaf size increases from the sunlit canopy to the lower strata of tropical forests, although there may be some tendency for leaf width to decrease slightly in moving from the lowest tree stratum to the forest floor (Brown 1919). Average leaf area for trees in the Mucambo rain forest studied by Cain et al. (1956) near Belem in Brazil ranges from 56.7 cm^2 for canopy species to 85.8 cm^2 for understory species; a similar trend is seen for the leaflets of species with compound leaves (figure 4). Finally, large leaves are known to characterize certain genera that colonize gaps and clearings in

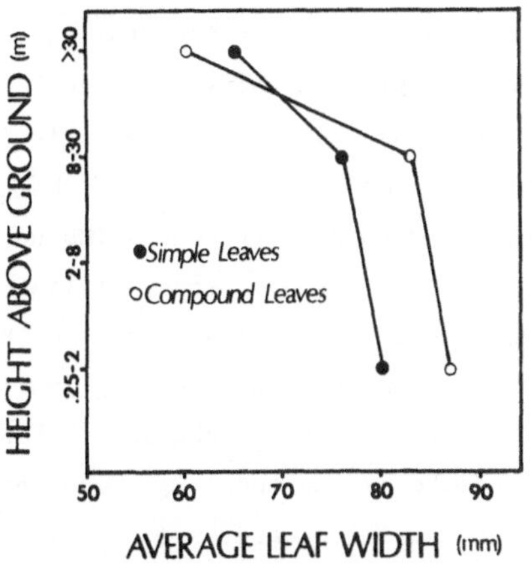

Figure 4. Average leaf width ($\bar{w}$) for species in different tree strata of the Mucambo rain forest. Species with simple and compound leaves plotted separately; $\bar{w}$ based on mean area of leaves and leaflets given by Givnish (1978) from data of Cain et al. (1957). Only three species occur in the 2-8 m stratum and are omitted.

tropical moist forests of South America, Africa, and Southeast Asia (e.g., *Cecropia, Musanga, Macaranga, Ochroma*), and many forest trees have much larger leaves as seedlings and saplings than as adults (e.g., *Campnosperma auriculatum* in Malaysia) (Whitmore 1975, Hallé et al. 1978, Hall and Swaine 1981).

Various arguments have been advanced to account for ecological patterns in leaf size (see review by Givnish 1978a). Schimper (1898) suggested that plants in dry areas have small leaves to reduce their transpiring surface and overall water loss. Thoday (1931) rightly noted, however, that some plants with small leaves have so many of them that their total transpiring surface is greater than that of comparable plants with larger leaves, so that total leaf area need have nothing to do with the size of individual leaves. Maximov (1929, 1931) presented a more fundamental criticism, observing that mere minimization of water loss could hardly be adaptive. Water vapor and carbon dioxide diffuse along the same pathway between leaf and atmosphere, so that any attempt to block water loss would necessarily halt the uptake of carbon. The latter would prove fatal to an autotrophic plant, and Maximov concluded that some compromise must be made between photosynthesis and transpiration. The precise nature of this compromise was, however, left unclear.

Attempts to explain the paradoxical occurrence of small leaves among plants of moist but nutrient-poor sites have mainly centered on contentions that nutrient-poor soils are effectively dry. Schimper (1898) argued that humic acids in peat and other impoverished soils inhibit root absorption and expose plants to functional drought. However, Small (1972a,b) found that bog shrubs are rarely exposed to water shortage and usually maintain high leaf water potentials. Brünig (1970, 1971, 1976, 1983) contends that heath forests and kerangas on tropical podsols are more exposed to drought than adjacent sites on deeper richer soils because (i) the indigenous plants show many apparent water-conserving adaptations, and (ii) the sterile sites have a lower water-holding capacity and are more prone to water shortage, if only at multi-year intervals. The first part of this argument is obviously circular, and raises the fundamental question of whether the xeromorphic traits represent adaptations to nutrient poverty itself. The second is only partly correct because the capacity of the rooting zone varies little, and often exceeds that of dipterocarp forests on more fertile sites, over most of the soil catena studied. Brünig's argument appears hard to sustain for sites that receive abundant aseasonal rainfall, and depends on the assumption that the advantage conferred by small, thick leaves during rare droughts outweighs their presumed disadvantage during the intervening years. Grubb (1974, 1977) has concluded that a similar argument applied to montane rain forests is unlikely to be correct.

Walter (1973) suggests that reduction in leaf size occurs in response to drought and/or mineral poverty because these conditions inhibit cell growth and expansion, and thus impede leaf growth. However, this is an argument about proximal rather than ultimate causes — the question is why certain developmental responses should be favored by natural selection. Why should leaf expansion or ultimate size be reduced by drought or nutrient poverty, whereas root growth is enhanced by the same conditions? Presumably, these responses would be selectively favored if they enhance plant growth and competitive ability. Thus, the central question is how can leaf size influence growth and maximize competitive ability in a given environment?

Leaf or leaflet width can affect the growth of a plant with a given total leaf area through its effects on gas exchange and leaf temperature. As leaf width increases so does the average thickness of the boundary layer of still air about the leaf, thus increasing the resistance to diffusion and heat exchange and affecting transpiration, leaf temperature, and photosynthesis. Gates and his colleagues have summarized the principal effects (Gates 1965, Gates and Papian 1971, Taylor 1975, Gates 1980). Under sunny conditions and moderate to high stomatal resistances ($r_s > 1\text{-}2$ s cm^{-1}), large leaves are warmer and transpire at a higher rate per unit leaf area than do smaller ones, with the increased concentration of water vapor inside the leaf at higher leaf temperatures outweighing the effect on transpiration of a slight increase in diffusive resistance. Under shady conditions, or at stomatal resistances less than a critical value for wholly or partly sunlit conditions, increased leaf size decreases leaf temperature and transpiration (Gates and Papian 1971, Taylor 1972, Geller and Smith 1980).

Based on these effects and their implications for plant growth, three recent models have been advanced for the evaluation of leaf size. First, Taylor (1972, 1975) and Taylor and Sexton (1972) proposed that leaf size is adjusted to maintain leaf temperature near the optimum for photosynthesis and/or to prevent thermal damage. However, the thermal response of photosynthesis is itself open to selection and tends to peak at higher temperatures in warmer environments (Mooney et al. 1977, Slayter et al. 1978). Thus, if leaf size were shaped by some other mechanism, the thermal response might adapt to the resulting leaf temperature and not vice versa. Parkhurst and Loucks (1972) advanced a second model in which leaf size is adjusted to maximize water-use efficiency, the ratio of photosynthesis to transpiration, and thus to maximize carbon gain for a given amount of water loss. However, natural selection is unlikely to maximize this ratio because it attains its peak value as numerator and denominator approach zero (Givnish and Vermeij 1976). Further, in the absence of other constraints this model predicts infinitely large or infinitely small leaves and, like the Taylor model, does not account for bog xeromorphism. Givnish and Vermeij (1976) and Givnish (1978a, 1979) propose a third model in which leaf size is adjusted to maximize whole-plant net carbon gain, based on the balance between photosynthetic benefits and below-ground costs at moderate to high stomatal resistances. Here I briefly review that model and some of its predictions under constant, sunlit conditions (figure 5).

As leaf size increases, transpiration increases dramatically with the resulting rise in leaf temperature, reflecting the nearly exponential rise in water vapor concentration within the leaf with leaf temperature. As transpiration increases, so should the amount of energy directed to the roots, in order to maintain leaf hydrature and the functional integrity of photosynthetic enzymes and membranes. In this simple model, the cost of these roots is assumed to increase in rough proportion to transpiration, making the implicit assumption that leaf water potential is held constant as leaf size varies under a given set of conditions. A more

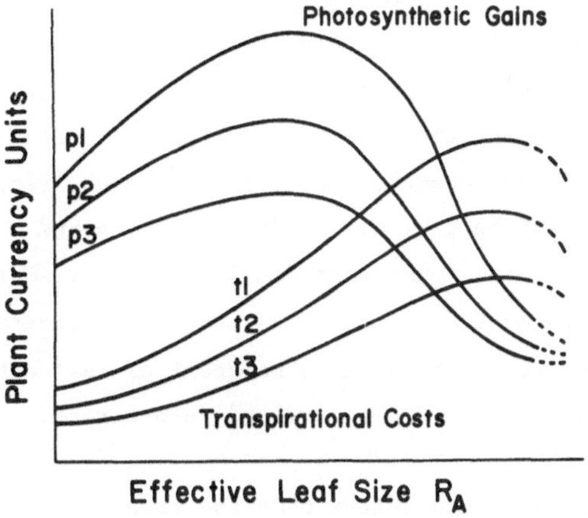

Figure 5. Benefit and cost curves of photosynthesis and transpiration for leaves in a sunny environment, as a function of effective leaf size. The different benefit curves represent expected photosynthesis in environments more (pl) or less (p3) sunny, warm, or rich in mineral nutrients. **Similarly, the set of cost curves indicates the** range of root costs associated with supplying transpirational losses in environments more (tl) or less (t3) dry or sunny. See Givnish (1979).

general and quantitative model would permit leaf water potential to vary as well, and would balance the energetic cost of roots with changes in plastid photosynthetic capacity (Givnish 1979). However, many of the qualitative tradeoffs are captured by assuming that transpirational costs are roughly proportional to the rate of transpiration. The constant of this proportionality should increase in drier sites as a result of lower soil moisture and hydraulic conductivity, resulting in decreased efficiency of water absorption by the roots.

As leaf size increases, it may also enhance photosynthesis by elevating leaf temperature and thus increasing the rate of carboxylation while only slightly raising overall diffusive resistance. Although the thermal response of most species tends to peak near temperatures characteristic of their habitat and decline at higher temperatures, I have argued elsewhere (Givnish and Vermeij 1976, Givnish 1978a, 1979) that such declines may not be a fundamental physiological constraint for a species or group of species, but the price of adaptation to a particular thermal regime — plants of warm habitats often have their peak photosynthetic rates at higher temperatures than do plants of cooler habitats (Mooney et al. 1978, Slatyer et al. 1978). Thus, the appropriate photosynthetic response to consider is not that of a single species grown under a single set of conditions, but the envelope of such responses to temperature for a given set of species. This envelope yields, by definition, the maximum photosynthetic rate at each temperature as a function of **leaf temperature, which rate would be favored in** species whose leaves operate at that temperature. This response envelope should increase over a wider range of leaf temperatures than the response of any single species, and may be monotonically increasing over a wide thermal range. In the remaining analysis, I will assume that the response envelope is monotonically increasing over reasonable environmental temperatures; if it is diatonic and peaks at some intermediate temperature, the predicted leaf size would shift so that leaf temperature is closer to the thermal optimum (Givnish and Vermeij 1976).

Thus, as leaf size increases, leaf temperature and photosynthesis should increase (figure 5). As leaf size continues to increase, the rate at which photosynthesis increases should decelerate as factors other than carboxylation limit the uptake of CO_2. The optimal leaf size is that which maximizes the differences between photosynthetic profits and transpirational costs (figure 5). Under drier conditions, the root cost of replacing a given water loss is greater, which should raise the cost curve multiplicatively and favor smaller leaves than under moister conditions (figure 5). Lower humidity favors the same result by increasing evaporation and steepening the cost curve, and by reducing the rate at which leaf temperature increases with leaf size, thus flattening the benefit curve and favoring smaller leaves.

This cost/benefit model can also be used to explain the trend toward small leaves on sterile soils. Soil poverty can reduce the concentration of nutrients that plants can economically sequester (Mooney and Gulmon 1979, Gulmon and Chu 1981). Medina (1970, 1971) showed that an enhanced supply of soil nitrogen results in a greater leaf N concentration and greater production of RuP_2 Carboxylase, and hence a higher rate of light-saturated photosynthesis. Similar results have been reported for a variety of other soil nutrients (Natr 1975). As soil fertility declines, and with it the concentration of photosynthetic enzymes, enzyme concentration itself may partly limit photosynthesis and reduce the extent to which changes in other limiting factors can increase photosynthesis (Givnish and Vermeij 1976). Indeed, the data of Gulmon and Chu (1981) show that a reduced N supply to *Diplacus aurantiacus* flattens its photosynthetic response to light.

Thus, as soil poverty increases, the extent to which leaf temperature can enhance photosynthesis should be reduced, flattening the benefit curve and favoring smaller leaves (figure 5). Thus, soil fertility itself can influence leaf size independent of any supposed correlations of soil fertility and moisture-holding capacity. The puzzling duality of morphological adaptations to drought and mineral poverty is thus explained by the cost/benefit model, and is now easy to understand in qualitative terms. In dry fertile sites, transpirational costs are absolutely high and increase at a high rate with leaf size, so that small leaves are favored. In moist but sterile sites, transpirational costs are relatively high compared with the photosynthetic benefits accruing from a given increase in leaf size, so that small leaves are again favored. Parallel conclusions apply to adaptations in leaf thickness (Givnish 1979), leaf reflectivity and inclination, and canopy aerodynamic roughness (see below). However, there are no grounds for believing that plants on sterile sites should show the same physiological tolerance for low leaf water potentials as do plants exposed to low soil water potentials on dry sites — to this extent, adaptations to drought and nutrient poverty are not exactly equivalent (Givnish 1979). Indeed, Corbett et al. (1979) and Peace and MacDonald (1980) found that small thick leaves of trees in montane rainforests and heath forests, respectively, are no more tolerant of desiccation than those of trees in lowland rain forests.

It should be emphasized that the predictions of the model do not depend on the absolute levels of photosynthesis and transpiration, but on the relative rates at which these change with leaf size. The absolute amount of transpiration is immaterial, only the rate at which photosynthesis varies for a given increment in transpirational costs is important. It is also important to recognize the nature of the cost of transpiration, as used here. If transpiration increases, the carbon cost envisaged is the construction and maintenance of roots adequate to maintain leaf water potential at a fixed level; no suggestion is made that the cost represents an increase in root respiration to supply an increased flow. In economic terms, the cost of transpiration is thus an average cost, not a marginal cost, and is analogous to the cost of, say, a telephone call. The price of a phone call does not reflect the minute marginal cost of electricity needed to make the call, but the cost of installing additional transmitting and switching capacity to permit such calls, averaged over the lifetime of the equipment. Similarly, the cost of an investment in transpiration is the amortized cost of constructing and maintaining roots sufficient to supply the added flow under the same conditions of leaf water potential. Marginal costs are important in nutrient uptake: Veen (1977, 1981) estimates a requirement of 36.8 mg O_2 for uptake of 1 milliequivalent of NO_3^- in maize, compared with 24.5 mg O_2 to grow 1 g of root and 0.77 mg O_2 per g root for daily maintenance respiration. The biological difference between passive and active uptake, and its implied difference between average and marginal costs, may allow a separation of the costs of transpiration and nutrient uptake and a unification of the approaches to below-ground costs taken by Givnish and Vermeij (1976) and Mooney and Gulmon (1979). To accomplish this, we will need measurements on fine root biomass, turnover rates, hydraulic conductivity, and respiration rates associated with root construction, maintenance, and nutrient uptake.

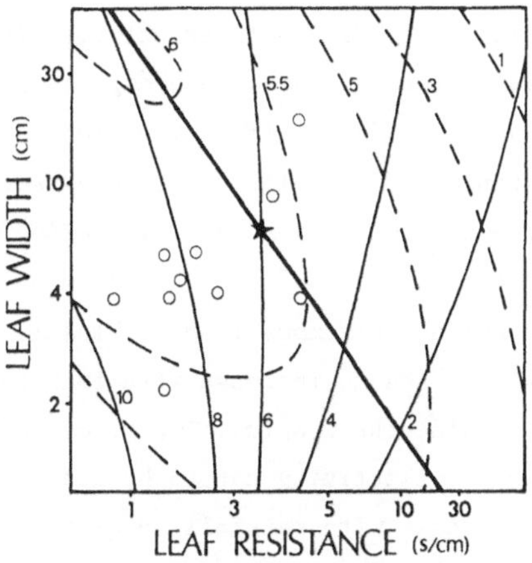

Figure 6. Predicted photosynthesis (--; x 10^{-8} g CO_2 cm^{-2} s^{-1}) and transpiration (—; x 10^{-6} g H_2O cm^{-2} s^{-1}) as a function of leaf size and stomatal resistance for typical summer conditions at the Michigan Biological Station. Circles indicate actual characteristics of each major tree species in the region (Taylor 1975). Solid curve represents locus of optimal leaf size and stomatal resistance as a function of the proportionality constant b (cost of transpiration); star indicates prediction for b = 0.005 g CO_2 g^{-1} H_2O.

One example should illustrate the potential use for such data and a quantitative approach in tests of the leaf size model or variants thereon. Figure 6 summarizes the dependence of photosynthesis and transpiration on leaf size and stomatal resistance modelled by Taylor (1975) for typical plants and conditions at the Michigan Biological station. The hollow circles represent the actual leaf characteristics of the dominant species. The leaf size and stomatal resistance that would maximize the difference between photosynthetic benefits and transpirational costs, vary along the curve shown, as the cost of transpiration varies from 0 to 0.012 g CO_2 g^{-1} H_2O s^{-1}. A value of b = 0.005 results from a simple back-of-the envelope calculation, assuming a hydraulic conductivity of 0.02 g H_2O cm^{-1} root day^{-1} MPa^{-1} (Greacen et al. 1976), a fine-root diameter of 0.1 cm and longevity of 2 weeks, a dry/fresh mass ratio of 0.2, a hydraulic head of 0.2 MPa, and rough equivalence of a gram of fixed CO_2 and a gram of plant tissue; respiration is ignored. The optimal leaf size and stomatal resistance predicted for b = 0.005 falls near the centroid of the leaves actually found in the environment. Whether this is purely the result of fortuitous number-juggling can only be determined by collecting actual data on root characteristics of plants in their native habitats.

2.1.2. Leaf inclination and reflectivity. Both of these traits affect the receipt of radiation by the leaf, its thermal budget and gas exchange, and thus the energetic balance between photosynthetic benefits and transpirational costs. Quantitative data on trends in these traits are sparse, but it is commonly recognized that leaf inclination and reflectivity increase toward dry, sunny, and/or nutrient-poor environments. Canopy leaves are usually held at a steep angle to the horizontal, either erect or pendent, whereas leaves of understory species are usually nearly horizontal (Walter 1973, Brünig 1976, 1983, Hall and Swaine 1981). Ashton and Brünig (1975) report that, in tropical gap succession, trees with large, more or less horizontal leaves are common mainly on moist fertile sites, whereas trees with smaller, steeply inclined leaves occur on drier or less fertile sites. Leaves of plants in heath forests, peat swamps, and bana tend to be steeply inclined (Brünig 1976, 1983, Medina et al. 1978, Sobrado and Medina 1979, Medina 1983). Many of these species also have high leaf reflectivity (short-wave albedo), such as *Shorea albida, Vatica brunigii* and *Xylopia coriifolia*, whereas others have glistening leaves with a reddish-green appearance, such as *Melanorrhoea inappendiculata* and *Whitodendron moultonianum*, all from heath forest on Borneo (Whitmore 1975). Finally, leaf reflectivity tends to increase with increasing acidity; outside the tropics, Ehleringer et al. (1981) found reflectivity caused by leaf pubescence to decrease with rainfall in a convergent manner among *Encelia* species of California and Chile.

Leaves can absorb, reflect, and transmit radiation over a wide range of wavelengths. Absorptance at different wavelengths has different physiological consequences, and four primary wavebands of interest can be identified: long-wave infrared, photosynthetically active radiation (PAR), the remainder of the visible spectrum plus short-wave infrared, and ultraviolet (Gates 1962, Robberecht 1980, Gates 1980). At ambient temperatures, leaves re-radiate in the long-wave infrared; because radiative emissivity is equal to absorptivity, many leaves have high absorption in the far infrared to allow radiative cooling (Gates and Benedict 1963) while leaving PAR interception unaffected. Similarly, many leaves have low absorption and high reflectance in the near infrared and non-PAR visible, perhaps because absorption at these wavelengths tends to increase heat load (Gates and Benedict 1963). Absorptance of ultraviolet light may be related to the avoidance of physiological damage, and tends to decrease toward high elevations in the tropics and elsewhere (Robberecht 1980).

The tradeoffs involved with leaf reflectivity in the visible spectrum (including PAR) are straightforward, and similar considerations would apply to leaf inclination from the horizontal (also see discussion by Mooney and Ehleringer 1978, Ehleringer and Mooney 1978, Forseth and Ehleringer 1979, Ehleringer and Forseth 1980). As reflectivity or inclination increase, leaf absorption of radiation decreases with consequent effects on photosynthetic benefits and transpirational costs. As light absorption increases, photosynthesis increases and then plateaus as other factors become limiting in a sunny environment (figure 7); if the air temperature is high, photosynthesis may decline at high absorption as leaf temperature exceeds the thermal optimum (cf. Medina et al. 1978, Ehleringer and Mooney 1978). In a shady environment, a higher absorptance is needed to allow photosynthesis to balance respiration., and photosynthesis rises more slowly with increasing absorption and is less likely to plateau. At the same time, the heat load on a leaf increases as light absorption rises and so do transpirational costs (figure 7). The curve for these costs should be higher and steeper in sunny or dry environments than in shady or moist environments. As before, the optimal absorption/reflectance should be that which maximizes the difference between the curves for photosynthetic benefits and transpirational costs. Thus, leaf reflectivity and/or inclination should increase with site aridity, and be greater in sunny habitats than shady habitats (figure 7;

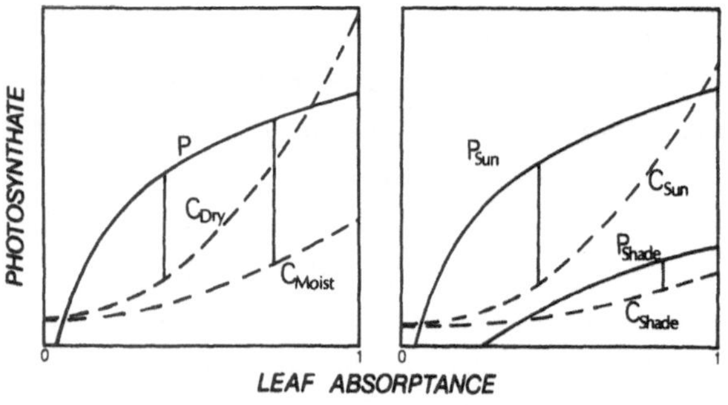

Figure 7. Hypothetical photosynthetic benefits and transpirational costs associated with changes in light absorption due to leaf reflectivity and inclination in relation to light intensity and moisture supply (see text).

cf. Mooney and Ehleringer 1978, Forseth and Ehleringer 1979). If soil infertility reduces the level of RuP_2 carboxylase and the consequent sensitivity of photosynthesis to light intensity as conditioned by leaf reflectance or inclination (see Medina 1970, 1971, Gulmon and Chu 1981, and section 2.1.1), mineral poverty will flatten the benefit curve and favor greater reflectance or inclination.

Although leaf reflectance and inclination have effects that are in many ways similar, there are two respects in which they differ. First, leaf reflectivity affects the interception of both direct-beam and diffuse radiation, whereas leaf inclination has its greatest effect on direct-beam radiation. More importantly, leaf inclination increases the leaf area that can be held over a given surface, though perhaps at the expense of greater light penetrance through the crown. (Such penetration could be beneficial in tall plants by allowing additional leaf layers to be held below the penumbras of the upper foliage Horn 1971).) Reflective leaves can be arranged more nearly horizontally to cast a denser shade and suppress competitors. It is perhaps of interest in this light to note that most *Eucalyptus* species have relatively broad, rather horizontal leaves in dense crowns while seedlings and young saplings, and less reflective but pendant leaves in open crowns as adults (e.g., Chippendale 1973). Low-growing mallee species in open, high-radiation vegetation often retain the juvenile foliage. Bloodwood eucalypts, typical of mesic forests, tend to bear their non-reflective foliage nearly horizontally.

Ehleringer and his colleagues (Ehleringer and Mooney 1978, Mooney and Ehleringer 1978, Ehleringer et al. 1981) have presented an elegant quantitative test for a cost/benefit model for the evolution of leaf pubescence and reflectivity in *Encelia*. The model departs somewhat from that just presented because of the means by which leaf reflectance is achieved. In desert *Encelia* species, increased leaf pubescence increases reflectivity and tends to increase photosynthesis by lowering leaf temperature toward the thermal optimum, and tends to decrease photosynthesis by decreasing chloroplast light interception and increasing diffusive resistance. Ehleringer and Mooney (1978) quantified these benefits and costs at the leaf level for various conditions and showed that the observed pattern of seasonal and geographic variation in leaf absorptance closely matched that expected. However, even this study is incomplete because the implicit below-ground costs associated with leaf traits are not considered. In particular, the predicted leaf absorptance depends almost entirely on stomatal conductance, and this presumably is set by tradeoffs between photosynthetic benefits and transpirational costs. Only an approach that incorporates below-ground costs can lead to quantitative predictions of trends in traits that influence gas exchange.

2.1.3. Leaf phenology. Several seasonal patterns of leaf activity are seen in trees of the humid tropics. Prominent functional types include evergreen, drought-deciduous, drought-green, and brevideciduous (Janzen 1970, Whitmore 1975, Hallé et al. 1978, Hall and Swaine 1981, Medina 1983). Within the evergreen class, several patterns of leaf expansion are found, ranging from almost continuous production in certain pioneer species (e.g., *Cecropia*) to discontinuous leaf flushes in many mature-phase species (Coley 1983). Drought-deciduous trees lose foliage during the dry season, with the effect of reducing not only transpiration but also respiration in the warm tropics (Janzen 1975). Such species may be obligately or facultatively deciduous, with the behavior of the latter depending on water availability through the dry season (DeOliveira and Labouriau 1961, Daubenmire 1972, Walter 1973, Frankie et al. 1974, Whitmore 1975, Medina 1983). In the Central American seasonal forests, species such as *Casearia arbora, Hura crepitans,* and *Sterculia apetala* are known to be facultatively deciduous, whereas species such as *Ceiba pentandra, Enterolobium cyclocarpum,* and *Spondias mombia* are obligately deciduous Medina 1983). In the understory of tropical deciduous forests, a few drought-green species such as *Jacquinia pungens* specialize on the seasonal window of light admitted by the canopy. Many species of the Malaysian rain forests are brevideciduous, deploying new leaves within a few days of shedding their previous crop (Holttum 1953, Whitmore 1975). Finally, a few species display a mixed or "manifold" leaf phenology, with different limbs in different states of leaf flush and leaf drop, as in the genus *Mangifera* (Koriba 1958).

Generally speaking, the incidence of deciduous species in the tropics increases with decreasing rainfall and increasing seasonality of rainfall, and is lower at higher elevations, on more sterile sites, and in lower tree strata (Beard 1955, Brünig 1970, Sarmiento 1972, Walter 1973, Whitmore 1975, Hall and Swaine 1981, Medina 1983). Among deciduous species of seasonal forests, leaf fall usually occurs during the dry season and new leaves are deployed around the beginning of the rainy season (Frankie et al. 1974 for New World; Koriba 1958, Fox 1972, for Old World).

The energetic costs and benefits of the deciduous and evergreen leaf phenologies have recently been reviewed by Chabot and Hicks (1982). In essence, the main tradeoffs associated with deciduousness involve the photosynthetic benefit of constructing a high-efficiency leaf adapted only to the favorable season vs. a leaf adapted to conditions year-round; the photosynthetic gain (or loss!) foregone during the unfavorable season; and the cost of leaf construction (see Monk 1966, Leigh 1972, Miller 1979, Miller and Stoner 1979). Basically, if the contrast in potential photosynthetic rate between the two seasons (weighted by their length) is low, and the cost of replacing foliage relative to those rates is high, evergreen leaves should be favored. Miller (1979) and Miller and Stoner (1979) present quantitative cost/benefit models that successfully predict the distribution of dominance by deciduous or evergreen species in areas of Mediterranean climate; such studies would be interesting and relatively easy to pursue along rainfall gradients in the tropics. Monk (1966) discussed the advantage of evergreen foliage in nutrient poor areas in terms of the reduced cost of nutrient acquisition for leaves that turn over less frequently. I would add that nutrient poverty may lower the level of photosynthetic enzymes present in leaves, reduce the sensitivity of photosynthesis to other limiting factors (see section 2.1.1), and thus reduce the contrast in photosynthetic rates likely to result between seasons, hence favoring evergreen leaves. The high concentration of leaf nitrogen and phosphorous in deciduous vs. evergreen woody species of the Venezuelan llanos (Cuenca 1976, Montes and Medina 1977, Medina 1983) is interesting in this regard and may represent divergence in phenology in response to differences in mineral absorption ability. On the other hand, Orians and Solbrig (1977) note that leaves must at least pay for the costs of their own construction, so that deciduous leaves with short lifespans <u>must</u> have higher photosynthetic rates than evergreen leaves and the differences in foliar nutrients may simply reflect this constraint.

Certain patterns of tropical leaf phenology are hard to reconcile with an approach based solely on the economics of gas exchange. Differences between continuous and discontinuous leaf production in pioneers vs. late-successional forest species seem best understood in terms of plant apparency and interactions with herbivores (Coley 1983). Perhaps the significance of the brevideciduous habit also lies along these lines. Another puzzling case is the production of new foliage a month or more before the end of the dry season by various trees and shrubs of the Venezuelan llanos (e.g., *Brysonima crassifolia, Palicourea rigida*). This would seem to be the worst possible time for leaf expansion on energetic grounds. Such woody plants usually have low coverage and little height growth once clear of the grass layer (Sarmiento 1983), so it is unlikely that a disproportionate advantage would accrue to plants that deploy early and overtop more conservative opponents. Sarmiento (1983) has made the interesting suggestion that leaf expansion may be timed to avoid the leaching of minerals by torrential rains from soft leaves that lack thick cell walls and a heavy cuticle. This is an important idea and bears further study.

2.2. Economics of support and supply

Because leaves of many different shapes can have the same effective size and impact on photosynthesis and transpiration, patterns in leaf shape cannot be explained solely in terms of the economics of gas exchange (Givnish 1979). An important additional consideration is the mechanical efficiency with which a photosynthetic surface composed of leaves of a given shape can be supported, supplied, and arranged. In the following sections I consider two constraints on leaf shape imposed by the efficient packing of foliage along branches, and by the functional differences between simple and compound leaves. Other aspects of the economics of support and supply for leaves, including the significance of parallel vs. pinnate venation and the basis for marginal dentition or lobation, are discussed elsewhere (Givnish 1978a, 1978b, 1979).

2.2.1. Leaf packing. Givnish (1979) concluded that the most efficient shape for the area supported and supplied by a leaf midrib should be roughly triangular, based on the constraints on leaf margin and midrib taper imposed by the need for mechanical stability and replacement of transpirational water loss. At the base of the leaf, the leaf margin should parallel the secondary veins and so form a basal angle of 180° when the secondaries are perpendicular, and a smaller basal angle when the angle between secondaries and midrib is acute.

Only a relatively few species have leaves of this shape. The majority are roughly oval to lanceolate in shape, taper proximally toward the base before the last secondary, and cannot be decomposed into triangular units supported by major veins (see Givnish 1979). Perhaps this is because the requirements for efficient leaf packing conflict with those for efficient support and supply within the leaf. Givnish's (1979) support-supply model implicitly assumes that a given leaf is essentially independent of, and relatively unaffected by, the shapes and positions of its neighbors. This assumption may approach reality when leaves are spaced far enough apart on a branch that they do not touch or shade one another, but this mode of leaf arrangement may be inefficient. It wastes space near the branch, so that greater lengths of branch are needed to accomodate a given amount of leaf surface; it also is an inefficient means of shading competitors.

Plants have evolved various ways to hold leaves to avoid self-shading and improve the density of shade they cast. Two modes of leaf arrangement are especially common and involve the packing of leaves in spirals or whorls about erect twigs (spiral phyllotaxis on orthotropic axes) or in planar arrays along more or less horizontal branches (distichous phyllotaxis on plagiotropic axes) (Horn 1971, 1975, Leigh 1972, 1975, Hallé et al. 1978). Spiral phyllotaxis on erect axes is common in sun plants, and distichous phyllotaxis on horizontal axes is common in shade plants (Leigh 1972, 1975, Hallé et al. 1978).

Although a triangular leaf may be most efficient when standing alone, it does not use space efficiently when packed in a planar, distichous

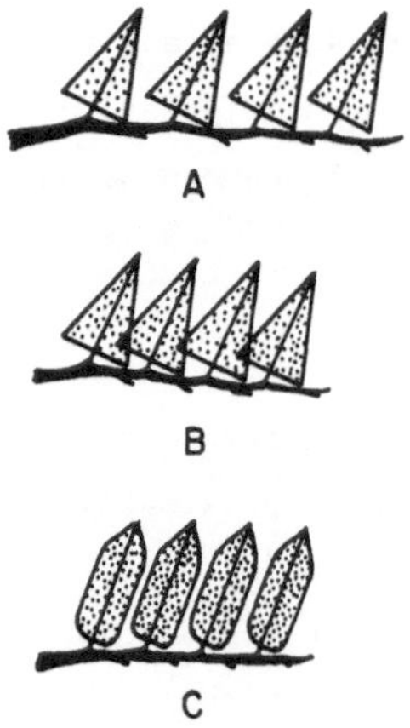

Figure 8. Packing of triangular leaves in a plane along a horizontal branch. (A) Arrangement of leaves so that they do not touch or overlap wastes much space along the branch. (B) Closer packing of similarly shaped leaves leads to overlap, shading, and potential wear and physical damage. (C) Efficient packing of triangular leaves in a plane along a horizontal branch.

array along a branch (figure 8). Even when such leaves are packed so closely that they touch, much of the space adjacent to the branch is not used, though it has been paid for in terms of woody tissue (Givnish 1979). If the leaves were more closely packed and the amount of branch invested per leaf reduced, triangular leaves would overlap. Such overlap is a respiratory drag for the shaded leaves and a potential source of abrasion for all leaves. On balance, plants having leaves without the zone of overlap would be favored. If the overlap is removed so that the leaves remain symmetric and the space between them is efficiently covered (allowing some space to permit convective and mechanical decoupling of neighboring leaves) the leaf arrangement in the lower part of figure 8 results. The limit to this process occurs either when the optimal effective leaf size is reached or when the leaves become too narrow to be supported effectively (Givnish 1979). Thus, leaves packed in planar arrays should be modified from a roughly triangular form to one in which the leaf margin roughly parallels the midrib over much of the leaf and tapers toward either end. The apical portion should remain roughly triangular.

Similar principles apply to the packing of leaves in spirals about erect twigs. In this case, however, the leaf bases must be packed radially to avoid self-shading from above, so that the margin of such leaves should taper outward from the leaf base for a certain distance, and then taper inward (figure 9). The narrowness of the leaf basal angle, and the position of the widest portion of the leaf, should depend on the effective number of interacting leaves being packed radially, and hence on the number of leaves per unit twig length and the solid angle from which light can strike the twigs by day. Basal angles should be lower in spiral phyllotaxis than in distichous phyllotaxis, provided that there are effectively more than two leaves being packed in a spiral. Basal angle and position of maximum width can also have implications for the economics of gas exchange of an entire erect axis or twig, even for leaves of constant effective size, as the following example suggests. Based on work by Woodson (1947), Wyatt and Antonovics (1981) recently studied a longitudinal cline in leaf shape in *Asclepias tuberosa* in the eastern and mid-

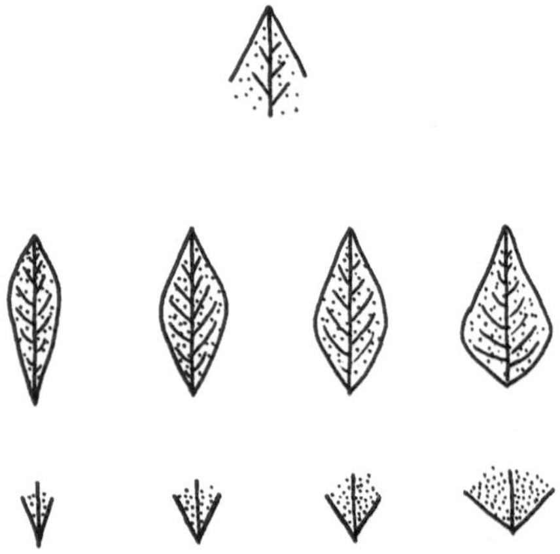

Figure 9. Leaf shape predicted for efficient packing in spirals about erect twigs; basal angle should decrease (left to right) as the effective number of leaves per unit twig length increases (see text).

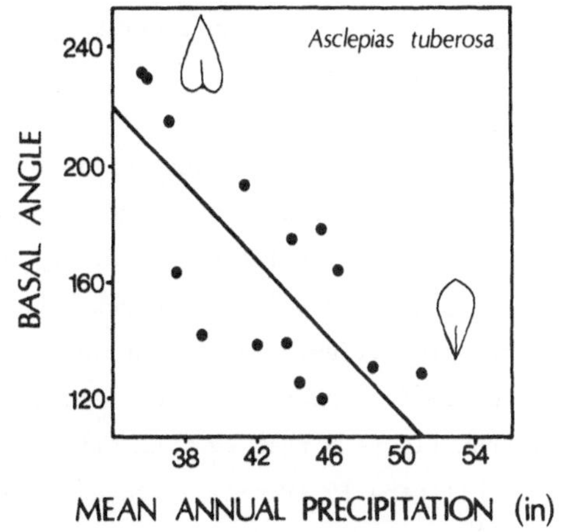

Figure 10. Cline in leaf basal angle vs. rainfall in *Asclepias tuberosa* (after Wyatt and Antonovics 1981).

western United States. Leaf shape in *A. tuberosa* ranges from obovate, with narrow basal angle and maximum leaf width beyond midleaf, to cordate, with broad basal angle and maximum width near leaf base. Leaf basal angle decreases and leaf shape shifts from cordate to obovate in moving along a west-to-east gradient of increasing rainfall (figure 10); the pattern is based partly on genetic differentiation and partly on developmental plasticity (Wyatt and Antonovics 1981). The pattern is puzzling because leaf width and effective size vary little over this gradient, so changes in leaf shape do not influence gas exchange at the level of individual leaves (Wyatt and Antonovics 1981, R. Wyatt, C. dePamphillis, personal communication). However, the observed differences in leaf shape can have profound effects on the photosynthesis and transpiration of the leaf assemblage on erect twigs (figure 11). For leaves of constant length, area, and maximum width, as the basal angle decreases and the point of maximum width moves away from the leaf base, the average amount of self-shading decreases. This is because most of the leaf surface is borne far from the vertical axis, where there is more space available in a given angular sector. Conversely, self-shading is nearly maximized when the point of maximum width is near the leaf base, since most of the leaf surface is packed into and shades the zone near the stem. Net photosynthesis, averaged over the leaf assemblage, should increase as self-shading decreases and absorbed PAR increases. This increase should be slight and plateau in strongly lit, open habitats with side lighting, and relatively more dramatic with less tendency to plateau in shaded habitats (figure 11). Transpiration should also increase as self-shading decreases and heat load increases; the increase in transpirational costs should be greater in dry habitats than in moist habitats (figure 11). Finally, the mechanical costs of supporting a given amount of leaf tissue should increase as the point of maximum width recedes from the leaf base, reflecting the increased length of lever arms (Givnish 1979); such costs must be added to transpirational costs to yield the total cost curve (figure 11). Thus, leaf basal angle should increase, and the distance from leaf base to point of maximum width should decrease, in moving toward sunny or dry environments. Thus, the model accounts for the relation of leaf shape to rainfall documented by Wyatt and Antonovics (1981), as well as their observation that obovate leaves are common in relatively dry but shaded sites in the Midwest. Such considerations, when applied in the context of other constraints, might be of general use in understanding

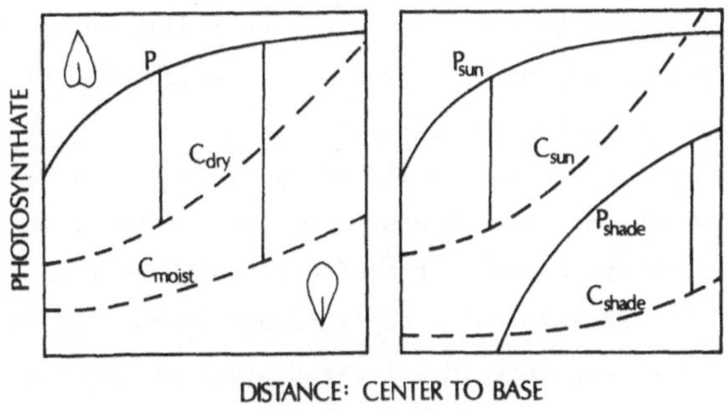

Figure 11. Hypothetical photosynthetic benefits, and combined mechanical and transpirational costs, for leaves of constant length, area, and effective size as a function of distance from leaf base to the point of maximum width (see text). Vertical bars indicate distance to base that maximizes net benefit.

Figure 12. (Upper) Efficient packing of bilaterally symmetric leaves in a planar array; note gap adjacent to proximal side of leaf base. (Middle) Increased efficiency of packing with asymmetric leaf bases in which an additional "half-leaf" has been added to the basal secondary vein on the proximal side. (Lower) Same, but half-leaf added to the distal side of the leaf base.

trends in the shape of leaves about erect axes.

As organs of energy capture, orthotropic axes with spiral phyllotaxis have an advantage in sunny environments because (i) erect lever arms are always more efficient and can support more leaf mass per unit twig mass; (ii) greater self-shading relative to distichy has relatively little effect on photosynthesis near saturation at high light intensity; and (iii) such self-shading decreases heat load and transpirational cost in a high radiation environment. Plagiotropic axes with distichous phyllotaxis should be favored in shady environments in spite of mechanical inefficiency, because (i) a low degree of self-shading has a large impact on net photosynthesis near the compensation point; and (ii) the additional transpirational costs imposed by direct exposure in a low-radiation environment are close to nil. As organs of growth, orthotropic axes should be favored in sun-adapted plants whose total carbon input would be most strongly influenced by vertical growth and attainment of the canopy, whereas plagiotropic axes should be favored in shade-adapted plants whose total input would be most strongly influenced by horizontal spread and increase in crown size.

There should be a special premium on close packing of leaves for plants growing near the leaf compensation point in dimly lit understories; Horn (1971, 1975) has discussed some compensating advantages of more diffuse leaf arrangements for plants in well-lit conditions. However, it should be clear from figure 8 that even an efficient packing of bilaterally symmetric leaves in a planar array will leave some areas (near the branch) uncovered. If the leaf base is symmetric, some space must remain adjacent to the proximal side of the leaf base (figure 12). Such an opening could be covered in at least two different ways in an asymmetric leaf base were to evolve. First, if tissue were added to the proximal side of the leaf base, perhaps through the development of tertiary veins on the basal secondaries on one side of the midrib, each gap would be covered by a leaf on the same side of the branch (figure 12, middle). Second, if a comparable area of tissue were added on the distal side of the leaf base, each gap would be covered by a leaf on the opposite side of the branch (figure 12, lower). As expected, asymmetric leaf bases are found almost entirely in species having plagiotropic branches (Hálle et al. 1978; examples include members of *Anisophyllea* (Rhizophoraceae), *Begonia* (Begoniaceae), *Columnea* (Gesneriaceae), *Celtis*, *Trema*, and *Ulmus* (Ulmaceae), and *Tilia* (Tiliaceae)). Asymmetric leaf or leaflet bases are common in species with determinate plagiotropic surfaces embodied as compound leaves (e.g., *Ptelea* (Rutaceae)) or determinate branches (e.g., *Phyllanthus* (Euphorbiaceae). In the common tropical tree *Trema*, the lower plagiotropic branches have distichously arranged leaves with asymmetric bases, while the upper orthotropic branches have spirally arranged leaves with symmetric bases (J. Sperry, personal communication).

Another constraint on close packing in distichous

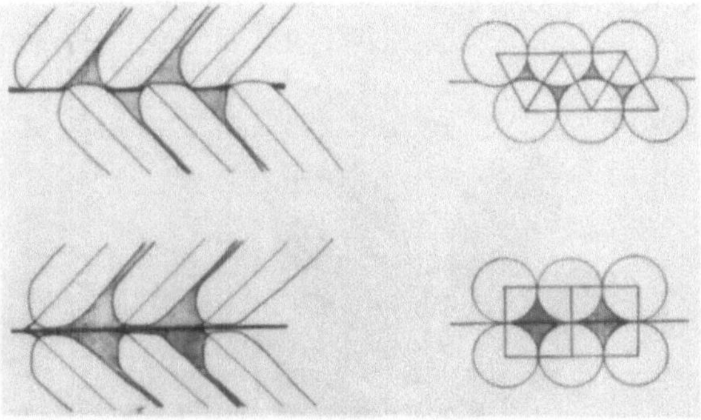

Figure 13. Packing of alternate vs. opposite leaves in a planar array (left) and packing of discs on triangular vs. rectangular grids (right). Note the reduction in uncovered space for close packing of leaf bases or discs on a triangular (alternate) grid.

arrays is whether the leaves are opposite or alternate. Just as packing circles on a triangular grid leaves 44% less uncovered space than packing circles on a square grid, so too with the distichous packing of alternate vs. opposite convex leaf bases along a horizontal branch — alternate leaves allow a far closer and more efficient packing of space near the branch (figure 13). Thus, one might expect a high incidence of alternate leaves in shade-adapted species of rain forest understories where efficient packing is strongly favored. Possession of alternate or opposite leaves is often a familial character (Cronquist 1981), and although most of the few families with opposite leaves are typically distributed in drier or more open habitats, there are several large tropical families that are common in rain forest understories (table 2). The principal examples include the Acanthaceae, Apocynaceae, Gesneriaceae, Guttiferae, Malpighiaceae, Melastomaceae, Monimiaceae, and Rubiaceae. However, many of these (Acanthaceae, Apocynaceae, Guttiferae, some Melastomaceae, Rubiaceae) usually have orthotropic shoots and so do not come under the correct argument. In other families, shade adapted species with opposite leaves and plagiotropic shoots often show a remarkable approach to an alternate leaf arrangement through anisophylly (figure 14). In the most common form of anisophylly, one member of a leaf pair is reduced in size and the position of the smaller leaf alternates from node to node. An efficient, quasi-alternate leaf mosaic results and is seen in many shade-adapted taxa with plagiotropic branches or shoots appressed to vertical surfaces, such as species of *Clidemia* and *Macrocentrum* of the Melastomaceae (Wurdack 1980), and *Gesneria*, *Pheidonophytocarpum*, and *Rhytaeniphyllum* of the Gesneriaceae (Skog 1976) in the New World Tropics. *Columnea* of the Gesneriaceae shows considerable infrageneric variation in anisophylly, shoot orientation and preferred habitat (Morley 1973, 1974) and thus provides useful material for a test of the model. In Jamaica, sun-adapted species with narrow leaves and pendulous epiphytic shoots (e.g., *C. microphylla*, *C. oerstediana*) or erect terrestrial shoots (e.g., *C. linearis*) are isophyllous or nearly so (figure 14). Species with somewhat broader, presumably shade-adapted leaves on plagiotropic shoots

Table 2. Dicotyledonous families characterized by opposite leaves (tabulated from Heywood 1978). Asterisk indicates family with modal distribution in tropical rain forests (see also Discussion).

*Acanthaceae	*Gesneriaceae
Aceraceae	Gomortegaceae
*Apocynaceae	Grubbiaceae
Asclepiadaceae	Guttiferae
*Austrobaileyaceae	Hippocastanaceae
Batidaceae	Labiatae
*Bignoniaceae	Loganiaceae
*Brunelliaceae	Loranthaceae
Callitrichiaceae	Lythraceae
Calycanthaceae	*Malpighiaceae
Caprifoliaceae	*Melastomaceae
Caryophyllaceae	*Monimiaceae
Casuarinaceae	Myrtaceae
Ceratophyllaceae	Oleaceae
*Chloranthaceae	Oliniaceae
Cistaceae	Pedaliaceae
Columelliaceae	Penaeaceae
Coriariaceae	Phrymaceae
Cornaceae	Punicaceae
*Cunoniaceae	*Quiinaceae
Dipsacaceae	Rhizophoraceae
Elatinaceae	*Rubiaceae
Eucryphiaceae	Salvadoraceae
Frankeniaceae	Sonneratiaceae
Garryaceae	Valerianaceae
Geissolomataceae	*Verbenaceae
Gentianaceae	Zygophyllaceae

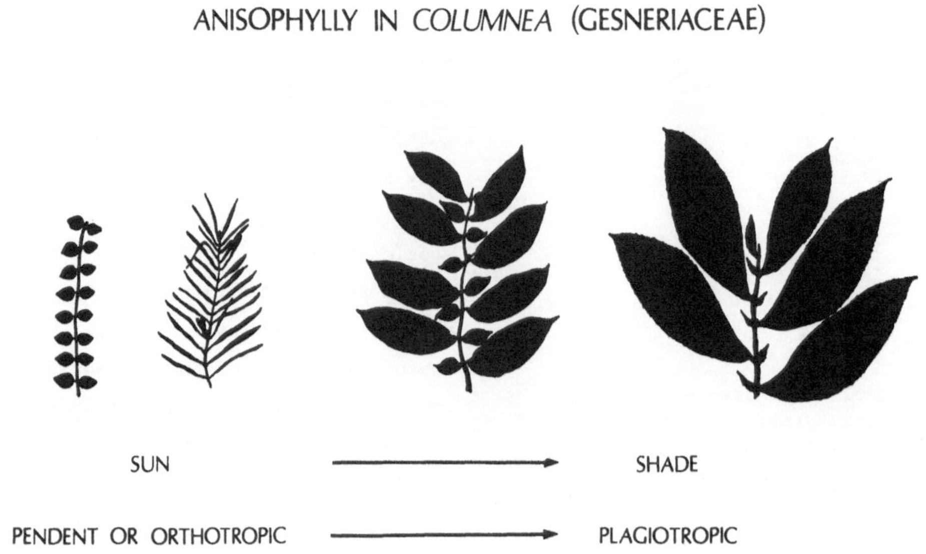

Figure 14. Anisophylly in *Columnea* (Gesneriaceae). Sun-adapted species with pendent or erect shoots (such as *C. microphylla* and *C. linearis*) are isophyllous; shade-adapted species with horizontal shoots and broader leaves (*C. harrisii, C sanguinea*) are markedly anisophyllous and approach an efficient **leaf mosaic of alternate leaves.**

(e.g., *C. harrisii, C. hispida, C. rutilans*) show moderate anisophylly, with the large leaf at each node being 2.5 to 3.5 times as long as the smaller one (Morley 1974). Finally, species with extremely large (> 20 cm length) and thin leaves on plagiotropic shoots (e.g., *C. purpurata, C. sanguinea*) display extreme heterophylly (figure 14), with the large leaf more than 6 times as long as the smaller neighbor. Morley (1973, 1974) interprets the trend toward anisophylly as a result of increased leaf size, suggesting that it imposes a need for reduction in size of one leaf at each node to aviod lamina overlap. The difficulty **with this argument is that it supplies no rationale** for why internode length should increase less rapidly than leaf size across species; indeed, the absolute internode length in large-leaved rain forest species like *C. sanguinea* far exceeds that of small-leaved sun epiphytes line *C. microphylla* (personal observation). However, the model presented here provides just such a rationale, invoking the premium on efficient leaf packing for plants of deep shade, and explaining why "alternate" anisophylly should be favored over other possible forms. It is interesting to note in this light that *Phainantha myrtilloides* (Melastomaceae), a climber with horizontal leaves that project from its vertical substrate, shows extreme anisophylly but always has the smaller leaf on the same side of the stem (Wurdack, personal communication). Study of cases like this, or the annual developmental shift from isophylly to extreme anisophylly in *Theligonum cynocranbe* (Heywood 1978), should cast additional light on the significance of anisophylly, alternate vs. opposite leaf arrangements, and other means of efficient leaf packing in plants with distichous phyllotaxis.

2.2.2. Compound leaves. The proportion of species with compound leaves tends to increase with decreasing rainfall in the lowland tropics and subtropics, being particularly common in thorn scrub, seasonal deciduous forest, savanna, and the upper stories of seasonal rain forest (Givnish 1978b). The significance of these trends may appear hard to fathom, because the only functional difference between simple leaves and compound leaves with leaflets of the same effective size and spacing is that the rachis on which the leaflets are borne is itself shed after the leaflets fall, whereas the twig bearing the leaves is often retained after they fall. Why should plants discard branchwork, in the form of leaf rachises, in which they have just invested considerable energy? How can compound leaves possibly be favored over simple leaves of the same size and spacing as their leaflets?

There are at least two potential advantages of such seemingly profligate behavior, however. First, in warm seasonally arid habitats that favor the deciduous habit, shedding the highest order branches (or rachises) is the most efficient means of reducing residual transpiration and respiration after leaf fall, since these branches have the highest surface-to-volume ratio, lowest suberization, and highest proportion of living cells (Givnish 1978a, 1978b, 1979). Second, among plants that colonize forest gaps (sensu Whitmore 1975, Hartshorn 1978, Bazzaz 1980) or other early successional habitats, compound leaves may be favored by virtue of the premium put on height competition for light. Such plants should branch rarely or not at all, at least initially, since branching diverts energy from the leader and may slow height growth. For this reason, and because the branches that are developed will have a short lifetime before they are shaded by new and higher branches, these branches should be made as inexpensively as possible. Compound leaves should be the ideal throwaway branches, given the lower energetic cost of parenchyma in rachises compared with woody tissue in twigs (Givnish 1978b).

These predictions are explored in detail elsewhere (Givnish 1978b, 1979). Compound leaves are indeed frequent in warm, seasonally arid habitats that favor the deciduous habit (see above), but not in dry habitats that favor evergreen leaves, such as the Mediterranean climate regions of California, Australia, and South Africa. Compound leaves are not simply a means to reduce effective leaf size through dissection of the photosynthetic surface, either. The trend toward effectively smaller leaves in moving from lowland rain forest to either montane (evergreen) rain forest or seasonal (deciduous) forest (see section 2.1.1) accompanies the predicted trends toward a higher incidence of simple leaves and of compound leaves, respectively (Brown 1919, Beard 1955, Givnish 1978b, 1979). The data of Cain et al. (1956) show the expected rise in the proportion of species with compound leaves toward the upper, more drought-prone, layers of Brazilian rain forest, even though the leaflets of the compound leaves have an average area similar to or greater than that of the simple leaves in each layer (figure 4). Furthermore, Givnish (1978b) has shown that the increased proportion of species with compound leaves in the canopy vs. understory of Jamaican dry forests results, as expected, from an increase in the proportion of species with deciduous compound leaves and a decrease in the proportion of those with evergreen simple leaves. Stowe and Brown (1981) have found an increase in the proportion of species with compound leaves with increasing seasonal aridity in the United States.

Species of gaps or early succession are often characterized by large compound leaves that may serve as throwaway branches. Hartshorn and Orians (personal communication) found that many of the tree species that invade forest gaps and openings at La Selva, Costa Rica, develop strong leaders with encircling spirals of compound leaves. Many species of early succession in the north temperate zone are also sparsely branched and have large, pithy compound leaves; indeed, most major woody taxa with compound leaves in the northeastern U.S. appear to colonize gaps (Givnish 1979). These include sumacs (*Rhus* and *Toxicodendron*), Kentucky coffee tree (*Gymnocladus dioica*), devil's walking stick (*Aralia spinosa*), Hercules' club (*Zanthoxylum clava-herculis*), and most hickories (*Carya*), ashes

(*Fraxinus*), walnuts (*Juglans*), and mountain ashes (*Sorbus*).

Palms (Arecaceae) and other plants with similar growth form in the Araliaceae, Cyatheaceae, and Dicksoniaceae would seem to be ideal gap colonists, with whorls of large compound leaves that concentrate growth on unbranched erect axes. Many palms are distributed in areas where unstable soils, battering winds, or repeated cycles of flooding and desiccation provide a chronically disturbed canopy (Givnish 1978b, 1979). Givnish (1978b) suggests that "feather" palms with long pinnate fronds should be favored in gap succession at various levels within relatively closed, moist forest, whereas "fan" palms with shorter palmate fronds should be favored in drier or more exposed habitats. The rationale is that the crown form of feather palms tends to minimize self-shading, and is less susceptible to overtopping because fronds with widely spaced leaflets sample radiation over a broad area. Fan palms have greater self-shading, which may be advantageous in sunny or dry habitats but disadvantageous in shady, moist areas (see section 2.2.1). In addition, their compact crown would be more susceptible to overtopping by a single branch of a competitor, but mechanically more efficient and resistant to wind in exposed situations (Givnish 1978b). Of the twelve subfamilies of palms (Tomlinson 1961, Moore 1973), nine have pinnate fronds, two (the Borassoideae and Sabaloideae) have palmate fronds, and one (the Lepidocaryoideae) has both types of frond. Interestingly, all but two genera of feather palms (*Butia* and *Phoenix*) have hypostomatous leaves, whereas more than half of all fan palm genera have stomata on both surfaces (Tomlinson 1961). Insofar as amphistomaty is characteristic of plants of sunlit habitats (Parkhurst 1980), the observed association between stomatal distribution and frond type provides tentative support for my hypothesis. Note as well that growth forms similar to feather palms occur in tree ferns (Cyatheaceae and Dicksoniaceae)

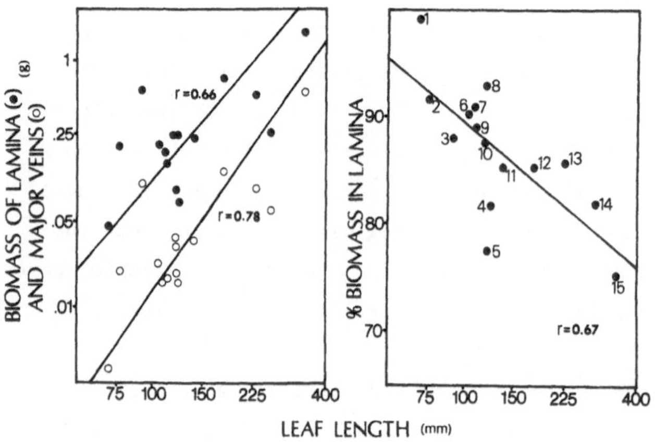

Figure 15. (Left) Dry mass of lamina and veins as a function of leaf length. (Right) Proportional allocation of biomass to lamina as a function of leaf length. Data are from (1) *Asimina triloba*, (2) *Betula lenta*, (3) *Ulmus rubra*, (4) *Kalmia latifolia*, (5) *Ostrya virginica*, (6) *Nyssa sylvatica*, (7) *Cornus florida*, (8) *Acer davidii*, (9) *Castanea dentata*, (10) *Lindera benzoin*, (11) *Prunus serotina*, (12) *Magnolia officinalis*, (13) *M. virginiana*, (14) *M. sprengeri* and (15) *M. macrophylla* (see text).

found typically in gaps in moist, tropical montane forests (Verdoorn 1938), and in pinnate-leaved species of Hawaiian *Cyanea* (Lobeliaceae) found typically in moist, shaded ravines in montane rain forests on Hawaii and Kauai (Carlquist 1970).

Rapidly growing, sparsely branched trees of tropical gap succession, such as *Cecropia* or *Macaranga*, lack compound leaves but may reap similar benefits by placing large, simple leaves on branch-like petioles, as the following argument suggests. Howland (1962) recognized that large leaves must contain proportionately more support tissue than smaller ones because they contain larger lever arms and thus are subject to great stress. He therefore concluded that it is more costly to build larger leaves. Indeed, Givnish's (1979) support-supply model predicts that support biomass in the midrib should increase as the cube of leaf length, whereas lamina biomass should increase as the square, for leaves of a given density. Data on 15 temperate woody species (figure 15) shows a trend close to that expected (principal axis regressions: $\ln y_v = 3.26 \ln x - 19.04$, $P < 0.01$; $\ln y_l = 2.67 \ln x - 14.95$, $P < 0.01$). The proportional allocation to lamina vs. veins > 0.25 mm diameter decreases sharply with leaf length ($y = -0.098 \ln x + 1.35$, $P < 0.01$), from about 95% of total dry mass in leaves 60 mm long to about 75% in leaves 400 mm long (figure 15). However, a direct application of Howland's (1962) argument to these data would be misleading and miss a point of fundamental signicance. A plant must construct support for its leaves both within and outside the foliage itself, and the woody tissue used to support leaves along stems is presumably more energetically expensive than the parenchyma used to support comparable areas of photosynthetic tissue within a leaf. Thus, the initial cost of construction may be lower for very large leaves, with woody twigs and branches being replaced by parenchymatous midribs, secondary veins, and/or rachises (Givnish 1979). However, over the long term this growth pattern may be inefficient because most of a plant's support tissue is shed when its leaves are. Such a plant would gain little energetic benefit through consolidation of new support tissue with old, and might lose much of its height during each leaf fall.

Thus, if a unit of leaf surface need not be replaced when it falls, and if its support is not needed for further growth, big leaves are mechanically cheaper than small leaves. On the other hand, if leaves are to be replaced or if their external support can be used to help subsidize future growth, smaller leaves should be cheaper. As discussed above, gap phase or early successional trees may represent plants in which large units of leaf surface and associated support often need not be replaced (Givnish 1979). In general, when height competition favors throwaway branches and the economics of gas exchange favors small effective leaf size, compound leaves should be favored. When gas exchange favors larger effective leaf size, large simple leaves should be favored. The restriction of large-leaved early successional plants to relatively moist, fertile sites in the humid tropics (cf. Ashton and Brünig 1975) accords with this view.

2.3. Economics of biotic interactions

Herbivores are thought to be a major selective influence on plants in the humid tropics, given the relative lack of abiotic controls on arthropod populations and the long periods of evolutionary time during which groups of plants, and potential enemies, could coevolve in mutual contact (Gilbert 1975, 1980, Janzen 1975, 1983, Janzen and Martin 1982). The study of the potential influence of herbivores on plant form and physiology is rapidly developing and may hold great significance for our understanding of tropical plant biology. Here, however, I will discuss only two questions that seem of particular importance in the context of this paper.

First, several authors have suggested that species with diverse chemical defenses may converge in leaf form to deter visually hunting specialist herbivores capable of detoxifying the compounds produced by each of their respective hosts (Gilbert 1975, 1980, Barlow and Wiens 1977, Rausher 1978, 1980, Wiens 1979, Givnish 1982). Such visual mimicry would make it difficult for a specialist herbivore to locate the few taxa whose metabolic poisons it can overcome. Rausher (1978, 1980) has provided support for this idea, showing that female *Battus philenor* butterflies locate their *Aristolochia* larval food plants based on leaf shape, with females having a more marked preference for an appropriate length/width ratio locating oviposition sites more rapidly. It remains to be seen whether broad-leaved *A. reticulata* is attacked more frequently in forest understories dominated by narrow leaved herbs, or whether *A. serpentaria* is attacked more frequently in areas with mainly broad-leaved herbs. Gilbert (1975) has presented the converse hypothesis that plant species sharing a similar chemical defense should diverge in leaf form to thwart related, visually hunting specialists. He suggests that the diversity of bizarre leaf shapes seen in sympatric *Passiflora* species results from coevolution with their group of visual specialist herbivores, the *Heliconius* butterflies.

These ideas are exciting, and it may be that visual mimicry is the proximal selective force favoring convergence in leaf form among certain taxa. However, this does not mean that the abiotic constraints on leaf form discussed in sections 2.1.1 - 2.2.2 are unimportant, even in cases where biotic pressures are clearly operating. If visual mimicry does occur, abiotic factors may determine which leaf forms should be converged upon because they otherwise increase competitive ability. For example, were visual mimicry to occur in rain forest understories, there is good reason to believe that species would converge on broad distichous leaves, rather than on the narrow, sclerophyllous leaf forms of *Eucalyptus*, *Casuarina*, and their mistletoes in semi-arid Australia (Barlow and Wiens 1977), or the grass-like foliage of orchids in sedgy bogs (Givnish 1982), of asters in salt marshes (e.g., *Aster spartinifolia*), or of prostrate *Protea* species in restionaceous fynbos (e.g., *P. lorea, P. restionifolia, P. scabra*). A central question for future research in the humid tropics is the relative importance of biotic and abiotic forces in shaping patterns of convergence in leaf form, although these forces are more likely to be complementary than mutually exclusive.

A second question of interest is whether the low level of leaf nitrogen in plants of sterile soils in the humid tropics (e.g., see Sobrado and Medina 1983) results from maximizing net plant growth when absorption costs are high (Mooney and Gulmon 1979), or from protecting foliage from insects when the nutrients in leaves are relatively expensive to replace (Janzen 1974). From the plant's viewpoint, there would seem to be less expensive means to suppress insect growth than by reducing leaf photosynthetic capacity, and tropical trees on sterile soils in Africa do have higher concentrations of leaf phenols and other N-poor toxins than trees on richer sites (McKey et al. 1978). However, the resolution of this question clearly will depend on the quantitative response of photosynthesis and insect growth to leaf nitrogen content, and the cost-effectiveness of defensive compounds, with the outcome likely to vary from species to species. For example, in *Diplacus aurantiacus*, the data of Gulmon and Chu (1980) and Lincoln et al. (1981) suggest that increasing leaf resin has a greater net benefit than decreasing leaf nitrogen at observed levels, in terms of its relative effects on insect survivorship and photosynthetic capacity. On the other hand, in *Eucalyptus* Morrow and Fox (1978) found that the concentration of essential oils had little effect on larval feeding and that leaf nitrogen content was the principal determinant of insect growth. Such approaches, bridging physiological ecology and plant plant-animal interactions, should be extremely valuable in the humid tropics, particularly when biotic costs and photosynthetic benefits are expressed in the same currency.

3. CANOPY LEVEL

Quantification of the geometric, mechanical, aerodynamic, and optical properties of plant canopies is a far more complex task than that for individual leaves. Perhaps as a consequence, there are very few hard data on the form and properties of tree crowns in tropical forests (but see Brünig 1970, 1971, Rollet 1974, Ashton 1978, Fisher 1978), most information being anecdotal or presented in the form of profile diagrams (e.g., see Davis and Richards 1933, 1934, Richards 1952, Ashton 1964, Whitmore 1975, Bourgeron and Guillaumet 1981, Hall and Swaine 1981). Thus, the discussion in the following sections is far more telegraphic and speculative than the corresponding sections on leaf form, and is aimed mainly at developing predictions for trends in crown form beyond those already alluded to in sections 2.1.2 - 2.2.2. I will briefly discuss three aspects of crown form (vertical leaf distribution, crown profile, and canopy aerodynamic roughness) with implications for the economics of gas exchange, and one aspect (single- vs. multi-stemmed growth) with implications for the economics of support and supply.

3.1. Economics of gas exchange

3.1.1. Vertical distribution of foliage. Horn (1971, 1975) has analyzed the photosynthetic costs and benefits associated with two basic kinds of vertical leaf arrangements in trees, which he calls monolayers and multilayers. Monolayers pack their leaves in a single shell, whereas multilayers scatter their leaves diffusely within a layer, but hold several layers of leaves. If the lower layers are held more than about 100 leaf (or lobe, or leaflet) diameters below the upper layers, they escape the penumbras cast by the upper leaves and each layer acts as a uniform density filter. Together with the nonlinear response of photosynthesis to light, this implies that a multilayer can hold far more productive leaf surface than the ground area its crown covers, even if its leaves are horizontal (Horn 1971). For example, consider a hypothetical multilayer growing in full sunlight, whose photosynthetic rate saturates at roughly 20% full sunlight and just balances respiration at 5% full sunlight, taking into account acclimatization to different levels of illumination. Then, if the multilayer scatters its leaves so as to cover about half the projected area of the plant crown, its uppermost layer will photosynthesize at the saturated rate over a leaf area just half that of a monolayer of the same size. Its second layer will be exposed to 50% full sunlight, photosynthesize at the full rate, and bring total photosynthesis up to that of the monolayer. A third layer would be exposed to 25% full sunlight, again photosynthesize at the maximum rate, increase total photosynthesis to 1.5 times that of the monolayer, and still transmit enough light to support additional lower layers. Horn (1971) concludes that total plant growth should be maximized by continuing this process until light levels reach the leaf compensation point, where the cost of adding a leaf just balances the energetic profit it earns.

The significant conclusions to be drawn from this model are (i) multilayers are more productive than monolayers under brightly lit conditions; (ii) monolayers are more productive in deep shade because they have less internal shading and no leaves operating below the compensation point; and thus (iii) multilayers (with small leaves and short leaf penumbras) should be favored in trees of early succession, and monolayers should be favored in trees that regenerate in the shade.

Horn's (1971) model works well for temperate pioneer species, but does not account for the megaphyllous, nearly monolayered trees of tropical early succession (e.g., *Ochroma*, *Musanga*). Some of the reasons for this apparent anomaly — such as high humidity, high rainfall, and potential nutrient flushes in tropical gaps (Bazzaz 1980) favoring large effective leaf size, as well as the advantages of throwaway branches — are discussed in sections 2.1.1 and 2.2.2.

One way in which Horn's model might usefully be modified would be to extend the notion of light compensation point, which is a familiar physiological concept and helps define the optimal number of leaf layers for a given set of conditions. The compensation point, as usually defined, is a poor measure of the net benefit of a leaf because it only accounts for the balance between photosynthesis and instantaneous leaf respiration. However, there are six additional energetic costs associated with photosynthesis that should increase the effective *ecological compensation point* at which total leaf benefits and costs just balance. These include (i) nighttime leaf respiration; (ii) effective daily cost of leaf construction, amortized over the lifetime of the leaf; (iii) marginal and average costs of roots, xylem, and phloem needed to supply an additional leaf (Raven 1976); (iv) marginal mechanical cost of supporting an additional leaf in a given position; and (v) expected loss of productivity due to herbivory or disease. Finally, the ecological compensation point must further be increased to account for the fact that, although a leaf may be operating above its compensation point as determined by the preceding five costs, its net photosynthesis may be so low that it would pay the plant to extract nutrients from that leaf, perhaps inefficiently and at energetic cost, and place them in a new, well-lit leaf (see Field 1981). This sixth energetic decrement in leaf productivity might best be considered an opportunity cost.

The result of incorporating these six costs into the Horn model is that the optimal number of leaf layers (and hence, leaf area index (LAI) for a given set of leaf inclinations) should decrease with habitat aridity, plant height, intensity of herbivory, and inefficiency of nutrient retranslocation from leaves. Put another way, the ecological compensation point should be lower in moist habitats, in short plants or herbs with less expensive mechanical tissue (Raven 1976, Givnish 1982), in infrequently browsed plants, and in plants with efficient nutrient retranslocation. These predictions have important implications for tree form and vegetation structure, but there are few data with which to test them as yet. Waring et al. (1978) demonstrate a regular increase in community LAI with moisture supply for coniferous forests in the Cascades and Siskiyou Mountains of Oregon, but the average LAI for tropical rain forests and seasonal forests reported by Schulze (1982) shows little systematic variation. More work is clearly needed in this area.

3.1.2. Canopy profile. Ecologists who construct profile diagrams have frequently remarked on apparent differences in the shape of tree crowns at different levels in the forest (e.g., Richards 1952). Emergents are said to have broad spreading crowns, trees of the continuous canopy below to have more rounded crowns, and subcanopy trees to have elongate crowns. Whether these differences in crown height/width ratio are real is difficult to judge, since most profile diagrams are drawn by an observer standing on the forest floor far below the canopy, so that perspective of the vertical dimension is foreshortened. If we assume that such differences are real, and that species at different levels are characterized by them, we still would not know whether these crown shapes are genetically determined or merely a result of environmental differences. For example, might not the increase in relative crown width with height be the result of populational thinning of a cohort of trees, with more space being made available to trees that survive later and later into the thinning process?

What are the implications of differences in crown shape for gas exchange? Leaving aside mechanical costs, which depend on the exact pattern of branching and growth by which a given crown is achieved (see section 2.2.2), the effects of crown shape on photosynthesis and transpiration may be similar to those caused by leaf inclination (see section 2.1.2). In regions where the sun passes close to the zenith, a plant with a hemispheroidal crown should experience lower light intensities and heat loads on leaves in the lower portions of its crown, even when growing in the open. This is not because leaves are assumed to lie tangent to a hypothetical crown surface. Rather, even if all leaves in the crown have the same orientation, those on the lower surface should receive less light because other portions of the crown will occlude direct-beam and diffuse radiation during part of the day. The upper portions of the crown are exposed more directly to the sun and its leaves may have higher photosynthetic rates, particularly among those just inside the crown surface. At the same time, the evaporative demand imposed by the greater heat load should increase transpirational costs in the upper portion of the crown. Thus, as moisture availability increases and the cost of transpiration decreases, crown shape should shift from cylindrical, strongly self-shading forms toward flat-topped canopies with maximal exposure (Givnish 1976; see Horn 1971, Brünig 1976). Along a gradient of increasing moisture supply in the central United States, crown shapes of the dominant tree genera do appear to shift from cylindrical forms in *Carya* and *Quercus* of semi-arid Ozark woodlands, to hemispherical forms in *Acer*, *Fagus* and *Tilia* of the mixed mesophytic forests, to flat-topped forms in *Fraxinus* and *Ulmus* in mesic swamp forests on ancient lake bottoms near the Great Lakes (Givnish 1976; see Küchler 1964 for vegetation types). The one elm species in the northcentral United States without a spreading, ulmaceous crown is the rock elm *Ulmus thomasii*, typically found on drier sites or on heavy soils. Other elms with more diffuse, rounded crowns include the winged elm (*U. alata*) and cedar elm (*U. crassifolia*), both typical of dry sites in the southeastern United States.

Obviously, the preceding argument is simplistic in that it is based on a single constraint. Nevertheless, it represents at least one of the selective pressures that may shape crown form when other factors, perhaps more important for trees in moist tropical forests, are taken into account. However, there is at least one glaring exception to the trends expected — the flat-topped *Acacia* trees of African semi-arid savannas. Walter (1973) states that the most important competitors of these plants may not be other trees, but grasses. In the tension zone between arid grassland and semi-arid woodland, grasses have an advantage in competing with trees for water, in that they have intensive, shallow root systems, can transpire at high rates during the growing season, and die back to the ground when soil moisture is exhausted. By contrast, trees have deeper and less intensive root systems, and have a permanent above-ground body that continues to respire and transpire after leaf fall (Walter 1973). Grasses could potentially outcompete trees by intercepting and transpiring most soil water before it reaches the trees' roots, leaving the trees high and dry after the rains cease. The trees' obvious advantage is that they can overtop and shade the grasses, whose vertical leaf surfaces are not well adapted to low light levels. The advantage of a flat crown may be that it maximizes the area shaded near a tree's roots, particularly at low latitudes, and may so suppress the growth of competing grasses that its own growth is enhanced, even at the cost of higher transpiration. However, a flat crown may have unexpected effects on transpiration as well, as the following section suggests.

3.1.3. Canopy roughness. As previously noted, Brünig (1970, 1971) found that the aerodynamic roughness of tree crowns and forest canopies tends to decrease in moving from lowland dipterocarp forests to heath forests in Borneo (figure 16), and he reports similar trends along gradients of decreasing soil fertility in the Rio Negro basin

Figure 16. Trends in forest stature and aerodynamic roughness of the forest canopy and individual tree crowns, along a gradient of decreasing soil fertility from lowland dipterocarp forests (left) to heath forests and kerangas in Borneo (after Brünig 1983).

(Brunig 1983). **The aerodynamic roughness of a** forest canopy or individual tree crown, as measured by the micrometeorological parameter z_o , determines the extent to which there is laminar air flow around plant crowns. Visually smooth canopies, with leaves packed toward the end of twigs that terminate on a common surface, tend to be aerodynamically smooth as well. The aerodynamic roughness of a forest canopy is a function of the roughness of individual crowns, their size, and their relative vertical position.

As Brünig (1970, 1971, 1976, 1983) himself states, increased canopy roughness should increase transpiration by permitting the movement of relatively dry air through the canopy. In addition, increased canopy roughness should increase light penetration into individual tree crowns (Brünig 1983), permit a multilayered leaf arrangement, and increase photosynthesis. Thus, as Brünig (1970, 1971) states, aerodynamically smooth canopies should be favored in dry areas because transpirational costs are high. However, to the extent that nutrient poverty reduces leaf nutrient levels, and thus reduces the maximum photosynthetic rate and its sensitivity to other limiting factors such as light (see section 2.1.1), a given increment in canopy roughness is likely to have a much smaller effect on photosynthetic benefits on sterile sites than on fertile sites. Thus, aerodynamically smooth canopies may represent an adaptation to shortages of either water or nutrients, and be yet another example of the natural duality of xeromorphism and peinomorphism (sensu Walter 1973). Thus, each of the characteristics Brünig considered to be adaptations to drought — small, thick leaves; high leaf albedo and inclination; aerodynamically smooth canopies — may actually represent direct adaptations to nutrient poverty. Of course, this need not be the case — wet sterile sites like bogs or peat swamps may favor xeromorphic plants because the costs of transpiration are actually high, whether as a result of root inefficiency under anoxic conditions, or because roots are restricted to superficial, oxidized soil layers and thus more exposed to short periods without rainfall (but see Small 1972a). In this regard, note that features that result in aerodynamically smooth canopies, such as the clustering of leaves toward the tips of branches, are also common in nutrient-poor montane rain forests. However, Buckley et al. (1980) found little physiological tolerance to desiccation in trees of montane rain forest, and Peace and MacDonald (1981) obtained similar results for heath forest.

Note also that high winds favor aerodynamically smooth canopies as a means of avoiding both desiccation and mechanical damage. Nolde (1941) and Horn (1971) suggest that the flat-topped, aerodynamically smooth canopies of savanna *Acacia* trees may actually serve to reduce transpiration, rather than increase it as suggested in the previous section. However, it should be recognized that canopy shape and aerodynamic roughness can vary somewhat independently, and that a smooth, steep-sided crown could serve to reduce transpiration even more than a smooth, flat crown. When canopy shape and aerodynamic roughness co-vary in **such a way that their effects on gas exchange** oppose each other, only a quantitative analysis can reveal the shape/roughness combination that would maximize growth.

3.2. Economics of support and supply

Hallé and Oldeman (1970) and Hallé et al. (1978) classify the growth patterns of tropical trees into 23 architectural models. These models are largely based on whether (i) plant has single or multiple trunks; (ii) trunks are branched or unbranched; (iii) inflorescences are terminal or lateral; (iv) branches are orthotropic or plagiotropic; and (v) leaf production is continuous or discontinous. The potential adaptive significance of variation involving points (ii), (iv), and (v) has been discussed in sections 2.1.1 - 2.2.2; (iii) involves reproductive biology and will not be addressed here. In this section the significance of variation in point (i) is analyzed, and implications drawn for the mechanical efficiency of crowns supported by a single (branching) stem or by multiple stems.

Such tree- and shrub-like growth forms involve a tradeoff between stem length and cost per unit length, at least if plant crowns are considered as static entities (see section 2.2.2, Givnish 1978b, and King 1981 for complications arising in dynamic systems). Multiple radial stems minimize the length of the support arm to each element of the plant crown. However, to ensure mechanical stability, the diameter of each stem should increase with the 3/2 power of its length, but less strongly with the mass of the crown element being supported (McMahon 1973, King 1981, Givnish 1982). Hence, coalescence of subparallel, radial stems into a single branching stem should be favored if the resulting increase in the total length of support arms is outweighed by the lower effective cost of such arms per unit length and crown element being supported. Low, relatively broad crowns should be supported on multiple stems because radii to different portions of the crown diverge strongly, so that the increase in total length of support arms would be relatively great. Similar considerations suggest that tall, relatively narrow crowns should be supported on a single, branching stem. Furthermore, among plants with a given crown width/height ratio, taller species should show less tendency toward the multi-stemmed habit, given the weaker dependence in tall plants of stem diameter on leaf mass vs. stem mass being supported (see Givnish 1982).

The predicted tendency for multi-stemmed woody plants ("shrubs") and single-stemmed woody plants ("trees") to be tall seems generally true, particularly in temperate areas. For example, among *Eucalyptus* species native to southwestern Australia there is a strong association between branching habit and typical crown height (table 3). Obligate mallee eucalypts, with multiple stems arising from a common rootstock, usually have a maximum crown height of 5 m or less; tree eucalypts are typically much taller (table 3). Species with a facultative growth form, showing a tendency toward the mallee or tree habit but usually a mallee when short, have the expected intermediate crown heights (table 3). Similar trends can be seen inter- and intraspecifically in *Banksia* and other elements of the flora of southwestern Australia, with a sharp drop in the proportion of shrubby species above a crown height of 4 meters (Givnish, unpublished data).

Table 3. Growth forms of *Eucalyptus* species native to the Goldfields region of southwestern Australia. Entries represent the number of species with a specified growth form at a given crown height. Data compiled from Chippendale (1973).

Height (m)	Mallee	Mallee/ tree	Tree/ mallee	Tree
0 - 3	19			
3 - 6	24	11	3	1
6 - 9		9	8	3
9 - 12	1	6	4	7
12 - 15				4
15 - 20		1	2	1
20 - 25				8
25 - 30				1

However, in rain forests the incidence of shrubby growth forms among woody plants < 4 m tall appears to be much less, with dominance by sparsely branched trees and (at lower abundance, particularly in South America) unbranched treelets or *schopfbäumchen* (Schimper 1898, Richards 1952, Whitmore 1975, Croat 1978, Hall and Swaine 1981). For example, a tabulation of data on growth forms of self-supporting woody plants native to Wet Evergreen Forest in Ghana (table 4) shows that only 6 of 95 species less than 10 meters tall are obligate shrubs. An additional 17 species are facultative shrubs or small trees, and 3 species are characterized by Leeuwenberg's model (Hallé et al. 1978), a rather shrub-like growth form with no main trunk above the first branchpoint. The

Table 4. Growth forms of erect woody plants native to Wet Evergreen Forest in Ghana. Data compiled from descriptions by Hall and Swaine (1980).

Growth form	No. of species
Tall trees (30+ m)	48
Medium trees (10-30 m)	72
Small trees (2-10 m)	
Branched	35
Unbranched	4
Pygmy trees (0-2 m)	
Branched*	19, 3L
Unbranched	11
Shrubs (0-5 m, branched at base)	6
Shrubs/small trees (0-10 m)	17

*L = Leeuwenberg's model (see text)

69 species less than 10 meters tall are trees, including 11 unbranched treelets and 19 branched pygmy trees less than 2 meters in height (table 4). Similarly, there are few multi-stemmed shrubs in the understories of rain forests in New Caledonia (Givnish, unpublished data) and on Barro Colorado Island in Panama (T. Croat, personal communication). In contrast, most woody taxa of low stature in moist temperate forest are shrubs (e.g., most species of *Cornus*, *Euonymus*, *Lindera*, *Kalmia*, *Ribes*, *Rubus*, *Rhododendron*, *Vaccinium*, and *Viburnum* in the northeastern U. S.).

The reasons for this anomaly are obscure. Shrubs differ from trees in having more meristems active, more potential points for stem regeneration, so that they may be favored in open habitats in which plant crowns are frequently destroyed, as in fire-swept chaparral (Miller and Stoner 1979), eucalypt forest, or Australian sand heath. Even if stem regeneration were a more important selective force favoring shrubby growth than differences in stem cost, however, it would be difficult to explain the multi-stemmed habit in short woody plants of closed temperate forests. Why should understory woody plants be shrubby in temperate forests and tree-like in tropical forests? Three possibilities come to mind, none particularly persuasive. First, an unbranched or sparsely branched treelet with throw-away, branchlike leaves may shed vines well and so escape a major source of potential mortality in tropical forests (Janzen 1975, Putz 1980). Second, a treelet with large leaves may be better adapted than a shrub with smaller leaves and axes for growth in dimly lit understories, by having a greater proportion of energy devoted to leaf vs. stem mass per node, and/or by improving height growth (see section 2.2.2). Third, the large leaf size favored in rain forest understories may impose a need for thick axes on which the leaves can be borne (see Hallé et al. 1978), and may thus favor sparsely branched growth forms. The functional significance of the sparsely branched treelet vs. understory shrub clearly requires further study.

3.3. Economics of biotic interactions

The possibility that certain aspects of crown form may represent adaptations to deter herbivores has received little discussion and will not be discussed at length. Attention is called to the suggestion by Janzen and Martin (1982) that certain aspects of crown form in trees of seasonal forests in Central America, notably trunk and branch spines, represent adaptations to browsing by large, extinct terrestrial herbivores. Large vertebrate grazers and browsers have low biomass in tropical rain forests, since most foliage is beyond the reach of terrestrial forms (Eisenberg 1980); the role of arboreal vertebrate folivores in shaping the evolution of plant crowns has received no attention (e.g., see Montgomery 1978). Invertebrate herbivores operate at a different spatial scale, and are probably met most effectively by mechanical, chemical, or biotic defenses at the leaf level.

Vertebrate pollinators and seed dispersers may play a more important role than herbivores in shaping crown form in tropical trees. Faegri and van der Pijl (1966, 1969) suggest that bat-pollinated or bat-dispersed trees may have canopies reorganized to permit nocturnal visitation by these large animals. Proposed adaptations for bat pollination and/or seed dispersal include the open structure of pagoda-like crowns (as in *Terminalia catappa*), long flower and/or fruit stalks hanging below the canopy (e.g., *Parkia*), and cauliflory and flagelliflory (e.g., *Ficus*, *Lansium*). Eggeling (1955) and Osmaston (1965) have described reorganization of the crown of pistillate trees of the bat-dispersed *Chlorophora excelsa*.

4. DISCUSSION

It is not possible in a single paper to analyze all aspects of adaptation in leaf form and canopy structure, even for those traits typical of tropical moist forests. Important traits not discussed here include (i) leaf drip tips (figure 1) and their potential function in more rapid drying of the leaf surface after rains (Dean and Smith 1979) or reduction of splash erosion below understory plants that most frequently possess elongate apices (Richards 1952, Williamson 1981); (ii) trunk buttresses and their potential function as tensile support elements (Henwood 1973); (iii) leaf traits that may increase light absorption by understory herbs, such as reddish leaf undersides that reflect unabsorbed photons of PAR back through the mesophyll (Lee et al. 1979), iridescent leaf surfaces that increase absorption of red light (Lee and Lowry 1975), and velvety leaf surfaces with lens-shaped epidermal cells that focus light on chloroplasts in the mesophyll (Schimper 1898); (iv) bright coloration of flushes of young foliage (Richards 1952); (v) stomatal clusters and stomatal domes in understory species (Skog 1976); and (vi) stilt roots.

Vines have also been ignored in this paper, even though they are a characteristic element of tropical rain forest, particularly as large woody lianas that invade the forest canopy and overgrow portions of tree crowns (Richards 1952). Such lianas are most abundant and diverse in tropical rain forests, and become progressively less important toward higher latitudes, drier climates, and sterile soils like those found in heath forests (Schenck 1892, Richards 1952); they tend to be sun- rather than shade-adapted. Croat (1978) reports that lianas encompass 13% of all species found on Barro Colorado Island; they comprise roughly a quarter of all woody angiosperms in that rain forest habitat. Biologists since Darwin (1865) have agreed that vines obtain a competitive advantage by being able to allocate less energy to support tissue and more to productive new leaves. It is thus clear why vines should be favored in areas supporting closed forests, but it is not clear why they should be less important in drier or nutrient-poor sites. Lianas tend to have xylem elements with large diameters, and this may render them more susceptible to cavitation (Zimmermann 1978). A deeper reason may be that in areas where plants must allocate much energy to roots for each allocation to leaves, relatively little energy will be diverted into support tissue and adoption of the vine habit can yield little advantage. If we use Monsi's (1968) model of exponential growth, this can be seen easily. The rate of change of leaf mass L is given by $dL/dt = fPL$, where f is the fraction of photosynthate allocated to leaves, and P is the net photosynthetic rate per unit leaf mass. If for each unit of energy allocated to leaves, a vine must allocate a fraction a to roots and a fraction b_v (b_t for trees), then the ratio of exponential growth coefficients of vine and tree is just $(1 + a + b_t)/(1 + a + b_v)$. Clearly, since $b_t > b_v$, this ratio and the vine's advantage will decrease as a and the allocation to roots in drier or more nutrient-poor environments increases. Vines have received little attention, and intensive study may yield important insights into the ecology and evolution of tropical plants. For example, Croat (1978) notes that over 50% of the climbing plants on Barro Colorado Island have opposite leaves and branching. He suggests this may be an adaptation to prevent falls from a tree crotch or other lodging point, by providing a broad base of support. This may help explain the unexpected occurrence of opposite leaves in certain typical families of rain forests, such as the Apocynaceae, Bignoniaceae, portions of the Celastraceae and Combretaceae, Loganiaceae, and Malpighiaceae (see table 2).

The preceding analysis should indicate three points of general importance. First, trends in leaf traits that influence the balance between photosynthesis and transpiration can only be understood by taking into account below-ground costs of water and nutrient uptake. Second, such an analysis suggests that many xeromorphic adaptations shown by plants in moist but nutrient-poor sites are actually adaptations to nutrient poverty itself, rather than to water shortages postulated by Brünig and others. Brünig's explanation may prove correct, but the argument advanced here indicates that the functional significance of bog xeromorphism should be reconsidered. Finally, although most models in this paper are qualitative in nature, it should be clear that a convincing test of their predictions must be quantitative. Only a quantitative model can offer, as well, a test of the central assumption that natural selection favors plants whose form tends to maximize whole-plant carbon gain.

5. ACKNOWLEDGEMENTS

It is a pleasure to thank L. Skog and J. Wurdack of the Smithsonian Institution, and R. Howard, P. B. Tomlinson, C. Wood, and especially P. Stevens of Harvard University for useful comments and data on anisophylly. J. Sperry checked the phyllotaxis and leaf symmetry of Trema in the field. S. Bartz and E. Burkhardt rendered invaluable technical assistance.

6. REFERENCES

Anderson JAR (1961) The structure and development of the peat swamps of Sarawak and Brunei, J. Trop. Geog. 18, 7-16.

Ashton, PS (1964) Ecological studies in the mixed dipterocarp forests of Brunei State, Oxford For. Mem. 25, 1-75.

Ashton PS (1978) Crown characteristics of tropical trees. In Tomlinson PB and Zimmermann MH, eds. Tropical trees as living systems, Cambridge, Cambridge Univ. Press.

Ashton PS and Brünig EF (1975) The variation of tropical moist forest in relation to environmental factors and its relevance to land-use planning, Mitt. Bundesforschungan. Forst-Holzwirt. 109, 60-86.

Barlow BA and Wiens D (1977) Host-parasite resemblance in Australian mistletoes: the case for cryptic mimicry, Evol. 31, 69-84.

Bazzaz FA and Pickett STA (1980) Physiological ecology of tropical succession: a comparative review, Ann. Rev. Ecol. Syst. 11, 287-310.

Beadle NCW (1962) Soil phosphate and the delimitation of plant communities in eastern Australia, Ecol. 35, 370-375.

Beadle NCW (1966) Soil phosphate and its role in molding segments of the Australian flora and vegetation, with special reference to xeromorphy and scleromorphy, Ecol. 47, 992-1007.

Beard JS (1944) Climax vegetation in tropical America, Ecol. 25, 127-158.

Beard JS (1955) The classification of tropical American vegetation-types, Ecol. 36, 89-100.

Brown WH (1919) Vegetation of the Philippine mountains, Manila, Phil. Bur. Sci., Dept. Agr. & Nat. Res.

Brünig EF (1970) Stand structure, physiognomy and environmental factors in some lowland forests in Sarawak, Trop. Ecol. 11, 26-43.

Brünig EF (1971) On the ecological significance of drought in the equatorial wet evergreen (rain) forests of Sarawak (Borneo). In Flendley JR, ed. The water relations of Malesian forests, Hull, Univ. Hull Press.

Brünig EF (1976) Tree form in relation to environmental conditions: an ecological viewpoint. In Cannel MGR and Last FT, eds. Tree physiology and yield improvement, London, Academic.

Brünig EF (1983) Vegetation structure and growth. In Golley FB, ed. Tropical rain forest ecosystems: structure and function, Amsterdam, Elsevier.

Buckley RC, Corlett RT and Grubb PJ (1980) Are the xeromorphic trees of tropical upper montane rain forests drought resistant?, Biotr. 12, 124-136.

Cain SA, de Oliviera Castro G, Pires JM and da Silva NT (1956) Application of some phytosociological techniques to Brazilian rain forest, Am. J. Bot. 43, 911-941.

Carlquist S (1970) Hawaii: a natural history, New York, Natural History Press.

Chabot BF and Hicks DF (1982) The ecology of leaf life spans, Ann. Rev. Ecol. Syst. 13, 229-259.

Chippendale GM (1973) Eucalypts of the Western Australia goldfields (and the adjacent wheatbelt), Canberra, Austral. Govt. Publ.

Cohen D (1970) The expected efficiency of water utilization in plants under different competitive and selective regimes, Isr. J. Bot. 19, 50-54.

Coley PD (1983) Herbivory and defensive characteristics of tree species in a lowland tropical forest, Ecol. Monog. 53, 209-233.

Cowling RM and Campbell BM (1980) Convergence in vegetation structure in the Mediterranean communities of California, Chile, and South Africa, Vegetatio 43, 191-198.
Croat TB (1978) Flora of Barro Colorado Island, Stanford, Stanford Univ. Press.
Cronquist A (1981) An integrated system of classification of flowering plants, New York, Columbia Univ. Press.
Cuenca G (1976) Balance nutricional de algunas leñosas de dos ecosistemas contrastantes: bosque nublado y bosque deciduo. Thesis, Univ. Central de Venezuela, Caracas.
Darwin C (1865) On the movements and habits of climbing plants, London, Longman and Green.
Daubenmire R (1972) Phenology and other characteristics of tropical semi-deciduous forest in northwestern Costa Rica, J. Ecol. 60, 147-170.
David TAW and Richards PW (1933) The vegetation of Moraballi Creek, British Guiana: an ecological study of a limited area of tropical rain forest, I. J. Ecol. 21, 350-384.
David TAW and Richards PW (1934) The vegetation of Moraballi Creek, British Guiana: an ecological study of a limited area of tropical rain forest, II. J. Ecol. 22, 106-155.
Dean JM and Smith AP (1979) Behavioral and morphological adaptations of a tropical plant to high rainfall, Biotr. 10, 152-154.
De Oliviera JGB and Labouriau LG (1961) Transpiração de algumas plantas de caatinga aclimatas do Jardim Botanico do Rio de Janeiro. I. Comportamento de *Caesalpinia pyramidalis* Tull., de *Zizyphus joazeiro* Mart., de *Jatropha phyllacantha* Muell. Arg. e de *Spondias mombin* Arruda, An. Acad. Bras. Ciênc. 33, 351-373.
Dolph GE and Dilcher DL (1980a) Variation in leaf size with respect to climate in Costa Rica, Biotr. 12, 91-99.
Dolph GE and Dlicher DL (1980b) Variation in leaf size with respect to climate in the tropics of the western hemisphere, Bull. Torr. Bot. Club 107, 154-162.
Eggeling WJ (1955) The relationship between crown form and sex in *Chlorophora excelsa*, Emp. For. Rev. 34, 294.
Ehleringer J and Forseth I (1980) Solar tracking by plants, Science 210, 1094-1098.
Ehleringer JR and Mooney HA (1978) Leaf hairs: effects on physiological activity and adaptive value to a desert shrub, Oecol. 37, 183-200.
Ehleringer JR, Mooney HA, Gulmon SL and Rundel PW (1981) Parallel evolution of leaf pubescence in *Encelia* in coastal deserts of North and South America, Oecol. 49, 38-41.
Eisenberg JF (1980) The density and biomass of tropical mammals. In Soulé ME and Wilcox BA, eds. Conservation biology, Sunderland, Sinauer.
Faegri K and van der Pijl L (1966) Principles of pollination biology, Oxford, Pergamon.
Faegri K and van der Pijl L (1969) Principles of dispersal in higher plants, New York, Springer.
Ferri MG (1961) Problems of water relations of some Brazilian vegetation types, with special consideration of the concepts of xeromorphy and xerophytism. In Plant-water relationships in arid and semi-arid conditions, New York, UNESCO.
Field C (1981) Leaf age effects on the carbon gain of individual leaves in relation to microsite. In Margaris NS and Mooney HA, eds. Components of productivity of Mediterranean-climate regions: basic and applied aspects, The Hague, Dr. Junk.
Forseth I and Ehleringer JR (1979) Solar tracking response to drought in a desert annual, Oecol. 44, 159-164.
Fox JED (1972) The natural vegetation of Sabah and natural regeneration of the dipterocarp forests. Ph.D. thesis, University of Wales, Cardiff.
Frankie GW, Baker HG and Opler PA (1974) Comparative phenological studies of trees in tropical lowland wet and dry forest sites of Costa Rica, J. Ecol. 62, 881-919.
Gates DM (1962) Energy exchange in the biosphere, New York, Harper & Row.
Gates DM (1965) Energy, plants and ecology, Ecol. 46, 1-13.
Gates DM (1980) Biophysical ecology, New York, Springer.
Gates DM and Benedict CM (1963) Convection phenomena from plants in still air, Am. J. Bot. 50, 563-573.
Gates DM and Papian LE (1971) An atlas of leaf energy budgets, New York, Academic.
Geller GN and Smith WK (1980) Leaf and environmental parameters influencing transpiration: theory and field measurements, Oecol. 46, 308-313.
Gentry AH (1969) A comparison of some leaf characteristics of tropical dry forest and tropical wet forest in Costa Rica, Turrialba 19, 419-428.
Gilbert LE (1975) Ecological consequences of a coevolved mutualism between butterflies and plants. In Gilbert LE and Raven PH, eds. Coevolution of animals and plants, Austin, Univ. Texas Press.
Gilbert LE (1980) Food web organization and the conservation of neotropical diversity. In Soulé ME and Wilcox BA, eds. Conservation biology, Sunderland, Sinauer.
Givnish TJ (1976) Leaf form in relation to environment. Ph.D. thesis, Princeton University, Princeton.
Givnish TJ (1978a) Ecological aspects of plant morphology: leaf form in relation to environment, Acta Biotheor. 27(7), 83-142.
Givnish TJ (1978b) On the adaptive significance of compound leaves, with particular reference to tropical trees. In Tomlinson PB and Zimmermann MH, eds. Tropical trees as living systems, New York, Cambridge Univ. Press.
Givnish TJ (1979) On the adaptive significance of leaf form. In Solbrig OT, Jain S, Johnson GB and Raven PH, eds. Topics in plant population biology, New York, Columbia Univ. Press.
Givnish TJ (1982) On the adaptive significance of leaf height in forest herbs, Am. Nat. 120, 353-381.
Givnish TJ and Vermeij GJ (1976) Sizes and shapes of liane leaves, Am. Nat. 100, 743-778.

Greacen EL, Ponsana P and Barley KP (1976) Resistance to water flow in the roots of cereals. In Lange OL, Kappen L and Schulze E.-D., eds. Water and plant life: problems and modern approaches, New York, Springer.

Grubb PJ (1974) Factors controlling the distribution of forest types on tropical mountains — new facts and a new perspective. In Flenley JR, ed. Altitudinal zonation of forests in Malesia, Hull, Univ. Hull Press.

Grubb PJ (1977) Control of forest growth and distribution on wet tropical mountains, Ann. Rev. Ecol. Syst. 8, 83-107.

Grubb PJ, Lloyd JR, Pennington TD and Whitmore TC (1963) A comparison of montane and lowland rain forest in Ecuador. I. Forest structure, physiognomy, and floristics, J. Ecol. 51, 567-602.

Gulmon SL and Chu CC (1981) The effect of light and nitrogen on photosynthesis, leaf characteristics, and dry matter allocation in the chaparral shrub, *Diplacus auranticus*, Oecol. 49, 207-212.

Hall JB and Swaine MD (1981) Distribution and ecology of vascular plants in a tropical rain forest, The Hague, Dr. Junk.

Hallé F and Oldeman RAA (1970) Essai sur l'architecture et la dynamique de croissance des arbres tropicaux, Paris, Masson.

Hallé F, Oldeman RAA and Tomlinson PB (1978) Tropical trees and forests, New York, Cambridge Univ. Press.

Henwood K (1973) A structural model of forces in buttressed tropical rain forest trees. Biotr. 5, 83-93.

Holttum RE (1953) Evolutionary trends in an equatorial climate, Symp. Soc. Exp. Biol. 7, 154-173.

Hartshorn, GS (1978) Tree falls and tropical forest dynamics. In Tomlinson PB and Zimmerman MH, eds. Tropical trees as living systems, New York, Cambridge Univ. Press.

Hedberg O (1951) Vegetation belts of the East African mountains, Sv. Bot. Tidsk. 44, 140-202.

Heywood VH (1978) Flowering plants of the world, New York, Mayflower Books.

Horn HS (1971) The adaptive geometry of trees, Princeton, Princeton Univ. Press.

Horn HS (1975) Forest succession, Sci. Amer. 232 (5), 90-98.

Howard RA (1969) The ecology of an elfin forest in Puerto Rico. 8. Studies of stem growth and form and of leaf structure, J. Arnold Arbor. 50, 225-267

Howland HC (1962) Structural, hydraulic, and "economic" aspects of leaf venation and shape. In Bernard EE and Kare MR, eds. Biological prototypes and synthetic systems, Ithaca, Cornell Univ. Press.

Janzen DH (1970) *Jacquinia pungens*, a heliophile from the understory of tropical deciduous forest, Biotr. 2, 112-119.

Janzen DH (1975) Ecology of plants in the tropics, London, Edward Arnold.

Janzen DH (1983) Food webs: who eats what, why, how and with what effects in a tropical forest? In Golley FB, ed. Tropical rain forest ecosystems: structure and function, New York, Springer.

Janzen DH and Martin PS (1982) Neotropical anachronisms: the fruits the gomphotheres ate. Science 214, 19-27.

King D (1981) Tree dimensions: maximizing the rate of height growth in dense stands, Oecol. 51, 351-356.

Koriba K (1958) On the periodicity of tree growth in the tropics with reference to the mode of branching, the leaf fall and the formation of the resting bud, Gard. Bull. 17, 11-81.

Küchler AW (1964) Potential natural vegetation of the coterminus United States, New York, Amer. Geogr. Soc. Publ. 146.

Lee DW and Lowry JB (1975) Physical basis and ecological significance of iridescence in blue plants, Nature 254, 50-51.

Lee DW, Lowry JB and Stone BC (1979) Abaxial anthocyanin layer in leaves of tropical rain forest plants: enhancement of light capture in deep shade, Biotr. 11, 70-79.

Leigh EG (1972) The golden section and spiral leaf arrangement. In Deevey ES, ed. Growth by intussusception, Hamden CO, Archon.

Leigh EG (1975) Structure and climate in tropical rain forests, Ann. Rev. Ecol. Syst. 6, 67-86.

Maximov NA (1929) The plant in relation to water, London, George Allen & Unwin.

Maximov NA (1931) The physiological significance of the xeromorphic structure of plants, J. Ecol. 19, 272-282.

McMahon TA (1973) Size and shape in biology, Science 179, 1201-1204.

Medina E (1970) Relationships between nitrogen level, photosynthetic capacity, and carboxydismutase activity in *Atriplex patula* leaves, Carn. Inst. Yrbk. 69, 655-662.

Medina E (1971) Effect of nitrogen supply and light intensity during growth on the photosynthetic capacity and carboxydismutase activity of *Atriplex patula* ssp. *hastata*, Carn. Inst. Yrbk. 70, 551-559.

Medina E (1983) Adaptations of tropical trees to moisture stress. In Golley FB, ed. Tropical rain forest ecosystems: structure and function, New York, Springer.

Medina E, Sobrado M and Herrara R (1978) Significance of leaf orientation for leaf temperature in an Amazonian sclerophyll vegetation, Radiat. Environ. Biophys. 15, 131-140.

Miller PC (1979) Quantitative plant ecology. In Horn D, ed. Analysis of ecological systems, Columbus, Ohio State Univ. Press.

Miller PC and Stoner WH (1979) Canopy structure and environmental interactions. In Solbrig OT, Jain S, Johnson GB and Raven PH, eds. Topics in plant population biology, New York, Columbia Univ. Press.

Monk CD (1966) An ecological significance of evergreenness, Ecol. 47, 504-509.

Monsi M (1968) Mathematical models of plant communities. In Eckhardt F, ed. Functioning of terrestrial ecosystems at the primary productivity level, Paris, UNESCO.

Montes R and Medina E (1977) Seasonal changes in nutrient content of leaves of savanna trees with different ecological behavior, Geo-Eco-Trop 1,

295-307.
Montgomery GG, ed. (1978) The ecology of arboreal folivores, Washington D.C., Smithsonian Inst.
Mooney HA and Ehleringer JR (1978) The carbon gain benefits of solar tracking in a desert annual, Plant Cell Environ. 1, 307-312.
Mooney HA and Gulmon SL (1979) Environmental and evolutionary constraints on the photosynthetic characteristics of higher plants. In Solbrig OT, Jain S, Johnson GB and Raven PH, eds. Topics in plant population biology, New York, Columbia Univ. Press.
Mooney HA, Ferrar PJ and Slatyer RO (1977) Photosynthetic capacity and carbon allocation patterns in diverse growth forms of *Eucalyptus*, Oecol. 36, 103-111.
Moore HE, Jr. (1973) The major groups of palms and their distribution, Gentes Herb. 11, 27-141.
Morley B (1973) Ecological factors of importance to *Columnea* taxonomy. In Heywood VH, ed. Taxonomy and ecology, New York, Academic.
Morley BC (1974) Notes on some critical characters in *Columnea* classification, Ann. Mo. Bot. Gard. 61, 514-525.
Nolde IV (1941) Zur Enstehung von Flachkronen bei tropisch-afrikanischen Bäumen, Kolonialforstl. Mitt. 3, 486-498.
Orians GH and Solbrig OT (1977) A cost-income model of leaves and roots with special reference to arid and semi-arid areas, Am. Nat. 111, 677-690.
Osmaston HA (1965) Pollen and seed dispersal in *Chlorophora* and *Parkia*, Commonw. Forest. Rev. 44, 97-105.
Parkhurst DF and Loucks OL (1972) Optimal leaf size in relation to environment, J. Ecol. 60, 505-537.
Peace WJH and MacDonald QD (1981) An investigation of the leaf anatomy, foliar mineral levels, and water relations of trees of a Sarawak forest, Biotr. 13, 100-119.
Putz FE (1980) Lianas vs. trees, Biotr. 12, 224-225.
Raunkiaer C (1934) The life forms of plants and statistical plant geography, London, Clarendon.
Rausher MD (1978) Search image for leaf shape in a butterfly, Science 200, 1071-1073.
Rausher MD (1980) Host abundance, juvenile survival, and oviposition preference in *Battus philenor*, Evol. 34, 343-355.
Raven JA (1976) The quantitative role of "dark" respiratory processes in heterotrophic and photolithotropic plant growth, Ann. Bot. 40, 537-562.
Richards PW (1952) The tropical rain forest: an ecological study, Cambridge, Cambridge Univ. Press.
Robberecht R, Caldwell MM and Billings WD (1980) Leaf ultraviolet optical properties along a latitudinal gradient in the arctic-alpine zone, Ecol. 61, 612-619.
Rollet B (1974) L'architecture des forêts denses humides sempervirentes de plaine, Nogent sur Marne, Centre Tech. Forest. Tropical.
Sarmiento G (1972) Ecological and floristic convergence between seasonal plant formations of tropical and subtropical South America, J. Ecol. 60, 367-410.
Sarmiento G (1983) Ecology of neotropical savannas, Cambridge, Harvard Univ. Press.
Schenck H (1892) Beiträge zur Biologie und Anatomie der Lianen. I. Beiträge zur Biologie der Lianen, Bot. Mitt. Trop. 4, 1-248.
Schimper AFW (1898) Pflanzengoographie auf physiologischer Grundlage, Jena, Fischer.
Schulze E-D (1982) Plant life forms and their carbon, water, and nutrient relations. In Lange OL, Nobel PS, Osmond CB and Ziegler H, eds. Encyclopedia of plant physiology (new series), Vol. 12B, New York, Springer.
Shields LM (1950) Leaf xeromorphy as related to physiological and structural influences, Bot. Rev. 16, 399-447.
Skog LE (1976) A study of the tribe Gesnerieae, with a revision of *Gesneria* (Gesneriaceae:Gesnerieae), Smithson. Contr. Bot. 29, 1-182.
Slatyer RO (1978) Altitudinal variation in the photosynthetic characteristics of snow gum, *Eucalyptus pauciflora* Sieb. ex Spreng. VII. Relationship between gradients of field temperature and photosynthetic temperature optima in the Snowy Mountains area, Austral. J. Bot. 26, 111-121.
Small E (1972a) Water relations of plants in raised *Sphagnum* peat bogs, Ecol. 53, 726-728.
Small E (1972b) Photosynthetic rates in relation to nitrogen recycling as an adaptation to nutrient deficiency in peat bog plants, Can. J. Bot. 50, 2227-2233.
Sobrado MA and Medina E (1980) General morphology, anatomical structure, and nutrient content of sclerophyllous leaves of the "bana" vegetation of Amazonia, Oecol. 45, 371-378.
Stowe LG and Brown JL (1981) A geographic perspective on the ecology of compound leaves, Evol. 35, 818-821.
Sugden AM (1982) The vegetation of the Serrania de Macuira, Guajira Colombia: a contrast of arid lowlands and an isolated cloud forest, J. Arnold Arb. 63, 1-30.
Tanner EVJ and Kapos V (1982) Leaf structure of Jamaican upper montane rain forest trees, Biotr. 14, 16-24.
Taylor SE (1971) Ecological implications of leaf morphology considered from the standpoint of energy relations and productivity. Ph.D. thesis, Washington Univ., St. Louis.
Taylor SE (1975) Optimal leaf form. In Gates DM and Schmerl RB, eds. Perspectives in biophysical ecology, New York, Springer.
Taylor SE and Sexton OJ (1972) Some implications of leaf tearing in Musaceae, Ecol. 53, 143-149.
Thoday D (1931) The significance of reduction in the size of leaves, J. Ecol. 14, 297-303.
Tomlinson PB (1961) Anatomy of the monocotyledons, Vol. II: Palmae, Oxford, Clarendon.
Veen BW (1977) The uptake of potassium, nitrate, water and oxygen by a maize root system in relation to its size, J. Exp. Bot. 28, 1389-1398.
Veen BW (1981) Relation between root respiration and root activity. In Brouwer R, Gasparíková O, Kolek J and Loughman BC, eds. Structure and

function of plant roots, The Hague, Dr. Junk.
Verdoorn F, ed. (1938) Manual of pteridology, The Hague, Martinus Nijhoff.
Volkens G (1887) Die flora des ägyptisch-arabischen Wüste auf Grundlage anatomisch-physiologischen Forschungen, Berlin.
Walter H (1973) Vegetation of the earth, New York, Springer.
Waring RH, Emmingham WH, Gholz HL and Grier CC (1978) Variation in maximum leaf area of coniferous forests in Oregon and its ecological significance, Forest Sci. 24, 131-140.
Webb LJ (1968) Environmental relationships of the structural types of Australian rain forest vegetation, Ecol. 49, 296-311.
Whitmore TC (1975) Tropical rain forests of the Far East, Oxford, Clarendon.
Wiens D (1979) Mimicry in plants. In Hecht MK, Steeres WC and Wallace B, ed. Evolutionary Biology, Volume 11, New York, Plenum.
Williamson GB (1981) Drip tips and splash erosion, Biotr. 13, 228-231.
Woodson RE (1947) Some dynamics of leaf variation in *Asclepias tuberosa*, Ann. Mo. Bot. Gard. 34, 353-432.
Wurdack JJ (1980) Flora of Ecuador, Stockholm, Swed. Natl. Res. Council.
Wyatt R and Antonovics J (1981) Butterflyweed re-revisited: spatial and temporal patterns of leaf shape variation in *Asclepias tuberosa*, Evol. 35, 529-542.
Zimmermann MH (1978) Structural requirements for optimal water conduction in tree stems. In Tomlinson PB and Zimmerman MH, eds. Tropical trees as living systems, New York, Cambridge Univ. Press.

LEAF ENERGY BALANCE IN THE WET LOWLAND TROPICS

N. Chiariello
(Department of Biology, University of Utah, Salt Lake City, UT 84112 USA)

ABSTRACT

The interplay between macroclimate and vegetation in the wet tropics creates leaf microclimates in which light intensity, humidity, air temperature, and wind velocity tend to be correlated. Along a vertical transect through the forest, this creates an increase in vapor pressure deficit (VPD) with increasing height. The correlated changes in environmental factors act to narrow the range of microclimates encountered along a vertical transect through the forest, making it possible to simulate leaf energy balance for "typical" microhabitats. Basing the simulations on leaves of the size most frequently encountered and with stomatal responses to light and VPD, leaf energy balance along a vertical gradient from forest floor to canopy can be summarized as follows:

As a result of high (>90%) humidities and low radiation in the understory, leaf temperatures are generally very close to air temperature. As radiation increases and humidity decreases, leaf overtemperatures rise to about 6°C, and transpiration rates increase. Simulations and several observations suggest that leaf temperatures near the top of the canopy or in clearings can exceed 40°C under moderately high radiation.

Leaf size has received considerable study in the wet tropics. Trends toward leaves larger than the predominant size (the mesophyll) occur with increasing moisture and increasing shade, but gap species often have large leaves. Optimization models have predicted that leaf size should be maximal in either the lowest or the intermediate strata. Simulations with varying leaf size and stomatal conductance suggest that leaf size has little effect on energy balance in understories and that interactions with conductance determine the consequences of leaf size in open conditions.

General trends in leaf energy balance for a number of sites and habitat types are not yet available. Further studies are needed of stomatal conductance and responses to VPD, leaf temperature relations, and absorptance properties.

1. INTRODUCTION

The application of biophysical principles for understanding energy exchange in leaves has become an essential link integrating studies of leaf adaptation. The approach is central in three respects. First, energy exchange is tightly coupled with carbon dioxide exchange in leaves, and is a major determinant of the variation in a leaf's photosynthetic rate as microclimate changes. Second, energy exchange processes define the coupling between the dynamics of two resources, carbon and water, and are thereby linked to root/shoot balance and whole plant allocation. Third, because the exchange of energy between a leaf and its environment is defined by fixed properties of interaction derived from physical principles, energy balance studies often yield insights that can be generalized to a whole community or a subset of the community. Leaves also interact with biotic agents (herbivores, other leaves), but the rules governing these interactions vary among species, among conspecific individuals, and through time for any species or individual. By virtue of the physical basis of energy exchange, energy balance analysis is uniquely powerful in comparative studies of leaf adaptation to different environments. The analysis allows us to predict whether temperature stress is likely to be a factor influencing leaf performance; it allows us to identify mechanisms by which leaves avoid thermal damage; and it allows us to compare these mechanisms with those observed in other environments.

Most studies of leaf energy balance have emphasized deserts and alpine sites, where abiotic factors are assumed to be important, if not dominant, selective forces. In these environments, leaves experience extremes of air temperature, evaporative demand, and irradiance. Understanding how these plants avoid desiccation and thermal damage is fundamental to explaining their persistence in these communities.

Leaf energy balance and related questions of leaf adaptation have been less studied in wet tropical ecosystems. One reason for this may be a lack of agreement about the nature of climatic stresses experienced by plants of the wet tropics. True rain forests, with no dry season, have been considered either free of climatic stresses (Hallé et al., 1978) or subject to climate-related nutrient limitation. Several studies have suggested that leaves of these forests should have adaptations for overcoming the low evaporative demand of the atmosphere, on the hypothesis that the high vapor pressure of the air restricts transpiration below values necessary for optimal nutrient uptake (Odum et al., 1970). This theory has been both supported (Leigh, 1975) and refuted (Grubb, 1977) in more recent discussions. Other studies have considered the optimization of leaf properties for maximizing carbon gain in relation to water loss (Parkhurst, Loucks, 1972; Givnish, Vermeij, 1976; Givnish, 1979), assuming that water loss is a cost. Still others have cautioned that optimization criteria based on water relations are premature in the wet tropics (Grace et al., 1980).

The diversity of conclusions on the importance of transpiration in the tropics provides a good indication of the state of tropical environmental physiology during the last decade. It also suggests that the importance of energy balance properties in explaining leaf characteristics in the wet tropics cannot be assumed. Heeding this caveat, the primary goal of this paper is to summarize the energy loads on leaves of the wet tropics in order to identify the likelihood of thermal damage or high water loss, and also general patterns of leaf temperature and transpiration, which can then be linked with studies of photosynthesis and allocation.

The discussion is divided into three sections. Section 2 describes the leaf microclimates encountered in the wet tropics, focusing on the vertical gradients in microclimate through the strata of a forest. Section 3 considers the properties of leaves important to energy balance, and loosely defines a "standard" leaf of the wet tropics, which is then used in Section 4 in energy balance simulations to predict general patterns of leaf temperature and transpiration in response to vertical changes in microclimate. Section 4 also reconsiders the application of leaf energy balance analysis in explaining trends in leaf size, which has been a primary focus of past energy balance studies in the wet tropics.

2. LEAF MICROCLIMATES

2.1. Vertical gradients in microclimate

Determining the problems of leaf energy balance posed by an environment requires an assessment of the environment the leaves experience, particularly the extreme conditions. This assessment entails microclimate descriptions consisting of instantaneous values of environmental parameters in the sites occupied by leaves. In principle, the parameters can be either measured or derived from a knowledge of the macroclimate and a model that predicts light attenuation through the canopy and air movement through the vegetation. Because microclimate models have not been widely tested for the wet tropics, this discussion will focus on measured environmental variables.

Suitable data on vertical microclimates have been gathered for several wet tropical forests. Comparisons between these data are often limited by the use of different meteorological instruments and by the limited number of conditions studied, but general trends are nonetheless apparent. Most striking among these trends is that environmental parameters covary through the strata of a wet tropical forest.

Vertical gradients vary in magnitude between different sites but are generally most pronounced around midday in the dry season (Fig.1) and somewhat less in the wet season. These gradients encompass the range of variation in forest microclimates under the most extreme conditions.

The most pronounced gradient is in light intensity. In some cases, less than one percent of the visible radiation and only five percent of the red and near infrared radiation reach the forest floor at midday (Evans, 1939; Allen et al., 1972; Pearcy, 1983) (Fig. 1a), though this varies somewhat between sites depending on forest structure (Walter, 1971; Chazdon, Fetcher, article in this volume). The sources and magnitude of longwave radiation also vary with position in the canopy since a leaf at the canopy top "sees" the sky but an understory leaf "sees" other leaves.

The vapor pressure of the air is generally high throughout the rain forest, but air vapor pressure may decrease with height above the forest floor (Aoki et al., 1975). Temperature also varies through the strata, with temperatures of 27-32°C just below the canopy top and 23-28°C in the understory (Evans, 1939; Cachan, Duval, 1963; Baynton et al., 1965; Aoki et al., 1975 (Fig. 1b). Air temperature differences between the understory and canopy top at midday may reach 6-7°C in some forests (Cachan, Duval, 1963; Aoki et al., 1975).

Increases in air temperature or decreases in vapor pressure with height result in decreases in relative humidity (Fig. 1c) and increases in saturation deficit (Fig. 1d) in moving from the understory to the canopy top (Evans, 1939; Cachan, Duval, 1963; Aoki et al., 1975). In the understory, relative humidity is generally greater than 70%, and often greater than 90%, but drops by as much as 30% through the vertical profile.

Wind speed is generally greater at the top of the canopy than in the understory (Fig. 1e). In some cases, wind speed apparently increases monotonically with height (Aoki et al., 1975). In others, wind speed reaches a local minimum near mid-canopy, where leaves are most densely packed (Baynton et al., 1965; Allen et al., 1972).

2.2. Expected profile of vapor pressure deficit

Covariance in the vertical gradients of microclimate parameters is most pronounced for

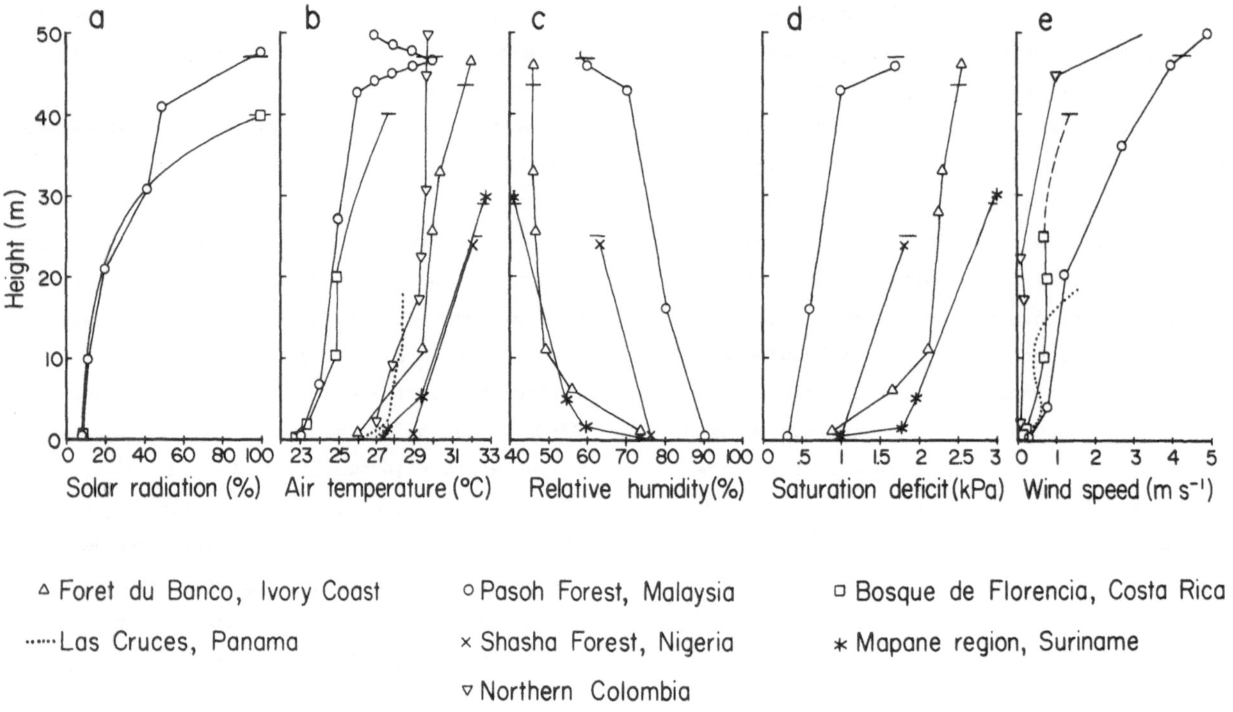

FIGURE 1. Vertical profiles of microenvironment near midday in the forests of the wet tropics. a) radiation (% intensity at the canopy top), b) air temperature, c) relative humidity, d) air saturation deficit, e) wind speed. Horizontal bars indicate the canopy height for studies reporting it. (Shasha forest data from Evans (1939); Las Cruces data from Hales (1949); Mapane region data from Schulz (1960); Banco forest data from Cachan, Duval (1963) and Cachan (1963); Northern Colombia data from Baynton et al. (1965a, 1965b); Bosque de Florencia data from Allen et al. (1972); and Pasoh forest data from Aoki et al. (1975)).

radiation, temperature, and humidity, which vary monotonically through the strata. The importance of the covariance is that each of the three parameters should tend to increase the vapor pressure deficit (VPD) between a leaf and the air as height above the forest floor increases. This is due to increases in the vapor content of leaves (due to increases in leaf temperature with increased radiation) and increases in the saturation deficit of the air, due to decreases in air vapor pressure and increases in air temperature.

The general trend toward drier, warmer, sunnier microclimates as we move up through the forest is repeated as we move out of the forest into gaps and clearings (Allee, 1926; Schulz, 1960; Grubb, Whitmore, 1966). In fact, clearings are likely to represent more extreme microclimates than the top of the canopy because wind speeds are lower in the gaps and because exposure of the soil to solar radiation tends to increase the temperature of nearby air.

The coupling between temperature, humidity, and light is a general consequence of the interplay between vegetation structure and macroclimate, and provides a clear framework for analyzing patterns in leaf energy balance. In the next section we examine general features of tropical leaves that are significant to energy balance properties, and then combine these features with the environmental trends already identified to describe general patterns in energy balance.

3. LEAF PROPERTIES

3.1. Leaf types

Compared with the incredibly high species diversity of wet tropical forests, most leaves are surprisingly similar in the widely surveyed characters that are important to energy balance. Exceptions are visually obvious, but most leaves, on the basis of species represented, are mesophylls (with the area of one side ranging from 2025 to 18225 mm^2), most are glabrous and hypostomatous, and many have drip tips (Cain et al., 1956; Webb, 1959; Grubb et al., 1963; Richards, 1964; Hall, Swaine, 1981).

3.2. Trends in leaf size

Departures from this predominant leaf type have generally been identified as trends along environmental gradients or as rough correlations with forest microenvironment. These studies have emphasized variation in leaf size. Several studies suggest that leaf size follows moisture gradients through the forest and through the strata of the vegetation at any one site. In a classification of Ghanaian forest types, Hall and Swaine (1981) found that macrophylls and megaphylls were present only in the wettest forests, and that microphylls became increasingly rare as moisture increased. Macrophylls are also most common in the wettest sub-formations of the Australian rain forest (Webb, 1959). In the Amazon basin, Cain et al. (1956) censused leaf size together with life form (sensu Raunkiaer, but subdivided into height classes) through the strata of a wet lowland forest. On a species basis, macrophylls were most important in lower strata, peaking at 50 percent of species in the geophyte class. However, species diversity also varied with life form and with stratum, resulting in a discrepancy between the greatest frequency of macrophylls and the largest number of macrophyllous species. The stratum with the highest representation of macrophylls was the geophyte class (one out of 2 species), but out of the total 23 macrophyllous species, over 40 percent were in the second tallest stratum. The only megaphyllous species was an epiphyte with unspecified leaf height.

Although these studies suggest relationships between leaf size and environment, they are concerned more with the overall structure and classification of vegetation, than with the specific microenvironments associated with various strata or leaf sizes. As a result, trends in leaf size are difficult to interpret and difficult to superimpose on the vertically structured microclimates. Parkhurst and Loucks (1972) used the data of Cain et al. (1956) to identify a trend towards increasing leaf size in moving from higher to lower strata, but the confidence intervals were wide. Thus, unless sample sizes are very large, even complete size class information does not yield significant trends in leaf size along environmental gradients.

Intraspecific comparisons in leaf size may be more interpretable, but few exist. Often, younger trees lower in the canopy have larger leaves than mature trees, and leaves of shade-grown trees are larger than leaves in the sun (Cain et al., 1956; Walter, 1971; Richards, 1964; Hall, Swaine, 1981).

Trends in leaf size also occur along altitudinal gradients. The proportion of trees with macrophylls is higher in lowland rainforest than in montane rainforest of Ecuador (Grubb et al., 1963). Ascending through the vegetation types of wet tropical mountains, the size of the dominant leaf-size class progressively declines, even within species that span adjacent formations (Grubb, 1977).

A final trend in leaf size is that pioneer species in tree-fall gaps generally have larger leaves than emergent trees (Ashton, 1978 in Bazzaz, Pickett, 1980). Some of the largest leaves of the wet tropics are found in clearings of the lowlands (Holdridge et al., 1971).

The available data suggest that against the vertical profile of forest microclimate, leaf size follows a general pattern of decreasing size with increasing height, but exceptions exist mainly in the lower strata. Also, the similarity in microclimate of gaps and treetops is not matched by a similarity in leaf size since many large leaves (e.g. Musaceae, Araceae) occur in clearings.

3.3. Absorptance

Patterns in leaf orientation and absorptance have received less study than leaf size, but several characteristics of leaves in the wet tropics suggest mechanisms to avoid high radiation. Steep leaf angles are characteristic of some species exposed to full sun and produce substantial reductions in leaf temperature (Medina et al., 1978). Upper canopy leaves generally are both steeply inclined and more sclerophyllous than leaves below them (Walter, 1971). The characteristic limpness of young leaves and shoots (Walter, 1971) as well as the midday wilting of many gap species (unpublished data) may have the same effect. Many of these leaves have a distinctive coloration and very low absorptance in the 200-800 nm waveband, which further reduces their radiation load (Lee, Lowry, 1980). Other leaves change their orientation diurnally, reducing their projected area along the solar beam (Ashton, 1978 in Bazzaz, Pickett, 1980).

3.4. Stomatal properties

Stomatal properties are also critical to leaf energy balance. In the handful of species now examined, stomatal conductance responds to light and VPD (or humidity) with varying sensitivities (Whitehead et al., 1981; Grace et al., 1982; Mooney et al., 1983) (Fig. 2). The VPD responses are generally similar to those observed in temperate plants (Hall et al., 1976). Leaves with this type of stomatal sensitivity are likely to have low stomatal conductance under

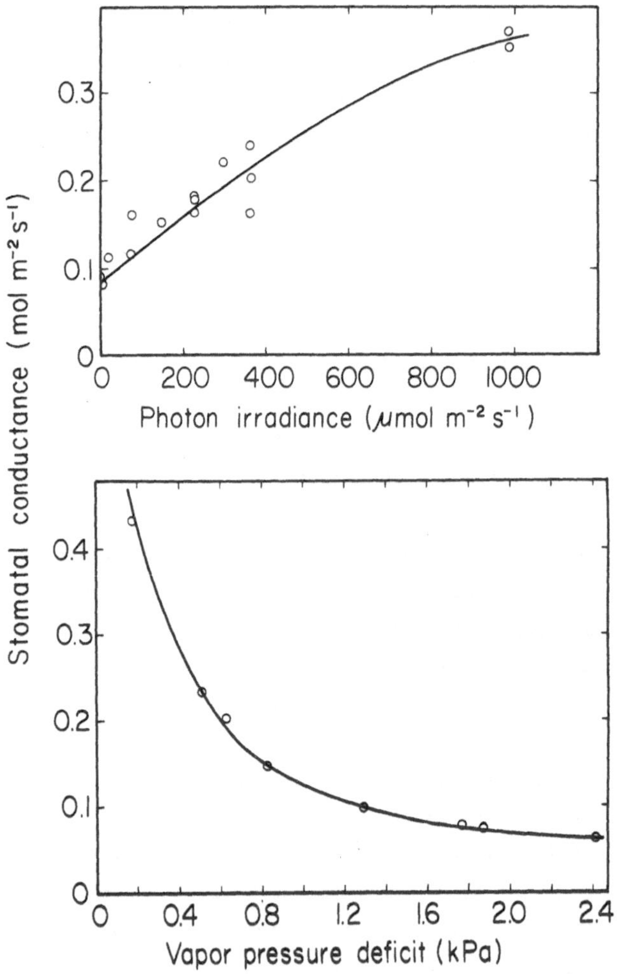

FIGURE 2. Stomatal conductance to water vapor in response to light intensity and vapor pressure deficit in *Piper auritum* (unpublished data of C. Field).

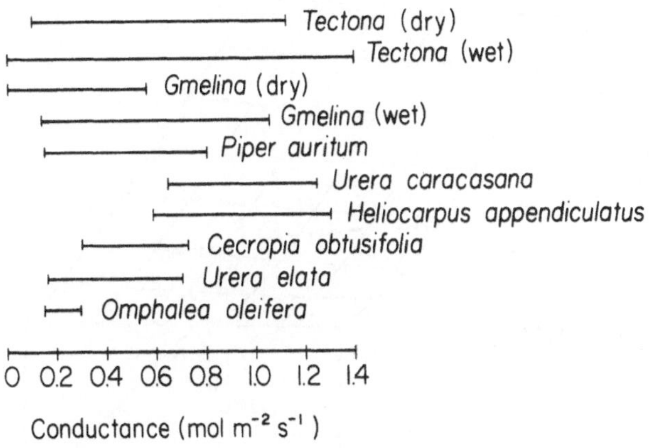

FIGURE 3. Range of stomatal conductance to water vapor in diurnal courses for several species. Tectona and Gmelina data from Grace et al. (1982) (wet season) and Whitehead et al. (1981) (dry season); others from unpublished studies at Los Tuxtlas, Veracruz, Mexico at the start of the rainy season.

the saturation deficits encountered at the top of the canopy (cf. Figs. 1d and 2). However, under some field conditions, conductances to water vapor as high as 1 mol m^{-2} sec^{-1} appear to be fairly common (Fig. 3). This value is high by comparison with plants of other habitats (Körner et al., 1979), possibly due to differences in ambient humidities rather than in stomatal characteristics. In species of various temperate environments, stomatal responses to VPD (Landsberg, Butler, 1980; Field et al., 1982) and high conductances (Geller, Smith, 1982) have important consequences for leaf energy balance. The high conductances in leaves of some tropical species may confer distinctive energy balance properties, but few species have been studied (Grace et al., 1980).

4. ENERGY BALANCE PREDICTIONS

4.1. Approach

To probe the effects of covarying trends in irradiance, humidity, and air temperatures on leaf temperature and transpiration, I employ a twofold approach. First I consider the effects of vertical changes in microclimate on a "standard" leaf, based on the predominant leaf type identified in Section 3. Then I review energy balance models of leaf size and, in a second set of simulations, examine the effects of leaf size and stomatal properties on energy balance at the two extremes of the vertical profile. The model for both simulations is based on the energy balance theory and equations developed by Penman (1948), Raschke (1956), and Gates (1962). In this model, the effects of temperature, pressure, and humidity on the properties of air are calculated from the subroutines of Tracy et al. (1978), which are based on the Smithsonian Meteorological Tables (List, 1971).

4.2. Simulations with a "standard" leaf

Constant environmental parameters for the simulations with a standard leaf include altitude, 10 m; wind speed, 1.0 m s^{-1}; longwave radiation, 400 W m^{-2}; and air temperature, 25°C. Constant leaf parameters include leaf absorptance to shortwave (0-3μm) radiation, 0.45; leaf characteristic dimension, 80 mm; and albedo of surrounding vegetation, 0.12. Incident shortwave radiation (10-700 W m^{-2}) and relative humidity (40-90%) vary independently. Stomatal conductance varies according to a stomatal response to light and VPD as described by Figure 2. The model was solved iteratively for both the leaf temperature satisfying the energy balance equation and the conductance consistent with the stomatal responses.

The simulations predict that leaf temperature increases with both increasing radiation and increasing humidity, but that it is more sensitive to radiation than to humidity (Fig. 4a). Increases in either variable cause leaf temperature to cross over from an undertemperature (leaf cooler than air) to an overtemperature (leaf warmer than air). The overtemperature reaches 6°C at the highest values of humidity and radiation, resulting in a leaf temperature of 31°C under the simulated conditions. Undertemperatures can occur under moderately low light (about 100 W m^{-2} absorbed shortwave radiation), but they are unlikely to occur in the understory due to the high relative humidity. In this simulation, the effect of the stomatal response to light and VPD is to reduce the sensitivity of leaf temperature to humidity, and to raise leaf temperature at low radiation levels. If conductance were constant, leaf temperature would increase more steeply with increases in humidity.

Increases in radiation cause transpiration to increase, but increases in humidity have the reverse effect (Fig. 4b). As a result, transpiration rates are predicted to be highest for the combination of high radiation and low humidity. The effect of the stomatal response is to reduce the sensitivity of transpiration to humidity, and to decrease transpiration at low radiation. With a constant conductance, the transpiration rate would respond more to the simulated range of humidity values than to the range in light. As a result of the stomatal response, leaf conductance to water vapor ranges from less than 0.1 (at low light and low humidity) to about 1.0 mol m^{-2} s^{-1} at moderate light and high humidity.

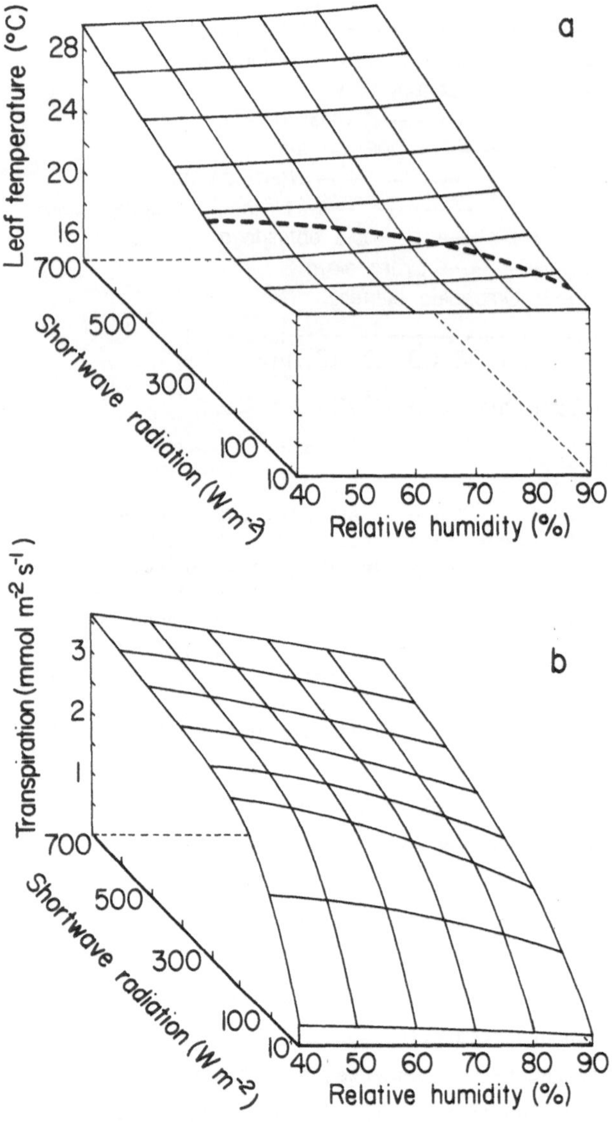

FIGURE 4. Results from energy balance simulations for "standard" leaves of the wet tropics. Constant parameters for the simulations are given in the text. Shortwave radiation refers to the incident radiation (0-3 μm). a) leaf temperature. The dashed line separates leaf overtemperatures (behind the dashed line) from leaf undertemperatures (in front). b) transpiration.

Predictions from the model generally agree with the limited available data on leaf energy balance. Under high radiation, predicted leaf overtemperatures can reach 6°C. Even higher overtemperatures would be predicted under conditions of lower wind speed, lower stomatal conductance, or higher radiation, or for leaf properties that have similar consequences (e.g. large leaves). Overtemperatures as high as 10 to 15°C, resulting in leaf temperatures of 46 to 48°C, have been observed in several species (Walter, 1971; Taylor, Sexton, 1972), suggesting that thermal stress may be an important factor affecting leaves under moderate to high radiation. Under low light and high humidity, the model predicts leaf temperatures near air temperature, as observed by Fetcher (1981).

Transpiration rates in the simulations range from about 0.2 mmol m^{-2} s^{-1} under low light and 90% relative humidity to about 2 mmol m^{-2} s^{-1} at high light and 60% relative humidity. This compares with a range of 0.24 to 2.0 mmol m^{-2} s^{-1} reported by Walter (1971), but the predicted minimum is lower than that measured in understories of a Puerto Rican montane forest (0.31-0.98 mmol m^{-2} s^{-1} (Odum et al., 1970)).

These simulations indicate that a typical leaf should generally be warmer than air under open conditions, and near or slightly above air temperature under more shaded conditions, especially if relative humidity varies with radiation. In the low-radiation understories, undertemperatures are prevented by the limited evaporative cooling at very high humidities, while under high radiation, undertemperatures are prevented by stomatal closure. If air temperature covaries with humidity and light, leaf temperatures exceeding 40°C are possible. The generality of this conclusion depends on the generality of stomatal responses to VPD. With constant high conductances, evaporative cooling reduces leaf temperature considerably at the top of the canopy (Fig. 5).

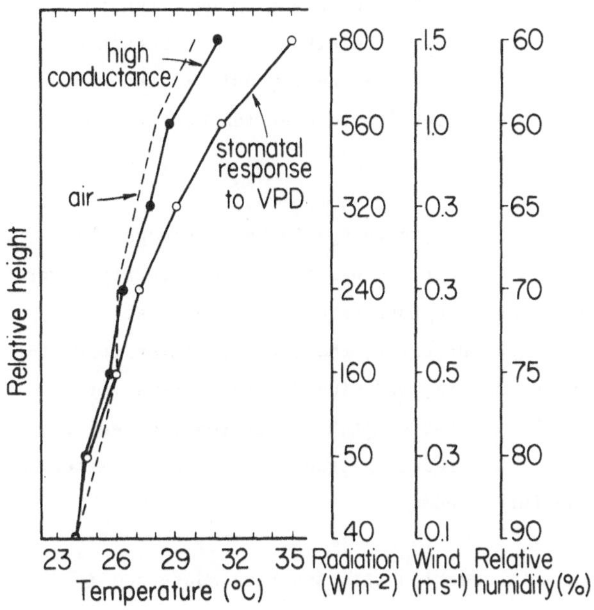

FIGURE 5. Air temperature and predicted leaf temperature through a vertical profile. Stomatal conductance is either high (1 mol m^{-2} s^{-1}) or varies with VPD according to Fig. 2. Leaf width is 80 mm, shortwave absorptance is 0.45, longwave radiation is 400 W m^{-2}. Air temperature, shortwave radiation, wind speed, and relative humidity vary with height as shown.

The frequency of temperature stress is a function of temporal variation in temperature, humidity, and radiation. In the understory, temperature and humidity are very stable, with small diurnal and seasonal fluctuations, but the radiation level can fluctuate widely over short time periods. Diurnal and seasonal fluctuations in air temperature, humidity, and radiation are largest at the canopy top and in clearings. Diurnal variation in temperature at the canopy top ranges from about 3 to 10°C, and relative humidity varies by up to 40% (Evans, 1939; Cachan, Duval, 1963; Schulz, 1960). Because microclimates result from the interaction between the vegetation and microclimate, the highest air temperatures and lowest humidities generally coincide with maximum radiation loads.

4.3. <u>Models of leaf size</u>

Several theoretical studies have attempted to explain observed trends in leaf size or to propose testable trends in leaf size by determining how leaf size should optimally change with microenvironment. Many of the studies of optimal leaf size have focused on the tropics (Parkhurst, Loucks, 1972; Taylor, 1975; Givnish, Vermeij, 1976; Givnish, 1979). The criterion for optimality in these studies is the maximization of carbon gain relative to transpirational losses, expressed as either a ratio (Parkhurst, Loucks, 1972; Taylor, 1975) or a difference (Givnish, Vermeij, 1976; Givnish,

1979). These criteria consider carbon dioxide exchange as well as energy exchange, but the assumptions about the effects of temperature on photosynthesis place the emphasis in the analyses on relationships between temperature and transpiration.

Leaf size affects energy balance through the effects of leaf dimensions on the boundary layer conductance to both water vapor and heat. As leaf size increases, the boundary layer conductance decreases, reducing the convective (or sensible) transfer of heat between the leaf and the air and reducing transpirational (or latent) transfer of heat.

The main question that has been addressed in optimality studies is, "What environmental conditions favor an increase in leaf size?" Parkhurst and Loucks (1972) found that increases in leaf size were favored by low radiation, warm temperatures, and low humidity. The effects of humidity were smaller than those of radiation and air temperature, implying that large leaves were most favored by warm, shady environments with low humidity as a secondary factor. Givnish and Vermeij (1976) predicted that low light would favor smaller leaves because large leaves would have increased leaf undertemperatures, a conclusion that depends on relatively low humidities in low-light microsites. They predicted that leaf size would be maximal at moderate light intensities and decrease at both extremes. Applied to a vertical transect through the rain forest, Parkhurst and Loucks' model predicts maximal leaf size in the understory, while Givnish and Vermeij's model predicts that leaves of the lowest strata should be smaller than those at intermediate heights. These predictions are all based on the assumption that leaf properties independent of leaf size (e.g. stomatal conductance) are fixed while leaf size varies.

4.4. Energy balance in large leaves with and without a VPD response

Other studies have demonstrated that stomatal properties influence the relationship between leaf size and energy balance (Gates, Papian, 1971; Taylor, 1975; Smith, Geller, 1980). Taylor (1975) showed that leaf size has a large effect on leaf temperature at low conductance (high stomatal resistance) but very little effect at high conductance. At low conductances, increased leaf size causes transpiration to increase, but at high conductances the relationship is reversed - transpiration decreases with increases in leaf size. As we have seen, stomatal conductance can be very high in some tropical species (Grace et al., 1982) (Fig. 3), and in some species with large leaves, a high stomatal conductance compensates for a low boundary layer conductance, resisting tendencies toward overheating (Grace et al., 1980). On the other hand, a stomatal response to VPD generally causes leaf temperature to increase with decreasing boundary layer conductance or increasing leaf size (Landsberg, Butler, 1980). Thus, it is not possible to generalize the effects of leaf size on leaf temperature without also specifying stomatal responses.

In evaluating the effects of leaf size, three factors are probably important. Environmental parameters covary through the forest strata. Maximum conductances may be high. And those species thus far examined have a stomatal response to VPD. To study the effects of these three factors, I designed simulations of leaf energy balance based on two types of microclimates - the sunny, open conditions found at the canopy top and in clearings, and the damp, dark understory. For each microclimate, I ask how leaf temperature and transpiration vary with leaf size and with either a constant conductance or a conductance set by the VPD response of stomata.

4.4.1. Exposed conditions. For a leaf in an environment of high irradiance, moderately high humidity, moderate wind and fixed high conductances, transpiration decreases monotonically with increases in leaf size. With fixed low conductances, transpiration is minimized by either very small or very large leaf sizes (Fig. 6a). If conductance is not fixed but responds to VPD, conductance and transpiration tend to decrease with increasing leaf size (Fig. 6a). Leaf temperature shows opposite trends. For any fixed conductance, leaf temperature increases as leaf size increases. A stomatal response to VPD and irradiance exaggerates the increase in leaf temperature with increasing leaf size (Fig. 6b).

These simulations indicate that leaves in sunny conditions are likely to have overtemperatures, and that large leaves are likely to have the highest overtemperatures. However, for leaves with high conductance or stomatal sensitivity to VPD, increases in leaf size reduce water loss.

4.4.2. Understories. In comparison with exposed microenvironments, a leaf in the understory receives little energy. Under these circumstances, the qualitative responses of transpiration and temperature to increases in leaf size are the same as in exposed conditions, but the magnitude of the response is very small (Fig. 7). The trends are the same, despite the lower radiation load, because increases in leaf size tend to decrease boundary layer conductance, independent of the radiation environment.

In very humid understories, decreases in transpiration resulting from increases in leaf size actually push leaves from undertemperatures to overtemperatures, which drives up the VPD and reduces stomatal conductance. However, because so little energy is received, the effects on leaf temperature and transpiration are very small. These simulations agree with Fetcher's

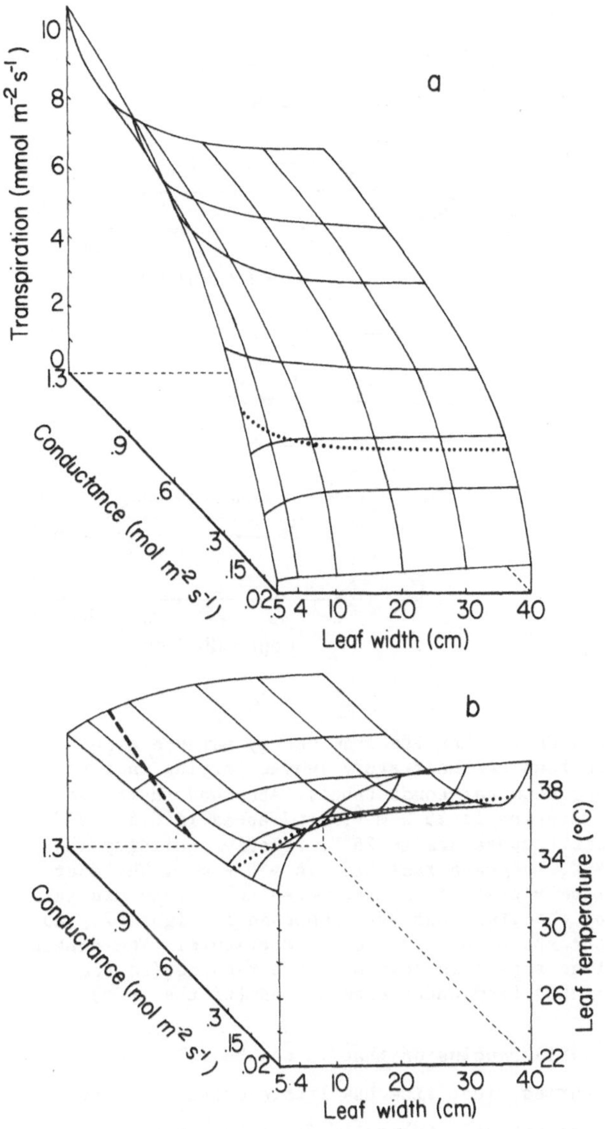

FIGURE 6. Results from energy balance simulations for canopy leaves varying in width and stomatal conductance. Absorbed shortwave radiation is 350 W m^{-2}, wind speed is 1.5 m s^{-1}, air temperature is 30°C, relative humidity is 60%, and longwave radiation is 440 W m^{-2}. The dotted line represents a leaf whose conductance is set by the light and VPD responses in Figure 2. a) transpiration. b) leaf temperature. The dashed line separates leaf overtemperatures (to the right) from undertemperatures (to the left).

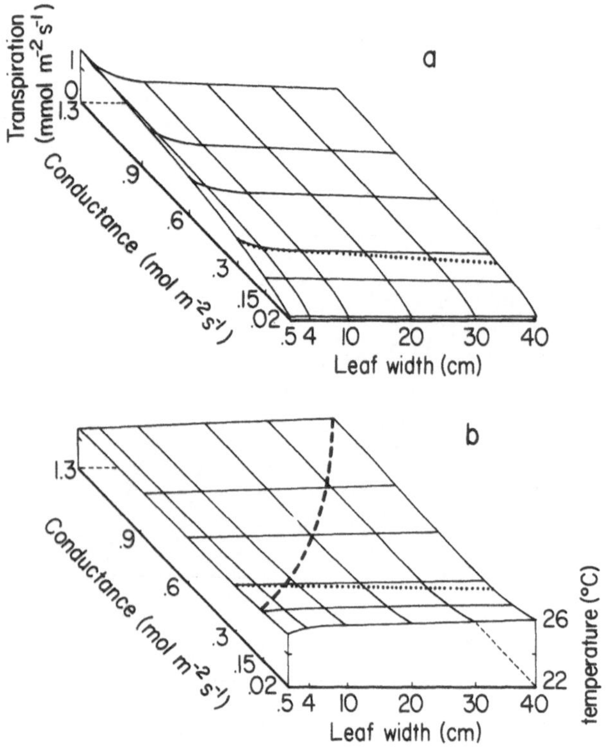

FIGURE 7. Results from energy balance simulations for understory leaves varying in width and stomatal conductance. Absorbed shortwave radiation is 25 W m^{-2}, wind speed is 0.5 m s^{-1}, air temperature is 25°C, relative humidity is 90%, longwave radiation is 430 W m^{-2}. The dotted line represents a leaf whose conductance is set by the light and VPD responses in Figure 2. a) transpiration. b) leaf temperature. The dashed line separates leaf overtemperatures (to the right) from undertemperatures (to the left).

(1981) conclusion that in the humid understories, leaf size has little effect on leaf temperature. Selection for other features influenced by leaf size could easily overwhelm selection for energy balance characteristics.

5. CONCLUSIONS AND AREAS FOR FURTHER STUDY

A number of studies have documented general characteristics of leaf microclimates in different strata of the forests of the wet tropics. I have emphasized that these forests show similar vertical trends in microclimate, but with respect to any one stratum, the forests appear to be quite different (Fig. 1). Humidity values differ the most; they are also often the most difficult to measure. To determine the basis of these differences, studies based on standardized instrumentation are needed. In addition, studies of canopy energy balance may help resolve discrepancies between forests.

The pronounced gradients in humidity and the differences between sites in humidity in the canopy have important implications for leaf energy balance. Because radiation and air temperature are highest in the top of the canopy, the low humidity results in either high transpiration (if conductance is high) or high leaf temperature (if conductance is low). Stomatal responses to humidity have been studied in only a few species, but all appear to reduce conductance as VPD increases. Considerably more work is needed on the stomatal properties of tropical species, and on the physiological consequences of high leaf temperatures in canopy or gap species with a stomatal response to VPD. For understory environments, further studies of energy balance should be designed to assess the significance of the small differences in leaf temperature and transpiration accessible through energy balance modification.

The effect of leaf size variation on energy balance depends on the value of stomatal conductance. I have emphasized this particular interaction between leaf characters because recent studies have provided the evidence that the interaction is of significance in the wet tropics. Other interactions may be equally important and need to be studied. Examples include the relationships among traits that determine convective properties, such as leaf size, sclerophylly, and orientation.

By comparison with studies of other ecosystems, relatively little is known about leaf absorptance of radiation in the wet tropics. This is surprising given the finding that both temperature and transpiration are more sensitive to radiation than to leaf size, and that significant overtemperatures are likely to occur under high radiation levels (Figs. 4, 5, 6).

Although leaf size has received considerable attention in the wet tropics, it remains poorly understood. Large leaves are found both in the understory, where they have little effect on energy balance, and in clearings, where they are likely to result in leaf temperatures over 40°C. Other interpretations of leaf size need further study.

ACKNOWLEDGMENTS

I am grateful to C. Field for many discussions of ideas related to this paper and for unpublished data (Fig. 2). H. A. Mooney, E. Medina, C. Vazquez Yanes, J. J. Landsberg and others provided helpful comments on an early draft and symposium presentation. N. Fetcher, H. Bollinger, and P. D. Coley provided several references. M. Martinez assisted with conductance surveys. B. J. Bannan assisted with the figures.

REFERENCES

Allee WC (1926) Measurement of environmental factors in the tropical rain-forest of Panama, Ecology 7, 273-302.

Allen LH Jr, Lemon E and Muller L (1972) Environment of a Costa Rican forest, Ecology 53, 102-111.

Aoki M, Yabuki K and Koyama H (1975) Micrometeorology of primary production of a tropical rain forest in West Malaysia, J. Agric. Meteor. (Tokyo) 31, 115-124.

Ashton PS (1978) Crown characteristics of tropical trees. In Tomlinson PB and Zimmerman MH eds. Cambridge, Cambridge University Press.

Baynton HW, Biggs WG, Hamilton HL Jr, Sherr PE and Worth JJB (1965) Wind structure in and above a tropical forest, J. Appl. Meteorol. 4, 670-675.

Baynton HW, Hamilton HL Jr, Sherr PE and Worth JJB (1965) Temperature structure in and above a tropical forest, Quart. J. Roy. Meteor. Soc. (London) 91, 225-232.

Bazzaz FA and Pickett STA (1980) Physiological ecology of tropical succession: a comparative review, Annu. Rev. Ecol. Syst.11, 287-310.

Cachan P (1963) Signification écologique des variations microclimatiques verticales dans la forêt sempervirente de basse Côte d'Ivoire, Annales Faculté Sciences Dakar 8, 89-155.

Cachan P and Duval J (1963) Variations microclimatiques verticales et saisonnières dans la forêt sempervirente de basse Côte d'Ivoire, Annales Faculté Sciences Dakar 8, 5-87.

Cain SA, Castro GM de Oliviera, Pires JM and da Silva NT (1956) Application of some phytosociological techniques to Brazilian rain forest, Am. J. Bot. 43, 911-941.

Evans GC (1939) Ecological studies on the rain forest of southern Nigeria II. The atmospheric environmental conditions, J. Ecol. 27, 436-482.

Fetcher N (1981) Leaf size and leaf temperature in tropical vines, Am. Nat. 117, 1011-1014.

Field C, Chiariello N and Williams WE (1982) Determinants of leaf temperature in California Mimulus species at diffrent altitudes, Oecologia 55, 414-420.

Gates DM (1962) Energy exchange and the biosphere. New York, Harper and Row.

Gates DM and Papian LE (1971) Atlas of energy budgets of plant leaves. New York, Academic Press.

Geller GN and Smith WK (1982) Influence of leaf size, orientation, and arrangement on temperature and transpiration in three high-elevation, large-leafed herbs, Oecologia 227-234).

Givnish T (1979) On the adaptive significance of leaf form. In Solbrig OT, Jain S, Johnson GB and Raven PH eds. Topics in plant population biology. New York, Columbia University Press.

Givnish TJ and Vermeij GJ (1976) Sizes and shapes of liane leaves, Am. Nat. 110, 743- 778.

Grace J, Fasehun FE and Dixon M (1980) Boundary layer conductance of the leaves of some tropical timber trees, Plant Cell Env. 3, 443-450.

Grace J, Okali DUU and Fasehun FE (1982) Stomatal conductance of two tropical trees during the wet season in Nigeria, J. Appl. Ecol. 19, 659-670.

Grubb PJ (1977) Control of forest growth and distribution on wet tropical mountains: with special reference to mineral nutrition, Annu. Rev. Ecol. Syst. 8, 83-107.

Grubb PJ and Whitmore TC (1966) A comparison of montane and lowland rain forest in Ecuador, J. Ecol. 54, 303-333.

Grubb PJ, Lloyd JR, Pennington TD and Whitmore TC (1963) A comparison of montane and lowland rain forest in Ecuador. I. The forest structure, physiognomy, and floristics, J. Ecol. 51, 567-601.

Hall AE, Schulze E-D and Lange OL (1976) Current perspectives of steady-state stomatal responses to environment. In Lange OL, Kappen I

and Schulze E-D, eds. Water and plant life: problems and modern approaches, pp. 169-188. Heidelberg, Springer.
Hall JB and Swaine MD (1981) Distribution and ecology of vascular plants in a tropical rain forest. The Hague, Dr W. Junk.
Hallé F, Oldeman RAA and Tomlinson PB (1978) Tropical trees and forest - an architectural analysis, New York, Springer-Verlag.
Holdridge LR, Grenke WC, Hatheway WH, Liang T and Tosi JA Jr (1971) Forest environments in tropical life zones, New York, Pergamon Press.
Körner C, Scheel JA and Bauer H (1979) Maximum leaf diffusive conductance in vascular plants, Photosynthetica 13, 45-82.
Landsberg JJ and Butler DR (1880) Stomatal response to humidity: implications for transpiration, Plant Cell Env. 3, 29-33.
Lee DW and Lowry JB (1980) Young-leaf anthocyanin and solar ultraviolet, Biotropica 12, 75-76.
Leigh EG Jr (1975) Structure and climate in tropical rain forest, Annu. Rev. Ecol. Syst. 6, 67-86.
List RJ (1971) Smithsonian Meteorological Tables. 6th rev. ed. Washington, DC, Smithsonian Institution Press.
Medina E, Sobrado M and Herrera R (1978) Significance of leaf orientation for leaf temperature in an Amazonian sclerophyll vegetation, Radiat. Environ. Biophys. 15, 131-140.
Mooney HA, Field C, Vazquez Yanes C and Chu C (1983) Environmental controls on stomatal conductance in a shrub of the humid tropics, Proc. Natl. Acad. Sci. USA 80, 1295-1297.
Odum HT, Lugo A, Clintron G and Jordan C (1970) Metabolism and evapotranspiration of some rain forest plants and soil. In Odum HT ed. A tropical rain forest, pp. I-103 - I-164. Oak Ridge, Tennessee, USAEC Division of Technical Information Extension.
Parkhurst DF and Loucks OL (1972) Optimal leaf size in relation to environment, J. Ecol. 60, 505-537.
Pearcy RW (1983) The light environment and growth of C3 and C4 species in the understory of a Hawaiian forest, Oecologia 58, 19-25.
Penman HL (1948) Natural evaporation from open water, bare soil, and grass, Proc. R. Soc. London A, 194:120-145.
Raschke K (1956) Uber die physikalischen Beziehungen zwischen Warmeubergangszahl, Strahlungsaustausch, Temperatur und Transpiration eines Blattes, Planta 48, 200-237.
Raunkiaer C (1934) The life-forms of plants and statistical plant geography, Oxford, Clarendon Press.
Richards PW (1964) The tropical rain forest. Cambridge, Cambridge University Press.
Schulz JP (1960) Ecological studies on rain forest in northern Suriname. Amsterdam, Noord. Hollandische, Uitg. Mij.
Smith WK and Geller GN (1980) Leaf and environmental parameters influencing transpiration: theory and field measurements, Oecologia 46, 308-313.
Taylor SE (1975) Optimal leaf form. In Gates DM and Schmerl RB, eds. Perspectives of biophysical ecology, pp. 75-86. New York, Springer-Verlag.
Taylor SE and Sexton OJ (1972) Some implications of leaf tearing in the Musaceae, Ecology 53, 143-149.
Tracy CR, Welch WR and Porter WP (1980) Properties of Air. Tech. Rep. No. 1. Laboratory for Biophysical Ecology, Madison Wisconsin.
Walter H (1971) Ecology of tropical and subtropical vegetation. Edinburgh, Oliver and Boyd.
Webb LJ (1959) A physiognomic classification of Australian rain forests, J. Ecol. 47, 551-570.
Whitehead D, Okali DUU and Fasehun FE (1981) Stomatal response to environmental variables in two tropical forest species during the dry season in Nigeria, J. Appl. Ecol. 18, 571-587.

TISSUE WATER DEFICITS AND PLANT GROWTH IN WET TROPICAL ENVIRONMENTS

ROBERT H. ROBICHAUX,[1] PHILIP W. RUNDEL,[2] LANI STEMMERMANN,[3] JOAN E. CANFIELD,[3] SUZANNE R. MORSE,[1] and W. EDWARD FRIEDMAN[1]
[1]Department of Botany, University of California, Berkeley, CA 94720;
[2]Laboratory of Biomedical and Environmental Sciences, University of California, Los Angeles, CA 90024;
[3]Department of Botany, University of Hawaii, Honolulu, HI 96822

ABSTRACT
Although the supply of water is usually abundant in lowland and montane wet tropical forests, plants growing in these environments may be exposed to moderate tissue water deficits on a fairly regular basis. In addition, these plants may experience severe water deficits every few years. Information on the nature of these water deficits, their physiological effects on growth, and the mechanisms by which they are tolerated in wet tropical plants is reviewed in the present chapter. Special attention is given to several recent studies involving wet forest species from Hawaii and Panama. These latter studies demonstrate the importance of variation in tissue elastic and osmotic properties as a means of promoting turgor maintenance in wet tropical plants.

1. INTRODUCTION

Lowland and montane tropical rain forests are among the wettest terrestrial habitats on the surface of the Earth (Walter, 1971). Annual rainfall in these habitats is often very high, with abundant precipitation occurring throughout most of the year. In extreme cases, such as on the island of Kauai in the Hawaiian archipelago, annual rainfall normally exceeds 12,300 mm/yr, and may exceed 15,000 mm/yr in particularly wet years (Mueller-Dombois et al., 1981). Yet, even in these extremely wet environments, dry periods of brief duration occur commonly on a diurnal and seasonal basis (Shreve, 1914; Walter, 1971; Grubb, 1977). In addition, drought years with unusually low precipitation may occur every ten to twenty years (Grubb, 1977; Buckley et al., 1980). As a consequence, plants growing in lowland and montane tropical wet forests may be exposed to moderate tissue water deficits on a fairly regular basis, and may experience severe deficits every few years.

The importance of tissue water deficits for plant growth in wet tropical environments can be inferred from correlations between habitat water availability and major phenological events, such as leaf shedding, flowering, and fruiting (Frankie et al., 1974; Opler et al., 1976; Alvim, Alvim, 1978; Augspurger, 1979; Reich, Borchert, 1982). For a lowland wet forest in Costa Rica, Frankie et al. (1974) observed that the period of maximal leaf shedding in the dominant tree species coincided with the first dry season during the year. In addition, Reich and Borchert (1982) observed that leaf shedding in a tropical tree species growing in a wet gallery forest in Costa Rica was correlated with onset of dry soil conditions. Alvim and Alvim (1978) demonstrated experimentally that leaf flushing and flowering in cultivated cacao plants were greatly enhanced when a period of high water availability followed a relatively extended period of low water availability. Such a relationship between water availability and flowering has also been observed for cultivated coffee plants (Alvim, 1960), for a lowland wet forest shrub species in Panama (Augspurger, 1979), and for two lowland wet forest tree species in Brazil (Alvim, Alvim, 1978).

Given that tissue water deficits may influence

plant growth in wet tropical environments, we address the following series of interrelated questions in the present chapter: (1) what levels of tissue water potential are experienced by plants growing in these environments, (2) how do the turgor and osmotic components of tissue water potential vary among these plants, (3) what effect does a decrease in tissue turgor pressure have on the physiological processes of these plants, and (4) what mechanisms are present in these plants that promote the maintenance of high turgor pressure as tissue water contents decrease?

2. TISSUE WATER DEFICITS AND PLANT GROWTH

2.1. A survey of the literature

Very few measurements of tissue water potential have been reported for plants growing in wet tropical environments (Fetcher, 1979; Robichaux, Pearcy, 1980; Oberbauer, 1982; Medina, 1983). Fetcher (1979) measured the water potentials of five tree species growing in a lowland wet forest in Panama. At the end of a prolonged dry season, diurnal water potentials varied between -0.5 and -2.5 MPa for most of the species, but reached values as low as -3.9 MPa in *Trichilia tuberculata* (= *T. cipo*). Water potentials increased significantly in all five species after the onset of the rainy season. Robichaux and Pearcy (1980) reported diurnal water potentials of -0.2 to -0.4 MPa during the wet season and -0.8 to -1.2 MPa during the dry season for an arborescent *Euphorbia* species growing in the understory of a mesic forest in Hawaii. Oberbauer (1982) measured diurnal water potentials as low as -1.7 MPa for the upper canopy leaves of *Pentaclethra macroloba*, a common tree species growing in the lowland wet forests of Costa Rica. Midday water potentials of the understory leaves of *P. macroloba* were 0.5 MPa higher. Finally, for three montane wet forest species in Puerto Rico, Medina (1983) reported diurnal water potentials of -0.1 to -1.0 MPa during an exceptionally clear day. These values suggest that plants growing in wet tropical environments experience tissue water potentials comparable to those of many woody plants growing in mesic temperate environments (Richter, 1976). As the data of Fetcher (1979) indicate, however, even plants in wet tropical environments may be subjected to severe tissue water deficits during prolonged dry periods.

The significance of these water deficits for plant growth depends in large part on the way in which the turgor and osmotic components of water potential vary among the plants growing in these environments. This stems from the fact that changes in turgor pressure appear to represent the principal means by which changes in plant water status are transduced into changes in plant growth (Hsiao et al., 1976; Turner, Jones, 1980). Unfortunately, there is very little information in the literature that documents how turgor pressure varies among plants growing in wet tropical environments. The only reports are those of Oberbauer (1982), Stemmermann (1983), and Ike and Thurtell (1981). As mentioned previously, Oberbauer (1982) analyzed the tissue water relations of the canopy tree species, *Pentaclethra macroloba*, growing in a lowland wet forest in Costa Rica. Values of tissue turgor pressure at maximal leaf hydration were 0.25 MPa higher in upper canopy leaves than in understory leaves. Presumably, this enabled the upper canopy leaves to maintain positive turgor pressures at lower tissue water potentials than the understory leaves. However, turgor pressures in leaves at all levels in the canopy approached zero during prolonged leaf exposure to direct solar radiation. Stemmermann (1983) analyzed the tissue water relations of several varieties of the Hawaiian tree species, *Metrosideros polymorpha*. Varieties of this species not only dominate the montane wet forests of Hawaii but also colonize recent lava flows and open bogs (Mueller-Dombois et al., 1981). Maximal values of tissue turgor pressure were 0.35 to 0.40 MPa higher in varieties of M. polymorpha that colonize exposed, dry substrates than in varieties that dominate the mature wet forests growing on older substrates. In addition, the former varieties were consistently able to maintain high and positive turgor pressures to lower tissue water contents than the latter varieties. The mechanisms promoting turgor maintenance in the drier site varieties of M. polymorpha appear to be similar to those described below for several Hawaiian species of Dubautia. In an experimental study with two cultivars of cassava (Manihot esculenta), Ike and Thurtell (1981) observed diurnal changes in tissue turgor pressure of 0.28 to 0.46 MPa for well-watered plants growing under controlled conditions in a growth chamber. The maximal value of tissue turgor pressure in these plants was 0.85 MPa. Several authors have measured the osmotic potentials of leaf extracts for plants growing in wet tropical environments (Blum, 1933; Walter, 1971; Medina, 1983). The extensive measurements of Walter (1971), for example, in a wet tropical forest in East Africa indicate that tissue osmotic potentials are lowest in canopy tree species, intermediate in lianous species, and highest in understory herbaceous species. These data imply that maximal tissue turgor pressures also vary among these plants, being highest in the canopy species and lowest in the understory species. However, these values must be interpreted cautiously, since sap extraction techniques are known to produce significant artifacts, such as dilution of symplasmic water with apoplasmic water (Tyree, Jarvis, 1982). In addition, it is not possible to evaluate whether the low tissue osmotic potentials in the canopy species reflect active solute accumulation or simple tissue desiccation. Unfortunately, these two processes have markedly different implications with respect to turgor maintenance by leaf tissue (Hsiao et al., 1976; Turner, Jones, 1980).

Given the limited number of turgor pressure measurements for plants growing in wet tropical environments, it is not surprising that very little information exists on the physiological effects of turgor pressure changes. Ike and Thurtell (1981) observed severe wilting in leaves of cassava when tissue water potentials declined to below -1.1 MPa. At these water potentials, tissue turgor pressures were near zero. In addition, the data of Oberbauer (1982) suggest that as tissue turgor pressure approached zero in leaves of Pentaclethra macroloba during prolonged leaf exposure to direct solar radiation, stomatal conductance decreased significantly. The limited data from these two studies thus suggest that the physiological processes of leaf expansion and stomatal opening in wet tropical plants may be markedly affected by significant changes in tissue turgor pressure. Such a pronounced turgor dependence would be consistent with the results of several recent studies involving plants from mesic temperate environments (Hsiao et al., 1976; Turner, Jones, 1980; Ludlow, 1980; Takami et al., 1981). A marked dependence of stomatal conductance on tissue turgor pressure also might account for the repeated observation that stomatal conductances and transpiration rates of wet tropical plants decline significantly as soil water availability and plant water status decrease to moderate levels (Okali, Dodoo, 1973; Fasehun, 1979; Fetcher, 1979; Buckley et al., 1980; Osonubi, Davies, 1980; Zobel, Liu, 1980; Peace, Macdonald, 1981).

If leaf expansion and stomatal opening in wet tropical plants depend intimately on tissue turgor pressure, then mechanisms promoting turgor maintenance should aid in the growth and survival of these plants under conditions of low moisture availability (Hsiao et al., 1976; Turner, Jones, 1980). As discussed by Tyree and Jarvis (1982), the two principal mechanisms promoting turgor maintenance in higher plants involve changes in tissue elastic and osmotic properties. A decrease in the bulk modulus of elasticity of the tissue causes a decrease in the magnitude of the change in turgor pressure for a given fractional change in tissue water content. A decrease in the osmotic potential of the tissue at full hydration causes an increase in the maximal value of tissue turgor pressure. This affects, in turn, the value of turgor pressure at tissue water contents below saturation. As illustrated in the following discussion, our recent work with several endemic Hawaiian species of Dubautia (Compositae) suggests that variation in these elastic and osmotic properties may have a significant influence on turgor maintenance in wet tropical plants.

2.2. Variation in the tissue water relations of Hawaiian *Dubautia* species

The twenty-one endemic Hawaiian species of Dubautia grow in a wide variety of habitats, including exposed lava, dry scrub, dry forest, mesic forest, wet forest, and bog (Carlquist, 1980; Carr, Kyhos, 1981). The variation in annual rainfall among these habitats is quite dramatic, ranging from less than 400 mm/yr in the dry scrub habitat to more than 12,300 mm/yr in the wet forest and bog habitats. Since these species also are very closely related, they provide an outstanding opportunity for examining the relationship between tissue water deficits and plant growth in a variety of tropical environments, including mesic and wet forests. Two of these species grow in contrasting habitats on the island of Hawaii (Robichaux, 1984). Dubautia scabra grows commonly in montane wet forest sites, where annual rainfall may exceed 5000 mm/yr. In contrast, D. ciliolata grows in dry scrub sites, where annual rainfall may be less than 500 mm/yr. At certain localities these two species may be found growing sympatrically. One such locality occurs at approximately 2000 m elevation on the slopes of Mauna Loa (Carr, Kyhos, 1981; Robichaux, 1984). At this locality, D. scabra and D. ciliolata are restricted to different lava flows, even though individuals of the two species may be found growing within a few meters of one another. Dubautia scabra is restricted to a 1935 lava flow, while D. ciliolata is restricted to an older, prehistoric lava flow. These two lava flows differ not only in age but also in physical structure (Robichaux, 1984). Annual rainfall at this site of sympatry is 800-1000 mm/yr.

The tissue water potentials experienced by these two species at this locality during a typical day differ significantly (Fig. 1). On every occasion that we have made measurements at this site, whether in winter or in summer, the midday water potentials of D. ciliolata have always been lower than those of D. scabra by at least 0.45 MPa. During particularly dry periods, this difference has been as great as 0.85 MPa.

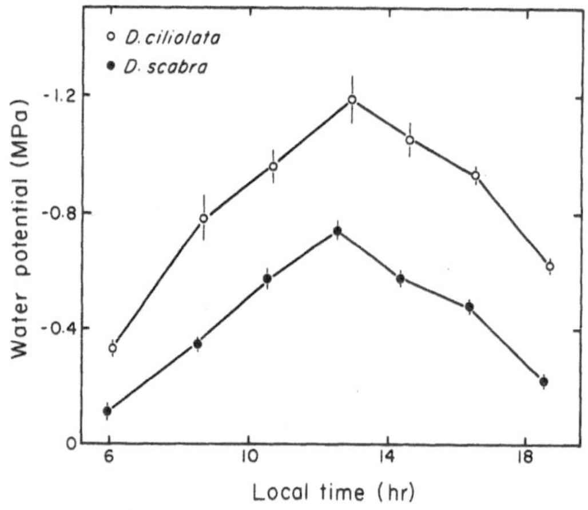

FIGURE 1. Diurnal water potentials for Dubautia ciliolata and D. scabra on 4 August 1982 near Puu Huluhulu, Saddle Road, Hawaii (Robichaux, 1984).

The tissue elastic and osmotic properties of these two species also differ significantly (Robichaux, 1984). At high tissue water contents, for example, D. ciliolata and D. scabra have bulk elastic moduli of 2.22 ± 0.48 MPa and 10.23 ± 1.75 MPa, respectively. In addition, the tissue osmotic potential at full hydration is -1.08 ± 0.04 MPa in D. ciliolata and -0.81 ± 0.04 MPa in D. scabra. The influence of these differences on the relative abilities of the two species to maintain high and positive turgor pressures as tissue water contents decrease is quite striking (Fig. 2). At a relative water content of 0.93, for example, the difference in tissue turgor pressure in these two species is 0.7 MPa. In addition, the relative water content at which turgor pressure reaches zero in D. ciliolata (0.75) is significantly lower than in D. scabra (0.88).

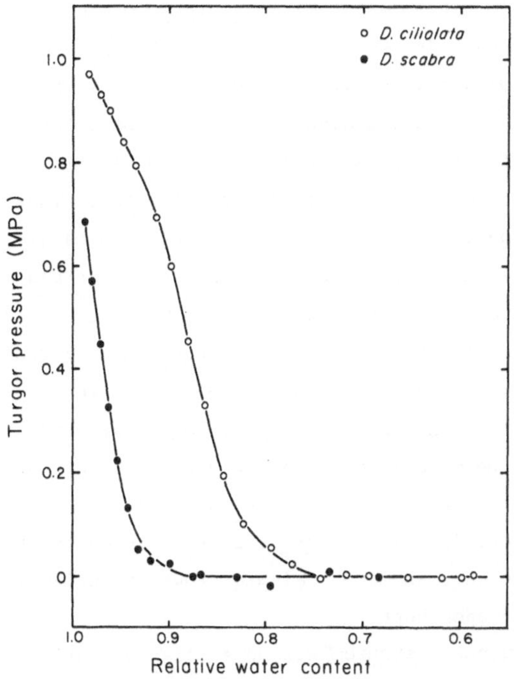

FIGURE 2. Relationship between tissue turgor pressure and tissue water content for Dubautia ciliolata and D. scabra. See Robichaux (1984) for a detailed discussion of the methods by which these relationships were obtained.

These differences in tissue elastic and osmotic properties appear to have a marked influence on diurnal turgor maintenance in these two species. To estimate diurnal turgor pressures in D. ciliolata and D. scabra, we first plot the data of Fig. 2 in a slightly modified form to obtain quantitative relationships between tissue water potential and tissue turgor pressure (Fig. 3). By combining the information in Fig. 3 with that in Fig. 1, we may then obtain a reasonably accurate estimate of diurnal changes in tissue turgor pressure in these two species (Fig. 4).

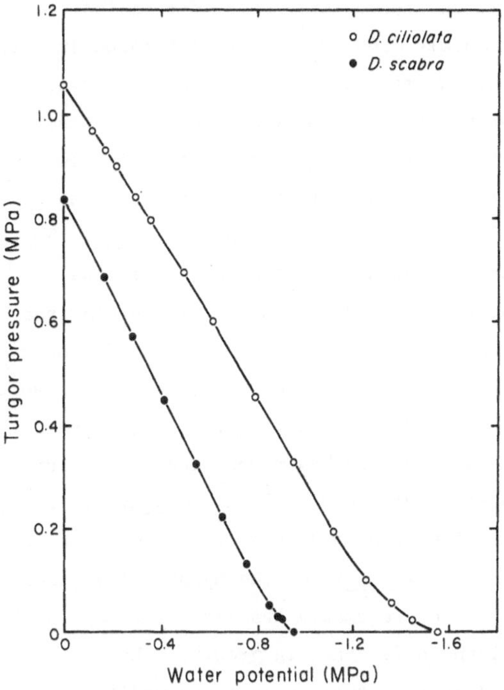

FIGURE 3. Relationship between tissue turgor pressure and tissue water potential for Dubautia ciliolata and D. scabra (Robichaux, 1984).

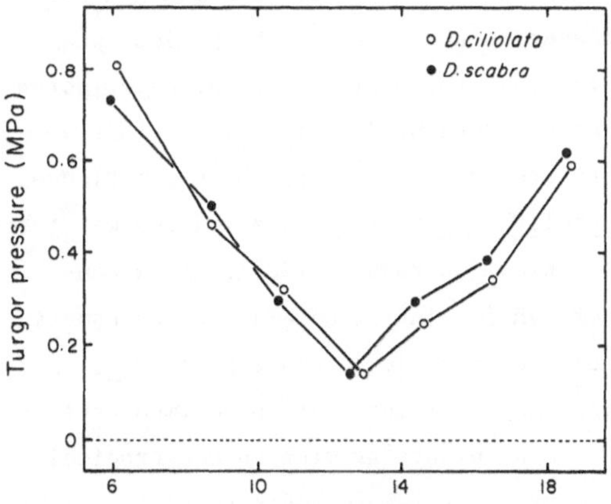

FIGURE 4. Estimated diurnal turgor pressures for Dubautia ciliolata and D. scabra on 4 August 1982 (Robichaux, 1984).

In marked contrast to their diurnal water potentials, the diurnal turgor pressures of D. ciliolata and D. scabra are very similar. Turgor pressures in both species decline from a maximal value of approximately 0.80 MPa in the morning to a minimal value of 0.14 MPa at midday. In addition, turgor pressures increase to approximately 0.60 MPa by late afternoon in both species. We may also calculate diurnal turgor pressures for a hypothetical individual that exhibits the tissue elastic and osmotic properties of D. scabra, yet that experiences the diurnal water potentials of D. ciliolata. Turgor pressures in such a hypothetical individual decline to 0 MPa by 1030 hr in the morning and ramain at 0 MPa until 1630 hr in the afternoon (Robichaux, 1984). In addition, turgor pressures in such an individual during the remainder of the day are several MPa lower than the actual turgor pressures experienced by D. ciliolata. This marked difference in diurnal turgor maintenance between D. ciliolata and the hypothetical individual would presumably translate into a significant difference in growth. These comparisons thus suggest that the modified elastic and osmotic properties of D. ciliolata may enable it to exploit a significantly drier environment than that occupied by D. scabra. While this comparison involves a montane wet forest species and a dry scrub species, we emphasize that the minimal diurnal water potentials experienced by D. ciliolata at this locality are very similar to those experienced by Pentaclethra macroloba in a lowland wet forest in Costa Rica (Oberbauer, 1982). As a consequence, shifts in tissue elastic and osmotic properties comparable to those seen in D. ciliolata may well turn out to be important in enabling many plants growing in wet tropical environments to tolerate moderate tissue water deficits. For example, the difference in osmotic potential at full hydration between D. ciliolata and D. scabra (0.27 MPa) is very similar to that between upper canopy and understory leaves of P. macroloba (0.25 MPa) (Oberbauer, 1982). This similarity correlates with the fact that the difference in the midday water potentials of the two Dubautia species is very similar to that of the upper canopy and understory leaves of P. macroloba (Oberbauer, 1982). A recent analysis by Stemmermann (1983) also suggests that differences in tissue elastic properties exist between varieties of the Hawaiian wet forest tree species, Metrosideros polymorpha. These differences, which are analogous to the difference between the two Dubautia species, may enable certain varieties of M. polymorpha to grow under moderately dry environmental conditions.

Preliminary work with two additional Hawaiian Dubautia species also illustrates the potential significance of variation in tissue osmotic properties (J.E. Canfield, unpublished). These two species grow in the Alakai Swamp on Kauai. This large, dissected, high-elevation plateau receives an average of 6,000-10,000 mm of precipitation per year. The vegetation of the Alakai is composed primarily of native wet forest, with a mosaic of open bogs scattered over the more level upland areas. Both of the Dubautia species are common in the Alakai. However, D. raillardioides is restricted to the wet forest, while D. paleata grows exclusively in the open bogs.

Preliminary measurements on a clear day at one locality in the Alakai indicate that the diurnal water potentials experienced by these two species may differ significantly (Fig. 5). Water potentials of the wet forest species, D. raillardioides, remain above -0.15 MPa throughout the day. In contrast, water potentials of the bog species, D. paleata, decrease from approximately -0.10 MPa in the morning to below -1.12 MPa in the afternoon. These differences in diurnal water potentials are paralleled by significant differences in the tissue osmotic properties of these two species.

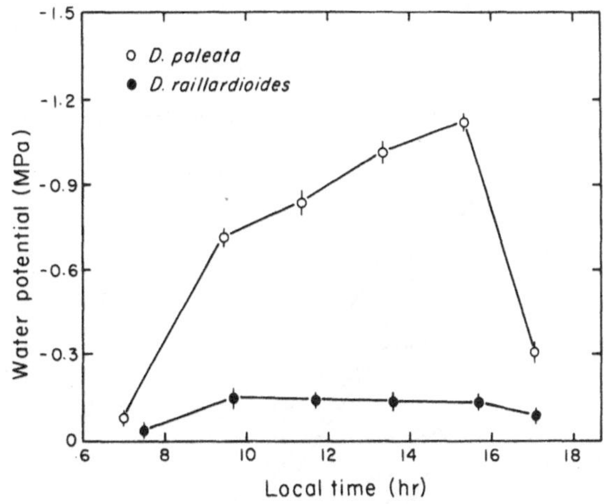

FIGURE 5. Diurnal water potentials for Dubautia paleata and D. raillardioides on 8 August 1981 at Lehua Makanoe bog, Alakai Swamp trail, Kauai (J.E. Canfield, unpublished).

The tissue osmotic potential at full hydration is -0.75 ± 0.05 MPa in D. raillardioides and is -1.20 ± 0.01 MPa in D. paleata. Since the tissue elastic properties of these two species are similar, this difference in osmotic potential at full hydration translates directly into a difference in the water potential at which tissue turgor pressure reaches zero. These water potentials are -0.95 and -1.35 MPa for D. raillardioides and D. paleata, respectively. Coupled with the data on diurnal water potentials, these latter data suggest that neither species experiences a condition of zero turgor during the day in its native habitat. However, were an individual with the osmotic and elastic properties of D. raillardioides to experience the diurnal water potentials of D. paleata, it would experience a condition of zero turgor for approximately 3 hr during the afternoon. As a consequence, the modified osmotic properties of D. paleata may enable it to exploit a more xeric environment than that occupied by D. raillardioides.

We are currently examining the extent of variation in these tissue elastic and osmotic properties for a large number of the other Dubautia species. This information should allow us to evaluate the role of these properties in promoting turgor maintenance in plants from a wide variety of tropical environments.

2.3. Tissue water relations of tropical forest species in Panama

Barro Colorado Island, a Smithsonian Institute research facility lying in Lake Gatun along the Panama Canal, is perhaps one of the most intensively studied areas of lowland tropical forest in the world. The vegetation on the island is classified as semi-evergreen rain forest, with a canopy height of up to 30 m and individual trees reaching 50 m in height (Knight, 1975). The mean annual precipitation is 2,750 mm/yr, but there is a pronounced dry season from mid-December until late April in which only about 180-260 mm of the annual total falls. Despite this regular seasonal drought, the great majority of the forest canopy is evergreen (Croat, 1978).

Soil moisture contents on Barro Colorado Island are typically 40-45% (by weight) during the wet season but drop steadily in the dry season to a low of about 25-30% in most years. Most streams on the island cease to flow at this latter time. The majority of precipitation during the dry season and early wet season goes into soil moisture storage. Even the first heavy rains in late April and early May produce less than 2% of their volume as runoff. Late in the rainy season, nearly 85% of the water falling as rain goes into runoff (Rubinoff, 1974).

A preliminary analysis of the components of tissue water relations in a series of important woody species on Barro Colorado Island suggests a pattern relating water relations to growth form (P.W. Rundel, unpublished). Saplings of canopy or subcanopy trees reach zero turgor pressure at significantly lower tissue water potentials than do understory shrubs. Trichilea tuberculata and Quararibea asterolepis, the two most numerous canopy trees in old-growth forests

TABLE 1. Maximal heights, tissue osmotic potentials at full hydration ($\overline{\pi}^o$), and tissue water potentials at zero turgor ($\psi_{P=0}$) in several woody species from Barro Colorado Island, Panama (P.W. Rundel, unpublished)

Species	Height (m)	$\overline{\pi}^o$ (MPa)	$\psi_{P=0}$ (MPa)
Canopy/Subcanopy Trees			
Trichilea tuberculata	30	-2.1	-3.1
Quararibea asterolepis	30	-2.5	-3.3
Hirtella triandra	20	-2.2	-3.1
Understory Shrubs			
Hybanthus prunifolius	3	-1.5	-2.3
Piper cordulatum	3	-1.8	-2.2
Faramea occidentalis	4	-1.1	-1.3
Gap Successional Trees			
Leuhea seemanii	30	-1.8	-2.4
Cordia alliodora	25	-2.1	-2.4

on the island, reach zero turgor at water potentials of -3.1 to -3.3 MPa (Table 1, Fig. 6). Hirtella triandra, an important subcanopy tree, reaches zero turgor in this same range of water potentials. Among the understory shrubs, however, zero turgor is reached at water potentials of -2.3 MPa or higher. In Faramea occidentalis, in particular, this point is reached at only -1.3 MPa. This difference between canopy trees and understory shrubs is maintained over a wide range of water potentials. At any given water potential the turgor pressure of canopy tree leaves is 0.4 - 1.2 MPa higher than that of understory shrub leaves. The osmotic potentials at full hydration range from -2.1 to -2.5 MPa in the three canopy and subcanopy species and from -1.1 to -1.8 MPa in the three understory shrubs (Table 1).

Although the saplings of the canopy and subcanopy species were growing in the same microenvironment as the similar-sized understory shrubs, the canopy and subcanopy species were inherently more drought-tolerant because of their tissue water relations. Indeed the osmotic potentials at full hydration in these trees are quite comparable to

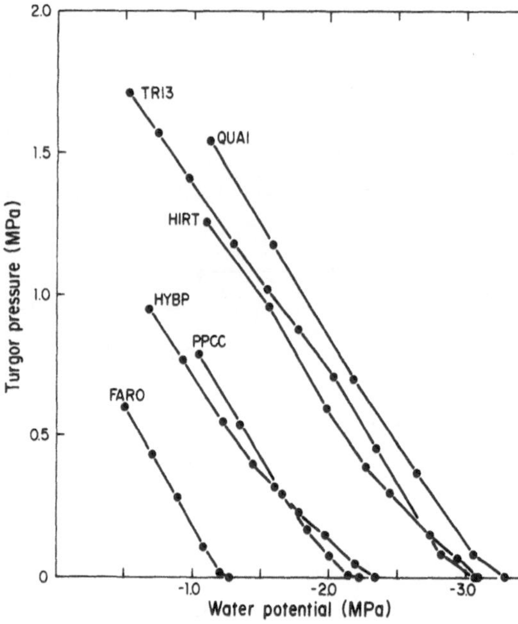

FIGURE 6. Relationship between tissue turgor pressure and tissue water potential for saplings of three species of canopy and subcanopy trees (Trichilea tuberculata, Quararibea asterolepis, and Hirtella triandra) and three species of understory shrubs (Hybanthus prunifolius, Piper cordulatum, and Faramea occidentalis) on Barro Colorado Island, Panama. Abbreviations are: TRI3 = T. tuberculata, QUA1 = Q. asterolepis, HIRT = H. triandra, HYBP = H. prunifolius, PPCC = P. cordulatum, and FARO = F. occidentalis (P.W. Rundel, unpublished).

those of many temperate forest trees, despite the wet forest environment on Barro Colorado Island. This genetic adaptation for drought tolerance in leaves of canopy tree species may relate to the greater transpirational stress experienced by leaves at the top of the canopy. As mentioned previously, Oberbauer (1982) has documented significant differences in the water potentials experienced by upper canopy and understory leaves of Pentaclethra macroloba in a non-seasonal, lowland wet forest in Costa Rica. Fetcher (1979) reported dawn and midday water potentials as low as -2.6 and -3.9 MPa, respectively, in *Trichilea tuberculata* at the end of the dry season on Barro Colorado Island. The dry season in the year of his study (1977) was relatively typical in severity and length. The subcanopy leaves used in his measurements apparently maintained a low leaf conductance to water vapor of about 0.5 mm/sec at midday, suggesting that the osmotic potential at full hydration was even lower than that found in leaves of sapling individuals in our study. With the onset of the rainy season, midday water potentials increased to -1.9 MPa.

The dry season of 1983 was the most severe on record for Barro Colorado Island. Only about 65 mm of precipitation fell from the end of November, 1982, to late April, 1983 (D. Windsor, personal communication). The severity of this drought was sufficient to impact the health of many forest species. Many species of understory shrubs were visibly affected by the drought in their loss of leaf turgor. Leaves that were normally horizontal in position were hanging vertically. Diurnal cycles of plant water potential measured in mid-March, 1983, indicated that leaves of *Faramea occidentalis*, *Psychotria marginata*, and *Psychotria furcata* remained at zero turgor throughout a 24-hour cycle. Leaves of two other common shrub species, *Hybanthus prunifolius* and *Piper cordulatum*, appeared to be above zero turgor only for very brief periods and had virtually no measurable leaf conductance at any time during the day. Saplings of many canopy tree species appeared to be tolerating the drought much better than the understory shrubs. These saplings showed little or no visible indications of wilting and commonly had low rates of leaf conductance in the early morning hours. Continued monitoring of these populations will be necessary to determine the extent of possible permanent damage resulting from these extreme drought conditions.

Successional trees colonizing large gaps on Barro Colorado Island exhibit tissue water relations that are somewhat intermediate between those of canopy trees and understory shrubs. Two common successional trees, *Leuhea seemanii* and *Cordia alliodora*, have osmotic potentials at full hydration of -1.8 and -2.1 MPa, respectively (Table 1). The water potential at zero turgor is -2.4 MPa for both species. Despite the severity of the drought in 1983, leaves of the gap colonizers remained under positive turgor pressure throughout the day and exhibited moderately high rates of leaf conductance to water vapor. Indeed, the diurnal cycle of leaf conductance during 1983 was not remarkably different from that measured in 1981 at the end of one of the wettest dry seasons on record (Fig. 7). This apparent paradox can be interpreted in terms of

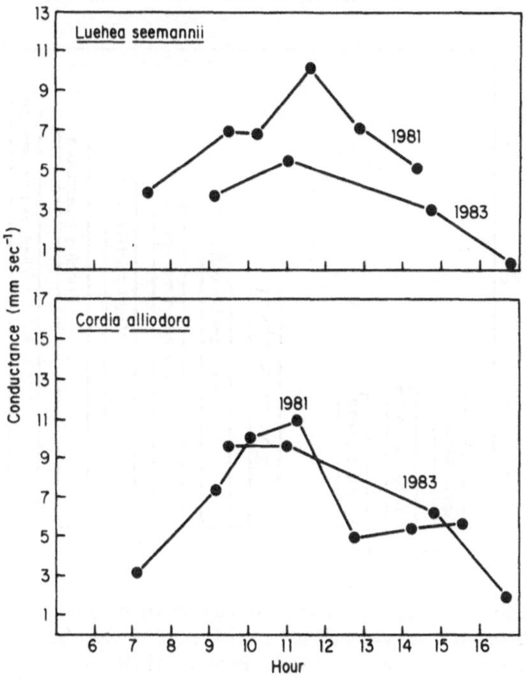

FIGURE 7. Diurnal cycle of leaf conductance to water vapor in gap successional trees at the end of the dry season in 1981 (a moderately wet dry season) and in 1983 (an extremely severe dry season) on Barro Colorado Island, Panama (P.W. Rundel, unpublished)

the soil moisture stress present in large, artificially maintained gaps in relation to the forest habitats. Predawn water potentials provide some indication of the soil moisture available for plant uptake. During March, 1983, predawn plant water potentials in the forest ranged from -1.9 to -3.4 MPa (Fig. 8). A leaf area index of about 6 m^2/m^2 or more was a major factor contributing to a loss of soil moisture through transpiration. In the large gap area, where a low leaf area index was maintained with regular clearing, the predawn water potentials in successional trees were consistently about -1.1 MPa. Thus, gap colonizers appear to be buffered against potential water stress by higher levels of soil moisture. The relationship between leaf area index and soil moisture stress during gap succession in wet tropical forests is clearly an important topic for expanded, future studies.

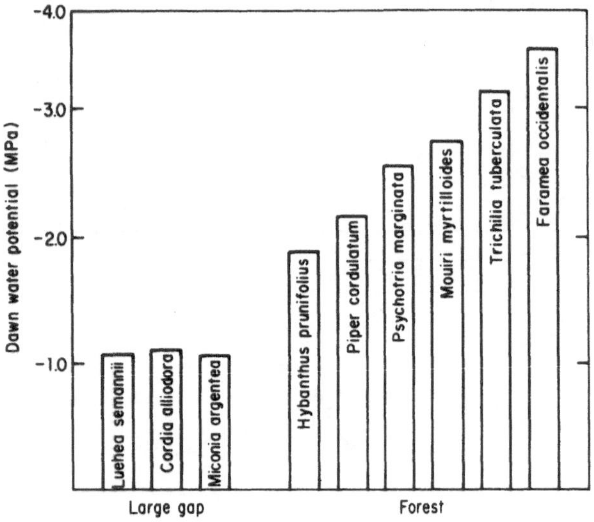

FIGURE 8. Predawn water potentials of species in forest and large gap habitats during March, 1983, on Barro Colorado Island, Panama (P.W. Rundel, unpublished).

3. CONCLUSION

While the supply of water is normally abundant in lowland and montane tropical wet forests, it is clear that plants growing in these environments may experience moderate tissue water deficits on a fairly regular basis. At the present time, we know very little about the nature of these deficits, their physiological effects on growth, and the mechanisms by which they are tolerated in wet tropical plants. Indeed, virtually any new research effort in this area will add significantly to our knowledge. Our only cautionary note in this regard is that care be given to the techniques that are used in studying the tissue water relations of these plants. Some excellent techniques are currently available that can be used even in remote areas in the tropics (Tyree, Jarvis, 1982).

Research on water deficits and plant growth in wet tropical environments may have significant implications for applied programs in forestry and agriculture (Hinckley et al., 1981). Unfortunately, the current rate of conversion of lowland wet tropical forests is exceedingly high (Myers, 1983). Obtaining information on plant responses to water stress may aid in predicting the long-term impact of this conversion, particularly with regard to the relative effects of differing types and levels of disturbance. In addition, such information may prove useful in the planning of reforestation programs.

ACKNOWLEDGEMENTS

The research of R. Robichaux, S. Morse, and W. Friedman is supported by NSF grant DEB 82-06411. The research of P. Rundel is supported by NSF grant BSR 82-14915. The research of L. Stemmermann and J. Canfield is supported by NSF grant DEB 79-10993 to D. Mueller-Dombois.

REFERENCES

Alvim PdeT (1960) Moisture stress as a requirement for flowering of coffee. Science 132, 354.

Alvim PdeT and Alvim R (1978) Relation of climate to growth periodicity in tropical trees. In Tomlinson PB and Zimmermann MH, eds. Tropical trees as living systems, pp. 445-464. Cambridge, Cambridge University Press.

Augspurger CK (1979) Irregular rain cues and the germination and seedling survival of a Panamanian shrub (Hybanthus prunifolius),

Oecologia 44, 53-59.
Blum G (1933) Osmotische Untersuchungen in Java, Ber. Schweiz. Bot. Ges. 42, 550-680.
Buckley RC, Corlett RT, and Grubb PJ (1980) Are the xeromorphic trees of tropical upper montane rain forests drought-resistant? Biotropica 12, 124-136.
Carlquist S (1980) Hawaii, a natural history, Lawai, Hawaii, Pacific Trop. Bot. Garden.
Carr GD and Kyhos DW (1981) Adaptive radiation in the Hawaiian silversword alliance (Compositae: Madiinae). I. Cytogenetics of spontaneous hybrids, Evolution 35, 543-556.
Croat TB (1978) The flora of Barro Colorado Island, Palo Alto, Stanford University Press.
Fasehun FE (1979) Effect of soil matric potential on leaf water potential, diffusive resistance, growth and development of Gmelina arborea L. seedlings, Biol. Plant. 21, 100-104.
Fetcher N (1979) Water relations of five tropical tree species on Barro Colorado Island, Panama, Oecologia 40, 229-233.
Frankie GW, Baker HG, and Opler PA (1974) Comparative phenological studies of trees in tropical wet and dry forests in the lowlands of Costa Rica, J. Ecol. 62, 881-919.
Grubb PJ (1977) Control of forest growth and distribution on wet tropical mountains: with special reference to mineral nutrition, Ann. Rev. Ecol. Syst. 8, 83-107.
Hinckley TM, Reich PB, and Morikawa Y (1981) Some aspects of the water relations of tropical tree species, XVII IUFRO Interdivisional Proceedings, pp. 91-96.
Hsiao TC, Acevedo E, Fereres E, and Henderson DW (1976) Water stress, growth, and osmotic adjustment, Phil. Trans. R. Soc. Lond. B. 273, 479-500.
Ike IF and Thurtell GW (1981) Water relations of cassava: water content, water, osmotic, and turgor potential relationships, Can. J. Bot. 59, 956-964.
Knight DH (1975) A phytosociological analysis of a species-rich tropical forest on Barro Colorado Island, Panama, Ecol. Monogr. 45, 259-284.
Ludlow MM (1980) Adaptive significance of stomatal responses to water stress. In Turner NC and Kramer PJ, eds. Adaptation of plants to water and high temperature stress, pp. 123-138. New York, Wiley & Sons.
Medina E (1983) Adaptations of tropical trees to moisture stress. In Golley FB, ed. Tropical rain forest ecosystems: structure and function, pp. 225-237. Amsterdam, Elsevier Scientific.
Mueller-Dombois D, Bridges KW, and Carson HL (1981) Island ecosystems: biological organization in selected Hawaiian communities, Stroudsburg, Penn., Hutchinson Ross.
Myers N (1983) Conversion rates in tropical moist forest. In Golley FB, ed. Tropical rain forest ecosystems: structure and function, pp. 289-300. Amsterdam, Elsevier Scientific.
Oberbauer SF (1982) Water relations of Pentaclethra macroloba, a wet tropical forest tree, Bull. Ecol. Soc. Amer. 63, 178.
Okali DUU and Dodoo G (1973) Seedling growth and transpiration of two West African mahogany species in relation to water stress in the root medium, J. Ecol. 61, 421-438.
Opler PA, Frankie GW, and Baker HG (1976) Rainfall as a factor in the release, timing, and synchronization of anthesis by tropical trees and shrubs, J. Biogeog. 3, 231-236.
Osonubi O and Davies WJ (1980) The influence of water stress on the photosynthetic performance and stomatal behaviour of tree seedlings subjected to variation in temperature and irradiance, Oecologia 45, 3-10.
Peace WJH and Macdonald FD (1981) An investigation of the leaf anatomy, foliar mineral levels, and water relations of trees of a Sarawak forest, Biotropica 13, 100-109.
Reich PB and Borchert R (1982) Phenology and ecophysiology of the tropical tree, Tabebuia neochrysantha (Bignoniaceae), Ecology 63, 294-299.
Richter H (1976) The water status in the plant - experimental evidence. In Lange OL, Kappen L, and Schulze E-D, eds. Water and plant life: problems and modern approaches, pp. 42-58. Berlin, Springer-Verlag.
Robichaux RH (1984) Variation in the tissue water relations of two sympatric Hawaiian Dubautia species and their hybrid recombinant, Oecologia (in review).
Robichaux RH and Pearcy RW (1980) Environmental characteristics, field water relations, and photosynthetic responses of C_4 Hawaiian Euphorbia species from contrasting habitats, Oecologia 47, 99-105.
Rubinoff RW (1974) Environmental monitoring and baseline data, Washington, D.C., Smithsonian Tropical Research Institute.
Scholander PF, Hammel HT, Bradstreet ED, and Hemmingsen EA (1965) Sap pressure in vascular plants, Science 148, 339-346.
Scholander PF, Hammel HT, Hemmingsen EA, and Bradstreet ED (1964) Hydrostatic pressure and osmotic potential in leaves of mangroves and some other plants, Proc. Natl. Acad. Sci. 52, 119-125.
Shreve F (1914) A montane rain forest: a contribution to the physiological plant geography of Jamaica, Washington, D.C., Carnegie Institution of Washington.
Stemmermann L (1983) Successional varieties of Metrosideros in Hawai'i, Pacific Science (in review).
Takami S, Turner NC, and Rawson HM (1981) Leaf expansion of four sunflower (Helianthus annuus L.) cultivars in relation to water deficits. I. Patterns during plant development, Plant, Cell, and Environment 4, 399-407.
Turner NC and Jones MM (1980) Turgor maintenance by osmotic adjustment: a review and evaluation. In Turner NC and Kramer PJ, eds. Adaptation of plants to water and high temperature stress, pp. 87-103. New York, Wiley & Sons.
Tyree MT (1981) The relationship between the bulk modulus of elasticity of a complex tissue

and the mean modulus of its cells, Ann. Bot. 47, 547-559.

Tyree MT and Hammel HT (1972) The measurement of the turgor pressure and the water relations of plants by the pressure-bomb technique, J. Exp. Bot. 23, 267-282.

Tyree MT and Jarvis PG (1982) Water in tissues and cells. In Lange OL, Nobel PS, Osmond CB, and Ziegler H, eds. Encyclopedia of plant physiology, new series, vol. 12B, physiological plant ecology II, pp. 35-77. Berlin, Springer-Verlag.

Tyree MT and Karamanos AJ (1981) Water stress as an ecological factor. In Grace J, Ford ED, and Jarvis PG, eds. Plants and their atmospheric environment, pp. 237-261. Oxford, Blackwell.

Tyree MT and Richter H (1981) Alternative methods of analysing water potential isotherms: some cautions and clarifications. I. The impact of non-ideality and of some experimental errors, J. Exp. Bot. 32, 643-653.

Tyree MT and Richter H (1982) Alternative methods of analysing water potential isotherms: some cautions and clarifications. II. Curvilinearity in water potential isotherms, Can. J. Bot. 60, 911-916.

Walter H (1971) Ecology of tropical and subtropical vegetation, New York, Van Nostrand Reinhold.

Zobel DB and Liu VT (1980) Leaf-conductance patterns of seven palms in a common environment, Bot. Gaz. 141, 283-289.

APPENDIX

The best available technique for measuring the tissue water relations of wet tropical plants is the pressure chamber technique. This technique was originally developed by Scholander et al. (1964, 1965) and has subsequently been refined by Tyree and co-workers (Tyree, Hammel, 1972; Tyree, 1981; Tyree, Karamanos, 1981; Tyree, Richter, 1981, 1982; Tyree, Jarvis, 1982). With this technique, it is possible to obtain a complete characterization of the turgor, osmotic, and elastic properties of intact, living tissue in the region between maximal tissue hydration and 50-60% of maximal tissue hydration. The technique is simple, requires minimal monetary investment in equipment, and may be performed even in remote areas in the tropics.

In the following brief discussion, we highlight the principal theoretical aspects of the pressure chamber technique. For a more detailed discussion, the reader is referred to the excellent recent review of Tyree and Jarvis (1982). For a discussion of the procedural aspects of the technique, the reader is referred to Robichaux (1984).

The water potential of the symplasm of a single cell (ψ) is equal to the sum of the turgor pressure (P) and the osmotic potential (π):

$$\psi = P + \pi. \qquad (1)$$

The values of P and π are presumed to be uniform throughout the symplasm, since it is unlikely that any significant pressure gradients can exist across the bounding membranes of the vacuole and cytoplasmic organelles (Tyree, Jarvis, 1982). However, the solutes contributing to the reduction in π in the various symplasmic compartments may be quite different (Tyree, Jarvis, 1982).

For a tissue, the water potential of the symplasm at equilibrium is

$$\psi = \overline{P} + \overline{\pi} \qquad (2)$$

where $\overline{P}$ and $\overline{\pi}$ are defined as weight-averaged values of tissue turgor pressure and tissue osmotic potential, respectively. In other words,

$$\overline{P} = \sum_{i=1}^{n} \frac{w_s^i}{W_s} P^i \qquad (3a)$$

and

$$\overline{\pi} = \sum_{i=1}^{n} \frac{w_s^i}{W_s} \pi^i \qquad (3b)$$

where P^i, π^i, and w_s^i are the turgor pressure, osmotic potential, and weight of symplasmic water, respectively, in the ith cell in the tissue, and W_s is the total weight of symplasmic water in the tissue (Tyree, Jarvis, 1982).

The value of the tissue osmotic potential may be approximated by

$$\overline{\pi} = - \frac{\phi \rho RTN}{W_s} \qquad (4)$$

where ϕ is an osmotic coefficient (to account for the non-ideality of solute behavior), ρ is the density of water in the symplasm, R is the universal gas constant, T is the Kelvin temperature, and N is the number of moles of solutes in the symplasm (ionic species counted separately) (Tyree, Jarvis, 1982).

As the water content of an initially saturated tissue decreases, tissue turgor pressure will decline and eventually will reach zero. If negative turgor pressures do not occur in the tissue, then at all lower water contents, $\psi = \overline{\pi}$. Hence,

$$\frac{1}{\psi} = \frac{1}{\overline{\pi}} = -\frac{W_s}{\phi\rho RTN} = \frac{W_a}{\phi\rho RTN} - \frac{R^*(W_s^o + W_a^o)}{\phi\rho RTN}. \quad (5)$$

where W_a is the weight of apoplasmic water in the tissue, W_s^o and W_a^o are the weights of symplasmic and apoplasmic water, respectively, in the tissue at saturation, and R^* is the relative water content of the tissue (Tyree, Jarvis, 1982). The third equality in Eq. (5) follows from the definition of R^*, since

$$R^* = \frac{W_s + W_a}{W_s^o + W_a^o} \quad (6)$$

and

$$W_s = R^*(W_s^o + W_a^o) - W_a. \quad (7)$$

If ϕ and W_a remain constant as R^* decreases below the point at which $\overline{P}$ reaches zero, then Eq. (5) describes a linear relationship between $1/\psi$ (or $1/\overline{\pi}$) and R^*. Hence, a plot of $1/\psi$ against R^* will yield a straight line in the region where $\overline{P} = 0$ (Fig. 9) (Tyree, Jarvis, 1982).

If ϕ and W_a also remain constant as R^* increases above the point at which $\overline{P}$ reaches zero, then Eq. (5) may be extrapolated to $R^* = 1$ (Fig. 9). At $R^* = 1$,

$$\frac{1}{\psi} = \frac{1}{\overline{\pi}} = -\frac{W_s^o}{\phi\rho RTN}, \quad (8)$$

which is the reciprocal of the tissue osmotic potential at full hydration ($\overline{\pi}^o$). From the extrapolation of Eq. (5), it is also possible to calculate $\overline{\pi}$ for any and all values of R^* between full hydration and the point at which $\overline{P}$ reaches zero. Given this calculated relationship between $\overline{\pi}$ and R^*, together with the measured relationship between ψ and R^*, it is then possible to calculate the relationship between $\overline{P}$ and R^* (Tyree, Jarvis, 1982). An example of the latter relationship is presented in Fig. 2, with the curve for D. ciliolata corresponding to the curve in Fig. 9.

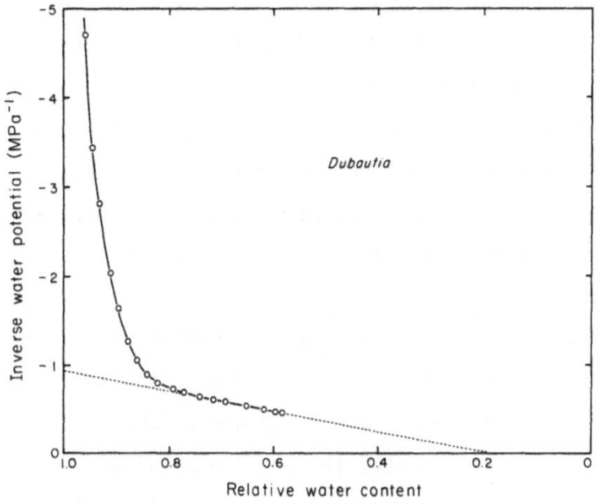

FIGURE 9. Relationship between the reciprocal of tissue water potential and tissue water content for *Dubautia ciliolata* (Robichaux, 1984).

The value of the bulk modulus of elasticity of the tissue may be calculated directly from the relationship between $\overline{P}$ and R^*. The weight-averaged bulk tissue elastic modulus ($\overline{E}$) may be defined as the change in tissue turgor pressure for a given fractional change in symplasmic water content. In other words,

$$\overline{E} = \frac{d\overline{P}}{dW_s} W_s \quad (9)$$

(Tyree, Jarvis, 1982). If W_a remains constant as W_s changes, then

$$\overline{E} = \frac{d\overline{P}}{dR^*} (R^* - R_a^*) \quad (10)$$

where R_a^* is the relative water content of the apoplasm (or the apoplasmic fraction). The value of R_a^* may be calculated from the original plot of $1/\psi$ against R^* by extrapolating Eq. (5) to $1/\psi = 0$ (Fig. 9) (Tyree, Jarvis, 1982). At this point,

$$R^* = R_a^* = \frac{W_a}{W_s^o + W_a^o} . \tag{11}$$

Three major assumptions of the pressure chamber technique are: (1) negative tissue turgor pressures do not occur (i.e., $\overline{P} \geqslant 0$), (2) as water is lost from the tissue, the concentration of solutes increases in an ideal fashion (i.e., ϕ = constant), and (3) all of the water lost from the tissue comes from the symplasm (i.e., W_a = constant = W_a^o). The consequences of violating these assumptions have recently been analyzed in detail by Tyree and Karamanos (1981), Tyree and Richter (1981, 1982), and Tyree and Jarvis (1982).

Several additional, potential sources of error with the pressure chamber technique include: (1) systematic and random errors in the measurement of ψ, (2) the existence of small internal disequilibria in ψ, (3) gradual changes in $\overline{\pi}$ over the time course of the measurements, and (4) plastic deformation of the cell walls over the time course of the measurements (Tyree, Karamanos, 1981).

PHOTOSYNTHETIC CHARACTERISTICS OF WET TROPICAL FOREST PLANTS

H. A. MOONEY (Stanford, California)
C. FIELD (Salt Lake City, Utah)
C. VÁZQUEZ-YÁNES (Mexico City, Mexico)

ABSTRACT

The leaves of forest trees of the wet tropics are generally large and live an average of about one year. They, however, vary greatly, dependent on site, in their average leaf specific weights and nitrogen contents, encompassing values found for leaves of plants inhabiting arid climates. Those factors which affect photosynthesis, either directly or indirectly, light, humidity, and CO_2 concentration, vary greatly from the top of the forest canopy to the forest floor. Those species which inhabit the understory, where the radiation level is only a few percent of that received in the open, utilize brief sunflecks to fix a large fraction of their daily carbon gain. They are able to respond quickly to an abrupt increase in radiation since stomata remain open even at very low light intensities. There is little information available on the photosynthetic responses of tropical trees to CO_2, humdity, temperature, and water potential.

There are intrinsic differences in the photosynthetic capacities of the various growth forms which inhabit humid tropical forests. Fast-growing gap species have the highest capacities and understory plants the lowest. In-progress studies, examining the photosynthetic characteristics of related species occupying different microhabitats will define those physiological properties which are site specific. Other studies comparing closely related species from a variety of climates, including the humid tropics, have shown that these latter species generally have lower photosynthetic capacities and lower nitrogen-use efficiencies than plants from drier climates.

Reports of similar net productivities of arid and tropical wet forests and of nutrient-rich and nutrient-poor tropical wet forests indicates that there may be large compensatory variation among communities in those components, both physical and biological, which determine net productivity. Detailed information is not yet available to assess this possibility.

1. INTRODUCTION

There have been a number of reviews in the past few years which have considered some aspects of the carbon balance of tropical rain forest plants (Bazzaz, Pickett, 1980; Mooney, et al., 1980; Jordan, 1983; Pearcy, Robichaux, 1983; Medina, Klinge, 1983). It is obvious from these that we still lack a comprehensive understanding of those factors which limit the productivity of lowland rain forest plants or of the specialized mechanisms which they may have evolved to gain carbon under the particular conditions prevailing in tropical wet climates. Although the new work appearing since these reviews has not been extensive it is nonetheless important to attempt another synthesis since there is a certain urgency for developing a framework for managing this vital but disappearing biosphere component. Here we concentrate principally on leaf photosynthesis. We first examine leaf properties, then the environment to which leaves are exposed in tropical rain forests, and then the kinds of photosynthetic responses found in the diverse leaf types found in these habitats. Finally we briefly review those factors controlling primary productivity of entire systems.

2. LEAF CHARACTERISTICS

Since leaves are the principal organs involved in photosynthesis we examine what is known to be special about those of plants of the tropical rain forest.

There is apparently a vast range of variability in leaf features of rain forest plants dependent primarily, it seems, on the nutrient status of the habitat (Table 1). Leaves of trees of extremely depauperate sites in the Amazon are not unlike those found in Mediterranean sclerophylls. That is, they are quite thick and have heavy cuticles. They have relatively high specific weights and very low nitrogen contents. They differ from Mediterranean sclerophylls primarily in their large size. In contrast, leaves of rain forest dominants from eutrophic regions are comparatively thin and they have a relatively high nitrogen content. It is thus difficult to generalize about "tropical rain forest" leaf types since they can vary to such a large degree.

Since nitrogen content is such a strong determinant of leaf photosynthetic capacity, it is of value to see what range this particular character spans for tropical rain forest plants. Mean values of leaf nitrogen content of

TABLE 1. Mean leaf properties of dominants of eutrophic and oligotrophic lowland tropical forests as compared to those of Californian sclerophylls.

	Tropical		Californian[4]
	Eutrophic (New Britain)[1]	Oligotrophic (Amazon)[2]	
Dominant leaf size (cm^2)	(mesophyll)	30.1	4.3
Thickness (μm)	230	457	-
Epidermal thickness (μm)			
Upper	3.8	17.8	8.6
Lower	2.2	10.5	6.8
Stomatal density (mm^{-2})	408	135	230
Specific leaf wt. ($g\ m^{-2}$)	(120)[3]	219	194
Nitrogen ($mg\ g^{-1}$)	20.8	8.4	10.1
Leaf lifetime (yr)	~1	~1	~2

1, Grubb, 1977; 2, Sobrado, Medina, 1980; 3, data from 1 for a eutrophic lower montane forest in Puerto Rico; 4, Thrower, Bradbury, 1977.

over 2% have been reported for at least two forests and values near 1.5% are common for plants on comparatively nutrient-rich volcanic soils (Table 2). Values less than one per cent nitrogen are found only in the very nutrient deficient sandy soils of the Amazon. These average values must be viewed with some caution, however, since nitrogen content varies greatly with leaf age--and the precise leaf age is not generally given in these survey studies.

In the lowland tropics leaf duration is generally only between 3 to 13 months (Medina, 1981). Evergreen leaves of other ecosystems generally live for multiple years. Frankie, et al., (1974) in their comprehensive study of the phenology of canopy and understory plants in the Costa Rican rain forest found that there are always some plants producing new leaf growth at all times of the year although there is a peak of activity during the driest months (Fig. 1). They found that some species have continuous flushing throughout the year, whereas others are discontinuous. In the overstory, in particular, some species drop their leaves for periods of 2 to 14 weeks before flushing a new set.

There are other characteristic features of leaves of tropical rain forest plants other than their generally large size which influence carbon gaining capacity either directly or indirectly. The general appearance of "drip tips" has been long noted, particularly on trees of the lowland tropical rain forests. These tips are thought to be important in reducing ion leaching from the leaves as well as making the leaf a less suitable microsite for the growth of pathogens and epiphyllae (Dean, Smith, 1978). Another rather distinctive feature which is found in certain understory herbs is the presence of a layer of anthocyanin just below the chlorenchyma which results in an enhancement

TABLE 2. Mean nitrogen contents and specific leaf areas of leaves of trees and shrubs of lowland tropical and lower montane tropical rain forests.

Locality	Soil type	Vegetation type	Species n	N (mg g^{-1})	Specific leaf weight (g m^{-2})	Ref
Amazon	Oxysol (laterite)	Mixed forest	7	12.7	135	1
Amazon	Sandy (podzol)	Tall caatinga	6	11.6	132	1
Amazon	Sandy (podzol)	Bana	14	7.4	213	1
Uganda	Metamorphic	Mixed forest	14	28.	-	2
Cameroon	Sandy	Mixed forest	14	17.	-	2
New Britain	Volcanic ash	Lowland forest	14	20.8	-	3
Puerto Rico	Volcanic	Lower montane	7	13.6	120	1
Puerto Rico	Andesite	Lower montane	28	16.2	-	3
New Guinea	Gabbritic	Lower montane	23	15.2	-	3
New Guinea	Gabbritic	Lower montane	12	13.2	-	3

1, Medina, 1981; 2, McKey et al., 1978; 3, Grubb, 1977.

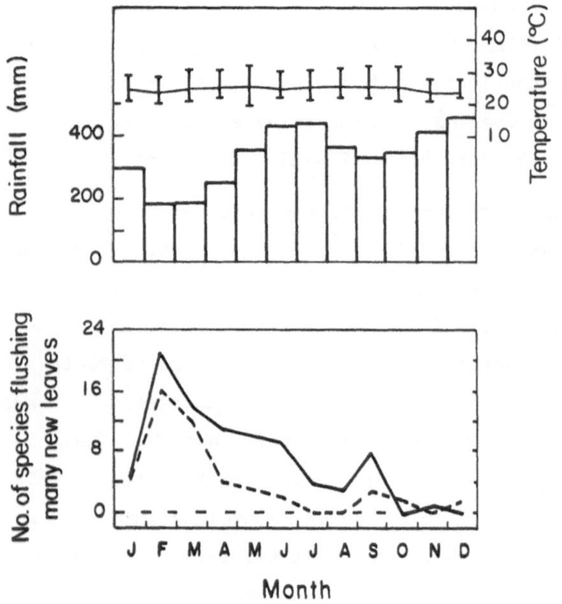

Fig. 1. Seasonal course of leaf flush in a tropical lowland rain forest in Costa Rica. Solid lines indicate overstory and dashed lines understory plants (from Frankie, et al., 1974).

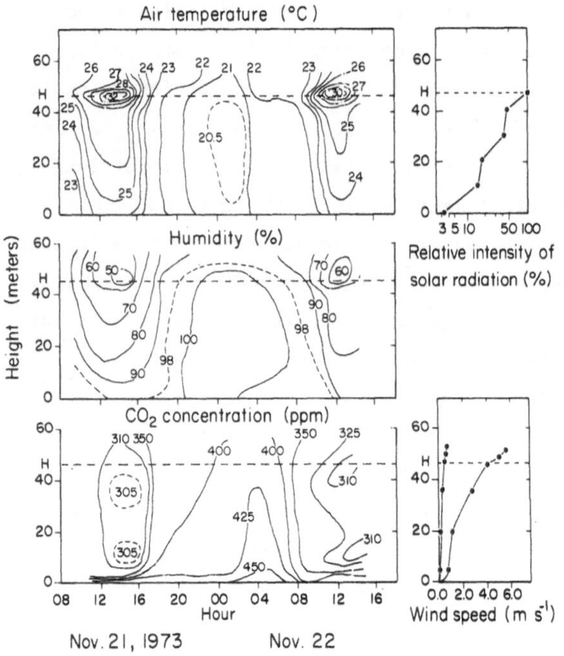

Fig. 2. Vertical profiles of temperature, humidity, CO_2, wind, and radiation in the Pasoh tropical rain forest in western Malaysia. H indicates the top of the canopy (from Aoki, et al. 1975)

of the capture of photosynthetically active radiation (Lee, et al., 1979).

3. THE VERTICAL PHOTOSYNTHETIC ENVIRONMENT

There is a vast difference in the environment to which leaves are exposed at the top of the canopy of a tropical rain forest versus that at the bottom, although measurements documenting these differences are few and are often for very short time periods only. Aoki, et al., (1975) give a particularly comprehensive set of measurements for a canopy of a Malaysian rain forest (Pasoh) which is summarized in part in Fig. 2. These measurements verify the widely recognized reduction in radiation through the canopy. Leaves of plants on the forest floor receive on average only a few percent of the total radiation received by leaves at the top of the canopy. As compared to the top of the canopy, the forest floor wind speeds are often quite low, humidity is high both during the day and night, diel temperature changes are small, and CO_2 concentrations can be relatively high. Since all of these factors can affect photosynthesis either directly or indirectly, one might expect quite different photosynthetic properties of leaves of the top of the canopy with those of the bottom. Some studies have shown, for certain factors at least, that this is indeed true. Unfortunately, comparative studies of responses of both upper and lower canopy leaves to all of the diverse factors affecting photosynthetic capacity have not yet been made.

3.1 Light environment and photosynthetic response

The large variation in the light environment of

a tropical forest is reflected in a large variation in the photosynthetic light response of the plants living in the various canopy levels. Plants of the understory have a very low photosynthetic capacity and light-saturated rates; whereas emergent trees are saturated at much higher light levels (Fig. 3). The whole forest probably is never light saturated. Differences in photosynthetic capacity and light requirement for photosynthesis for a variety of growth forms are shown in more detail in Table 3. The ground herbs have very low compensation points which is due to their low respiration rates. All forms which live in a high light environment have moderately high photosynthetic

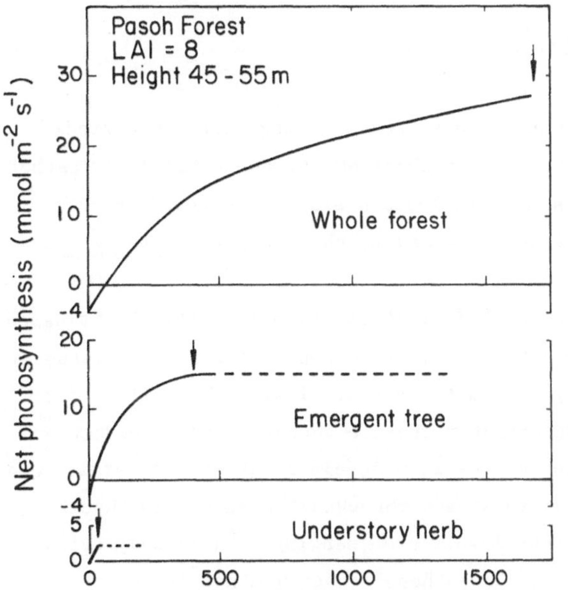

Fig. 3. Photosynthetic response to light of an understory herb, an emergent tree (Shorea leprosula), and of an entire Maylasian rain forest (Pasoh). Data for the entire forest from Aoki, et al., 1975 (1 cal cm^{-2} min^{-1} converted to the approximate sunlight equivalent of 1323 μmol m^{-2} s^{-1}). Data for Shorea and unidentified herb from Koyama, 1981). The Shorea rates were higher than that measured on other emergent trees and more similar to secondary forest species. Respiration rate also unusually high on Shorea, probably due to leaves being newly expanded.

capacities. These rates are comparable to or somewhat higher than sun leaves of woody plants from other climates, but lower than many herbaceous sun plants (Larcher, 1975).

The photosynthetic response of plants to the remarkably low light levels, which characterize the rain forest floor, has been studied in some detail (Bjӧrkman, et al., 1972; Pearcy, 1983; Pearcy, Calkin, 1983). Bjӧrkman and Ludlow (1972) found that the amount of photosynthetically active radiation (photon flux density, PFD, 400-700 nm wavelength band) on the bottom of a Queensland rain forest was exceedingly low (Fig. 4). Diffuse PFD amounted to only about 5 μmol m^{-2} s^{-1}; whereas on the top of the canopy over 2000 μmol m^{-2} s^{-1} was measured during midday.

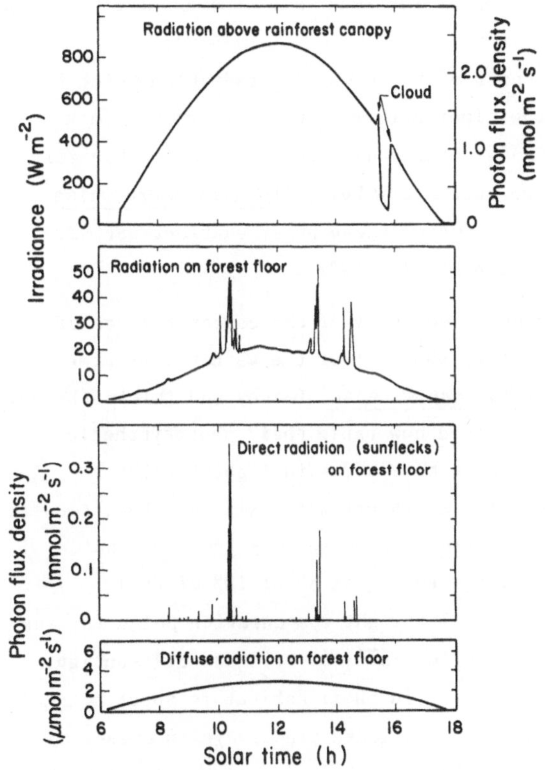

Fig. 4. Radiation components above and below a Queensland rain forest (from Bjӧrkman, Ludlow, 1972).

TABLE 3. Characteristic Differences in Photosynthetic Capacities of Leaves of Diverse Tropical Rain Forest Plants (from Koyama, 1981)

Growth Form	Maximum photosynthetic rate ~$\mu mol\ m^{-2}\ s^{-1}$	Light saturation ~$\mu mol\ m^{-2}\ s^{-1}$	Light compensation ~$\mu mol\ m^{-2}\ s^{-1}$
Upper canopy			
Sun type	12.6-18.9	250-370	12
Shade type	6.3-9.5	125-185	6-12
Lower canopy			
Shade type	4.4-5.0	125	6-12
Secondary forest	12.6-15.8	125-245	6-12
Climbers	12.6-18.9	125-245	6-12
Ground herbs	1.3-1.9	25-37	2.6-6

Determinations made at a chamber temperature of between 26-30C

Over half of the clear day radiation which fell on the floor occurred during brief sunfleck periods. Nearly half of the total carbon gain of the understory herb, Alocasia macrorrhiza occurred during these brief sunfleck periods (Björkman et al., 1972).

In a detailed study of the carbon balance of understory saplings of the C_3 Hawaiian tree, Claoxylon sandwicense, Pearcy and Calkin (1983), demonstrated how rapid their photosynthetic response is to changes in light level (Fig. 5). As is characteristic of shade leaves, those of Claoxylon have very low respiration rates and saturate at about 10% of full sunlight. The light compensation point is less than 2 $\mu mol\ m^{-2}\ s^{-1}$, or 0.1% of full sunlight. Diffuse light in their habitat is about 20 $\mu mol\ m^{-2}\ s^{-1}$ and sunflecks are always in excess of 200 $\mu mol\ m^{-2}\ s^{-1}$, the saturation point for this species. The leaves of Claoxylon attain light saturated rates during brief natural sunfleck periods (Fig. 5). Pearcy and Calkin estimate that photosynthesis during sunfleck periods account for about 60% of the total daily carbon gain. Sunflecks average between 10 to 30 minutes per day in the habitat of Claoxylon.

The efficient utilization of sunflecks is based on the very fast response of the photosynthetic apparatus to changing light levels (Fig. 5b). Importantly stomata increase their conductance to near maximum levels early in the morning and remain steady throughout midday even though light levels are changing dramatically (Fig. 5e). Thus these leaves are not limited by conductance when light flecks impinge upon them. The growth rate of individuals of Claoxylon in the understory is directly related to the potential duration of sunflecks it receives (Fig. 6).

Oberbauer (1983) has recently noted that even canopy species such as Pentaclethra macroloba (from Costa Rica) may have a relatively low light requirement. Growth of this species does

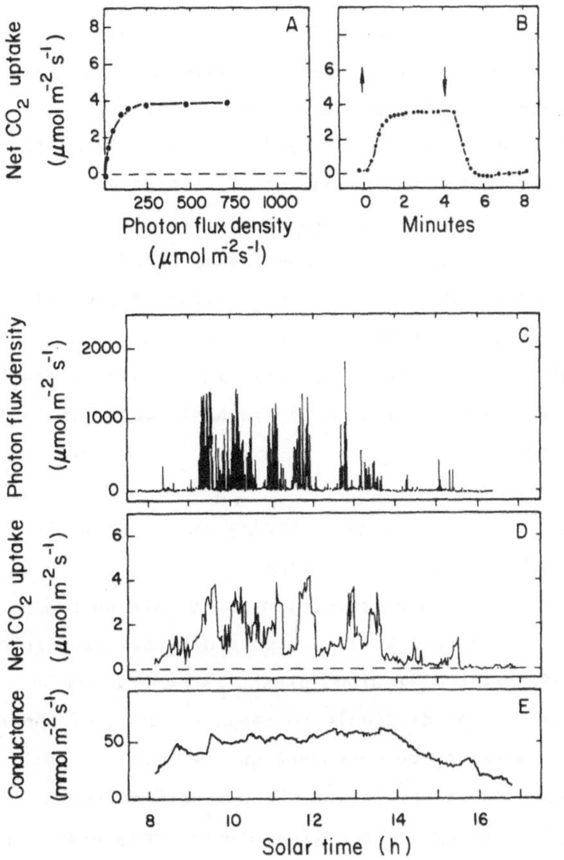

Fig. 5. Responses of in situ leaves of the C_3 Hawaiian tree Claoxylon sandwicense to light conditions. Measurements made on leaves of saplings growing in a shady understory habitat. A, photosynthetic light response curve, B, photosynthetic response to light on (arrow up) and light off (arrow down), C, photon flux density in the natural habitat, D, diurnal course of photosynthesis. and, E, conductance. Correspondence of light and photosynthesis displaced somewhat due to differing positions of light sensors and lags in photosynthesis measuring system, E, diurnal course of conductance (from Pearcy and Calkin, 1983).

not increase at levels above 20% of full sun. Light levels above this result in leaflet closure and reduced light interception. This canopy species showed low acclimation potential to different light regimes as predicted by Bazzaz and Pickett (1980).

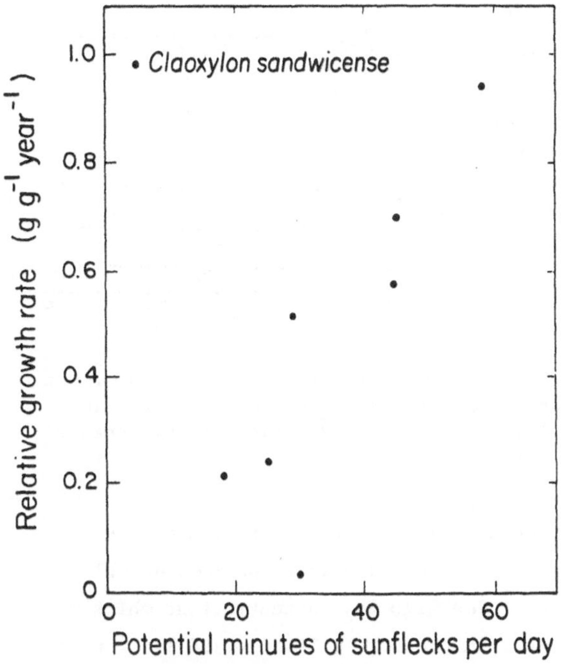

Fig. 6. Relative growth rates of Claoxylon as a function of microsite sunfleck duration (from Pearcy, 1983).

3.2. Humidity and photosynthetic responses

As noted above, there can be a considerable differential in the humidity environments of overstory and understory leaves. There has, however, been no study of the significance of these differences on the gas exchange responses of leaves occupying these dissimilar microsites. We have, however, reported the responses of leaves of one shrub species, Piper hispidum, which is found in the understory and in light gaps of the lowland wet tropics of Mexico (Mooney, et al., 1983). We found that leaves do not close their stomata fully in the dark at high humidities (Fig. 7). In the light, at humidities of 75%, the stomata have a conductance only 1/5th of that at 95% humidity. Further, in contrast to most plants, the stomata are not sensitive to CO_2 concentrations. Such a response would mean that

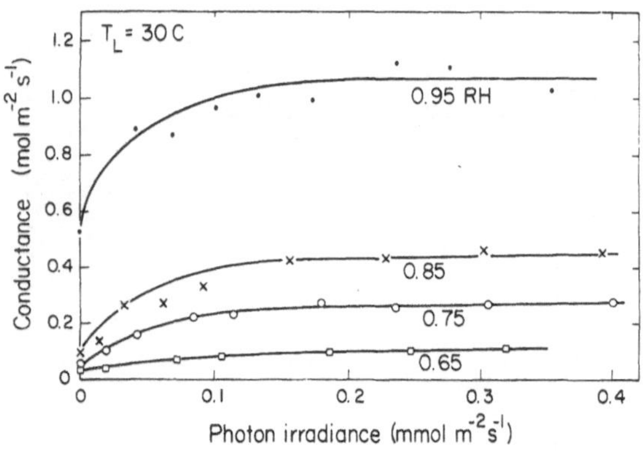

Fig. 7. Response of stomata of leaves of the tropical shrub Piper hispidum to radiation at different relative humidities (from Mooney et al., 1983).

in the understory, where high humidities prevail, the stomata would be open at all times. The high CO_2 concentrations which prevail on the forest floor would not act to close stomata as they might for CO_2 sensitive species. There would be little loss of water, however, because of the low water concentration gradient between leaf and air. There would, however, be a large potential gain in carbon since when a light fleck hits a leaf it would be able to immediately respond to its full photosynthetic capacity, unrestricted by the generally slow response of stomata to changing light levels. This indeed may explain the fast response to changing light levels noted above for Claoxylon.

We obviously need further studies on humidity responses of different life forms which experience different humidity regimes.

3.3. CO_2

There have been essentially no studies of the effect on CO_2 on the photosynthetic capacity of wet tropical plants.

3.4. Water potential and photosynthetic response

Another critical need is for information on the photosynthetic response of wet forest plants to moderate water stress, which must be frequent in the upper canopy, as well as in all parts of the vegetation during the brief drought periods which characterize even the wettest rain forests. Oberbauer (1983) has recently measured water potential components of leaves of Pentaclethra macroloba growing in the under canopy (<2m) to the upper canopy (>25m) and demonstrated compensations maintaining equal turgor in the dissimilar environments. No direct measures of the effects of differences in water potential on photosynthesis were made however.

4. AN EVOLUTIONARY APPROACH TO THE STUDY OF PHOTOSYNTHETIC ADAPTATION

From the above it is clear that data on the physical, chemical, and physiological properties of leaves of lowland plants is scanty and in many cases difficult to compare. Many of these properties are dependent on leaf age and yet this parameter is not often specified for a given study. The photosynthetic responses reported are for measurements obtained utilizing a variety of techniques (e.g., the study cited above by Koyama (1981) was done on excised branches, whereas that of Pearcy and Calkin (1983) was done on in situ leaves). The precise climatic and soil properties prevailing at a given study site may not be specified and thus cross-site literature comparisons are sometimes difficult to interpret.

4.1. Hawaiian comparisons

The recent studies of Pearcy and Robichaux and their co-workers have provided a new level of precision for interpreting which characteristics typify leaves of tropical wet habitats (Robichaux, Pearcy 1980 a,b, 1983; Pearcy, Robichaux, 1983 a,b; Pearcy, 1983, Pearcy, Calkin, 1983; Pearcy, et al., 1983). They have compared members of a number of genera which have representatives in widely different habitat types. Included in these comparisons have been both C_3 and C_4 species.

They recently compared the responses of pairs of species of Euphorbia (C_4) and Scaevola (C_3) originating from a variety of habitats extending from a wet forest (annual precipitation, 4000-6000, mm/yr) to a dry scrub (400-600 mm/yr) (Robichaux, Pearcy, 1983). Measurements made on leaves grown in a common environment showed that the wet forest species had similarly low rates when expressed on either a weight or area basis as compared to species pairs from drier habitats (Fig. 8). Nitrogen-use efficiency (CO_2 gained/N content in leaf) was also lowest in the wet site species although the site differences among the C_3 species was not great. Water-use efficiency (CO_2 gained/water lost), determined at a low but unspecified water vapor gradient, did not vary much between sites although it did between genera, as would be expected in comparisons of C_3 and C_4 species.

These comparisons are particularly instructive since it is thought that all of the species in each genus evolved from single species colonizers. There is remarkable convergence in the traits of the C_3 and C_4 species in the wet forest site (except for water-use efficiency). It should be noted, however, that C_4 species are rare in wet forest tropical habitats.

4.2. Mexican Pipers

A study of physiological divergence in the genus Piper has brought a somewhat different approach to the study of adaptation to tropical wet climates. This very large tropical genus (~2000 species) generally has many species representatives in any given tropical locality, each occupying somewhat different microsites

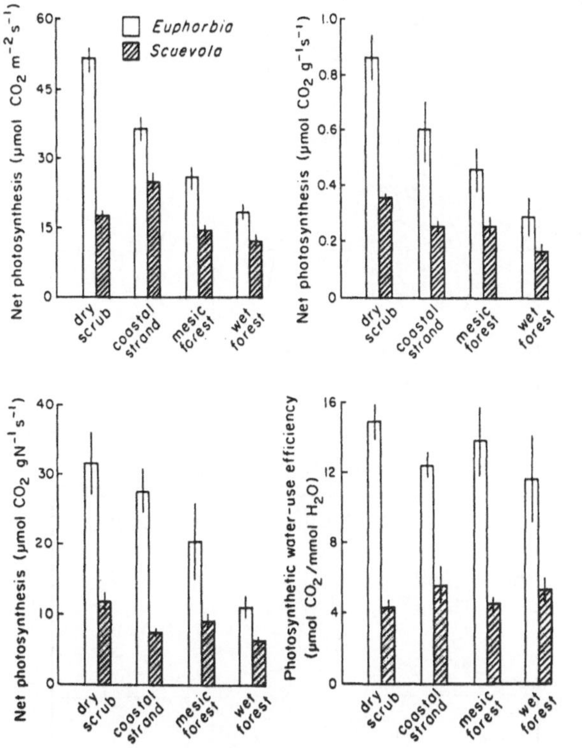

Fig. 8. Comparative photosynthetic characteristics of matched pairs of species of a C_4 genus, Euphorbia, and a C_3 genus, Scaevola, originating from diverse habitats in Hawaii. Measurements made on plants which were grown in a common environment (from Pearcy and Robichaux, 1983).

that are formed during the gap-filling process in the forest. In contrast then to the Hawaiian studies, one can view divergence within a given climatic type but in different microhabitats.

We have been studying those species occurring at the Los Tuxtlas Biological Preserve in the State of Veracruz, Mexico. Annual precipitation is over 4000 mm and the mean annual temperature is 27 C. We have examined leaf properties of nine species of this genus ranging from understory shrubs of the primary forest to successional herbs and shrubs.

Mean leaf size differs by over a factor of ten, extending from leaves less than 40 cm^2 to those which average nearly 500 cm^2 (Fig. 9). The

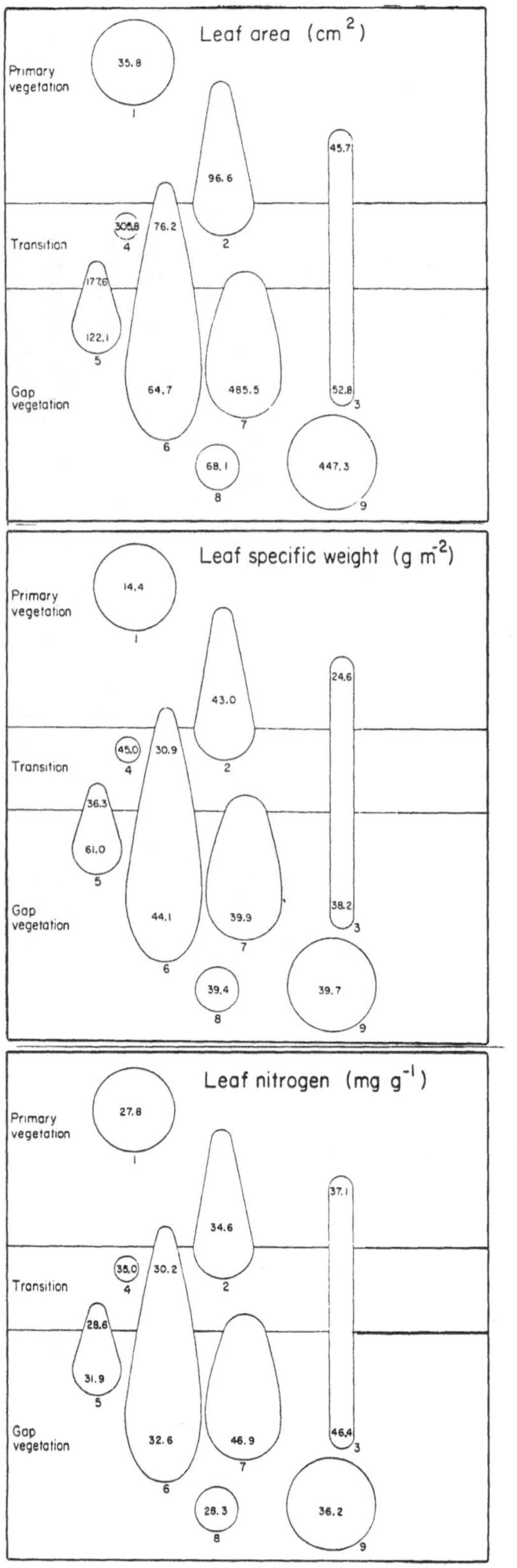

Fig. 9. Characteristics of the leaves of *Piper* species growing in the rain forest of Los Tuxtlas, Mexico. Size and positions of figures indicates relative abundance and distribution of the species in relation to the primary or disturbed vegetation. Numbers refer to the following species: 1, *Piper aequale*, 2, *P. lapathifolium*, 3, *P. amalago*, 4, *P. yzabalanum*, 5, *P. sanctum*, 6, *P. hispidum*, 7, *P. auritum*, 8, *P. aduncum*, 9, *P. umbellatum*.

largest leaved species grow in full sun; *P. umbellatum*, which is a ruderal herb, and *P. auritum*, a pioneer small tree. Leaf specific weight in all of the *Piper* species is rather low ranging from less than 15 g m^{-2} to just over 60 g m^{-2}. The higher values tend to be in those species occurring in sunny secondary or early gap vegetation. Leaves of species which occur both in the open vegetation and in the understory of the primary or transitional forest have leaves in the former habitat which are generally double the specific weights of those in the closed vegetation. The leaves of all of the species have relatively high nitrogen concentrations, although the highest values occur for leaves of sun species, or of plants growing in the sun of those species occuring in more than one habitat. The generally higher leaf specific weight of the sun leaves, and of the generally higher nitrogen concentration on a weight basis, results in a generally higher nitrogen content on an area basis for the sun types.

We have additionally made preliminary studies of the photosynthetic characteristics of four of these species; *P. hispidum*, a shrub species found in the gap vegetation and in the transitional forests, *P. auritum*, a small tree of the gap vegetation; *P. umbellatum*, a ruderal herb, and *P. amalago*, a shrub found in all habitats, both sun and shade.

The light saturated rates of these species are relatively low (100-300 nmol CO_2 g^{-1} s^{-1}; 3.4-8.5 μmol m^{-2} s^{-1}) (Table 4). This is somewhat surprising in view of their high leaf nitrogen contents. Two of the *Piper* species have

TABLE 4. Comparative photosynthetic characteristics of Mexican Piper species. These are values derived from single field determinations and are to be considered preliminary (unpublished data)

	Piper amalago shade	Piper amalago sun	P. hispidum	P. auritum	P. umbellatum
Mean maximum photosynthetic rate					
Area basis (μmol m^{-2} s^{-1})	3.4	4.8	6.5	6.2	8.5
Weight basis (μmol g^{-1} s^{-1})	120	201	184	111	296
N basis (μmol $molN^{-1}$ s^{-1})	55	72	106	47	150
Dark respiration (μmol m^{-2} s^{-1})	0.14	0.70	1.7	1.0	1.0
Light compensation (μmol m^{-2})	<5	12	20	12	15
Light saturation (μmol m^{-2})	300	300	300	400	300

relatively low instantaneous nitrogen-use efficiencies (less than 70 μmol CO^2 mol N^{-1} s^{-1} or 5 or less μmol CO^2 g N^{-1} s^{-2}). Efficiences this low are generally found on extreme sclerophylls (Field, Mooney, unpublished). At the present time we do not know the basis for this low efficiency. Since both the sun and shade leaves have low photosynthetic capacities and high nitrogen contents, this low NUE is probably not due to a disproportionate investment in light harvesting versus carbon reduction machinery as might be expected between sun and shade species. It may be that the low efficiency is due rather to a high proportion of leaf nitrogen being invested in N-based defensive compounds rather than in photosynthetic machinery. Piper species often contain alkaloids (Atal et al., 1975) the commonest of which (piperine) is insecticidal (Harvell, et al., 1943).

Both the Hawaiian and Mexican studies have confirmed the relatively low photosynthetic capacity of rain forest plants. They also indicate that some at least may have relatively poor instantaneous nitrogen-use efficiencies. Studies are needed of integrated nitrogen-use efficiency during the entire life of a leaf and of a plant.

5. PRODUCTIVITY

We will not review productivity studies in any great detail since there have recently been rather comprehensive overviews (Murphy, 1975; Brown, Lugo, 1982; Jordan, Herrera, 1981; Jordan, 1983).

The productivity of any stand is dependent on certain plant properties such as photosynthetic capacity of leaves, costs of making and maintaining tissues, leaf duration and amount, and allocation pattern of root over shoot. Important environmental limitations are radiation amount and duration, moisture, nutrient level and temperature. Except for nutrients in certain habitats, tropical wet climates have conditions which would promote high productivity, that is, high annual amounts of radiation, water generally not limiting in most of the canopy at least, and favorable temperatures for optimum metabolism through the day as well as year round.

As noted above, it is somewhat difficult to generalize about the leaf properties of tropical wet forests since they depend in large part on habitat nutrient status. Extensive data are not available on leaf duration in tropical forests. As stated earlier, Medina (1981) gives a figure of 3-13 months for lowland tropics. Understory plants may keep their leaves in excess of two years (Bentley, 1979). Grubb (1977) indicates a mean leaf life to 14-18 months in the upper montane rain forests. It is likely that the more nutrient deficient sites have plants with leaves of somewhat longer durations than the nutrient-rich habitats. Grubb also notes that leaf area index, for a variety of tropical forests, ranges between 5 to 10 with the higher values characterizing lowland forests. High leaf area indices and leaf durations of a full year are properties which can contribute to high annual productivity.

Because of the great diversity of tropical forest types, it is perhaps more instructive to look first in detail at a couple of forests rather than viewing averages in order to understand those factors limiting production. Jordan and Herrera (1981) have recently made such a comparison of the productivity of two tropical forests. One of these, the Amazonian caatinga, grows on nutrient poor podzolic soils, and the other, a Tabonuco forest, is a lower montane rain forest in Puerto Rico which grows on volcanically-derived soils (Table 5). Even though these forests grow on soils of very different nutrient levels, they have virtually identical annual primary productivities. Jordan and Herrera believe that the very specialized nutrient-conserving mechanisms of the plants of the nutrient-poor forest enable them to maintain this comparatively high productivity. Although the Amazon forest may have a high efficiency of nutrient use, it still has features which reduce productivity which are the result of low nutrient availability (higher leaf specific weight, lower leaf nitrogen, and greater allocation of carbon to roots rather than productive leaves). Thus, there must be other factors leading to this convergence in production other than nutrient conservation. Part of the explanation may lie in the differences in climate between these forests. The Puerto Rican forest, which has the potentially more productive structure, occurs at a latitude of nearly 20° N, in contrast to the Amazonian forest which is only 2 degrees north of the equator. Temperatures are cooler in the lower montane forest. Thus, it is possible that the similar productivities of the two forests are a result of compensating factors; the more favorable productive structure occurs in a climate more limiting to production.

Brown and Lugo (1982) have recently presented productivity data for a number of tropical forests. They show how total net productivity varies between 1 to 2 kg m^{-2} going from subtropical dry to tropical moist climates (Fig. 10). Whereas net productivity varies by only a factor of two among the various forest types Kira, 1975, gross productivity varies by a factor of six. The higher biomass supported by the wetter forests evidently results in a high respiration load. About 80% of the plant gross productivity is respired in the wet forests, whereas this figure falls to about 40 per cent in the subtropical dry forest, a value more typical of temperate deciduous forests (Larcher, 1975).

An important feature of tropical forests is the relatively fast closure time following disturbance, as contrasted with temperate forests (Fig. 11). This fast closure time is

TABLE 5. Productive characteristics of two tropical forests (all data from Jordan, Herrera, 1981, unless otherwise noted)

	Puerto Rico (Tabonuco)	Amazon (Tall Caatinga)
Latitude (degree)	18° 19'	1° 54'
Annual		
Precipitation (mm)	3450[1]	3000+[3]
Mean temperature (C)	23[1]	26.2[5]
Insolation (kcal cm^{-2})	101	?
Leaf Properties		
Mean nitrogen (mg g^{-1})	13.6[3](16.2)[4]	11.6[3]
Specific leaf wt (g m^{-2})	120[3]	132[3]
Leaf area index (m^{-2} m^{-2})	6.4[4]	5[6]
Biomass (kg m^{-2})		
Leaves	0.8	0.8
Stems + branches	19.0	15.8
Total aboveground	19.8	16.6
Roots	6.5	13.2
Total living	26.3	29.8
Root/shoot ratio	0.32	0.80
Annual productivity (g m^{-2})		
Wood	486	440
Litter	547	572
Total net	1033	1012

1, Brown, Lugo, 1982; 2, Jordan, 1971; 3, Medina, 1981; 4, Grubb, 1977; 5, Herrera, et al., 1978; 6, E. Medina, personal communication

due in part to the maintenance of year round photosynthetic gain in the tropical forests. Another important characteristic is the allocation pattern through time. Leaf biomass quickly attains an apparently optimum level and then remains at a constant amount with time, although it becomes a decreasing function of the total biomass (Fig. 12). This means that, in essence, all of the light resource is quickly captured by the forest with time and that competition for light results in increased biomass accumulation, but no further light harvesting tissue can be produced. This is also true to a certain extent with the nitrogen resource. Most of the accumulation of nitrogen occurs in the first few years (Fig. 14a).

In order to understand the productive nature of tropical forests we need more studies which give the full details of those aspects of the environment which affect productivity, as well as the details of the productive structure of the community. Unfortunately, only partial information is generally available, making comparisons difficult.

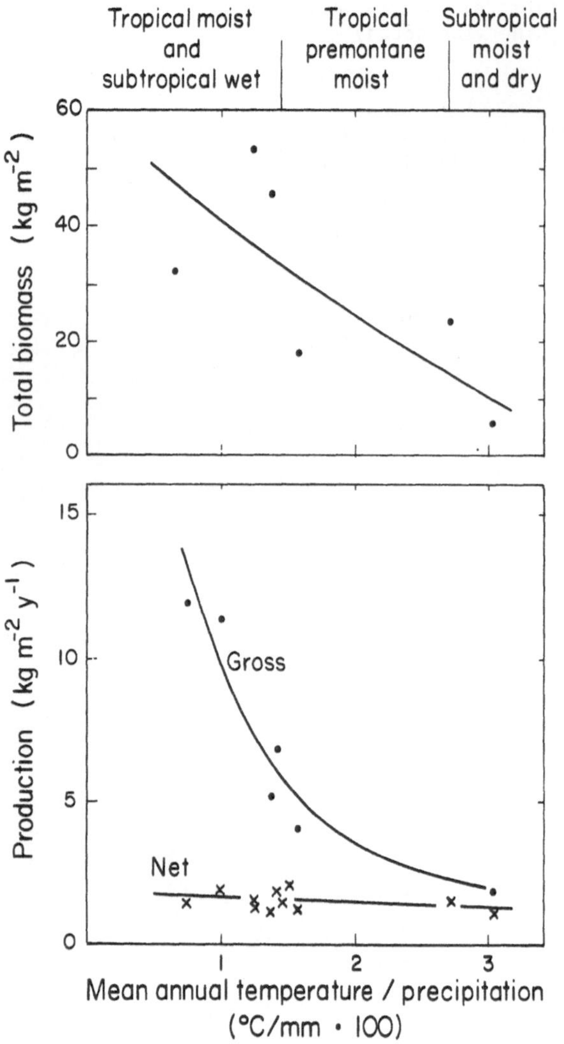

Fig. 10 Standing biomass (top) and gross and net annual productivity (bottom) of a variety of tropical vegetation types. Bottom scale is mean annual temperature (C)/mean annual precipitation (mm) times 100 (from Brown and Lugo, 1982).

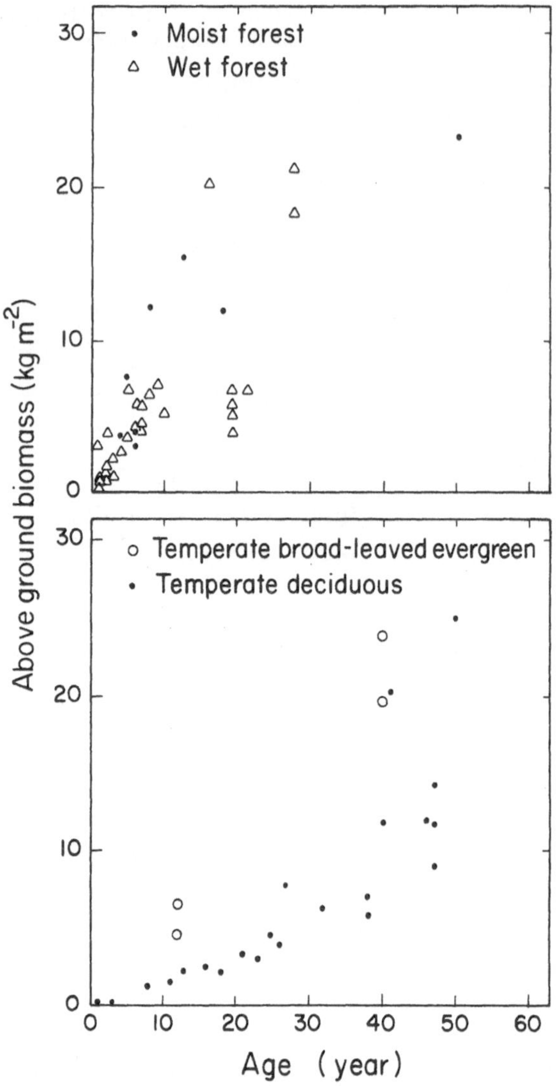

Fig. 11. Above-ground biomass versus stand age for tropical and temperate forests (from Brown and Lugo, 1982, after many sources).

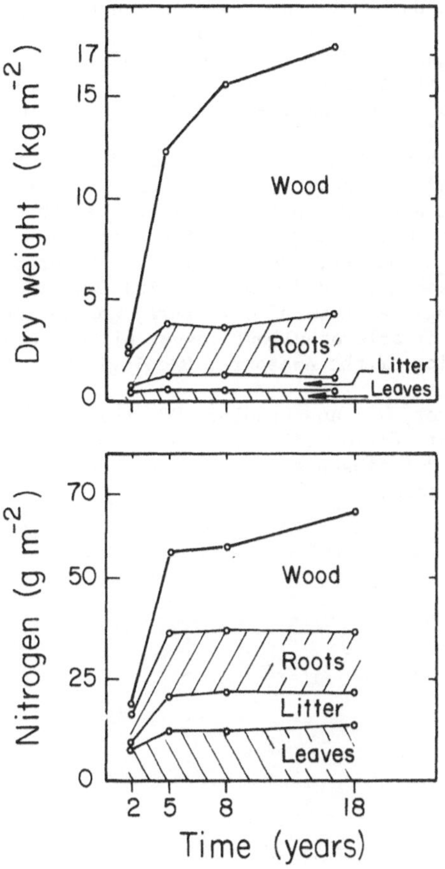

Fig. 12. Increase in dry weight (top) and nitrogen content of biomass (bottom) with time of a secondary forest following crop abandonment. (After Bartholomew, et al., 1953).

REFERENCES

Aoki M, Yabuki K, and Koyama H (1975) Micrometeorology and assessment of primary production of a tropical rain forest in West Malaysia, J. Agr. Met. 31, 115-124.

Atal CK, Dhar, KL, and J. Singh (1975). The chemistry of Indian Piper species Lloyda 38, 256-264.

Bartholomew WV, Meijer J, and Laudelout H (1953) Mineral nutrient immobilization under forest and grass fallow in the Yangambi (Belgian Congo) region, Publ. de l'Inst. Nati. d'Etudes Agron. Congo Belge, Serie Sci. 57, 1-27.

Bentley, BL (1979) Longevity of individual leaves in a tropical rain forest understory. Ann. Bot. 43, 119-121.

Bazzaz FA and Pickett STA (1980) The physiological ecology of tropical succession: a comparative review, Ann. Rev. Ecol. Syst. 11, 287-310.

Björkman O and Ludlow M (1972) Characterization of the light climate on the floor of a Queensland rain forest, Carnegie Inst. Wash. Yearbook. 71, 85-94.

Björkman O, Ludlow M, and Morrow PA (1972) Photosynthetic performance of two rain forest species in their native habitat and analysis of their gas exchange, Carnegie Inst. Wash. Yearbook. 71, 94-107.

Brown S, and Lugo A (1982) The storage and production of organic matter in tropical forests and their role in the global carbon cycle, Biotropica 14, 161-187.

Dean JM and Smith AP (1978) Behavioral and morphological adaptations of a tropical plant to high rainfall, Biotropica 10, 152-154.

Frankie GW, Baker HG, and Opler PA (1974) Comparative phenological studies of trees in tropical Wet and Dry forests in lowlands of Costa Rica, J. Ecol. 62, 881-919.

Grubb PJ (1977) Control of forest growth and distribution on wet tropical mountains; with special reference to mineral nutrition, Ann. Rev. Ecol. Syst. 8, 83-107.

Harvill EK, Hartzell A, and Arthur JM (1943) Toxicity of piperine solutions to houseflies, Cont. Boyce Thompson Inst. 13, 87-97.

Herrera R, Jordan CF, Klinge H, and Medina E (1978) Amazon ecosystems: their structure and functioning with particular emphasis on nutrients, Interciencia 3, 223-232.

Jordan CF (1971) Productivity of a tropical forest and its relation to a world pattern of energy storage, J. Ecol. 59, 127-142.

Jordan, CF (1983) Productivity of tropical rain forest ecosystems and the implications of their use on future wood and energy sources, In Golley FB, ed., Ecosystems of the world, Vol 14A, pp 117-136, Amsterdam, Elsevier Sci Publ Co.

Jordan CF and Herrera R (1981) Tropical rain forests; are nutrients really critical, Amer. Nat. 117, 167-180

Kira, T (1975) Primary production of forests, In Cooper JP, ed. Photosynthesis and productivity in different environments, pp 5-40, Cambridge, Cambridge Press.

Koyama H (1981) Photosynthetic rates in lowland rain forest trees of peninsular Malaysia, Jap. J. Ecol. 31, 361-369.

Larcher W (1975) Physiological plant ecology, Springer Verlag. Berlin, Heidelberg, NY. 252 p.

Lee, DW, Lowry JB, and Stone BC (1979) Abaxial anthocyanin layer in leaves of tropical rain forest plants; enhancer of light capture in deep shade, Biotropica 11, 70-77.

Medina E (1981) Nitrogen content, leaf structure and photosynthesis in higher plants, Report to the UNEP study group on photosynthesis and bioproductivity.

Medina E and Klinge H (1983) Tropical forests and tropical woodlands. In Lange OL, Nobel PS, Osmond CB Ziegler HJ, eds. Encyclopedia of plant physiology, new series, Vol. 12D Physiological plant ecology IV. Berlin, Springer-Verlag.

McKey D, Waterman P, Mbi C, Garlan J, and Struhsaker T (1978) Phenolic content of vegetation in two African rain forests; ecological implications, Science 202, 61-64.

Mooney HA and Chiariello N (1983) The study of plant function--the plant as a balanced system. In Dirzo R and Sarukhan J, eds., Darwin Centenary, Massachusetts, Sinauer Press.

Mooney HA, Björkman O, Hall AE, Medina E, and Tomlinson PB (1980) The study of physiological ecology of tropical plants-current status and needs, BioScience 30, 22-66.

Mooney HA, Field C, Vazquez-Yanes C, and Chu C (1983) Environmental controls on stomatal conductance in a shrub of the humid tropics, Proc. Nat. Acad. Sci. 80, 1295-1297.

Murphy PG (1975) Net primary productivity in tropical terrestrial ecosystems. In Lieth H, Whittaker RH, eds., Primary productivity of the biosphere, pp 218-231, New York, Springer Verlag

Oberbauer SF (1983) The ecophysiology of Pentaclethra macroloba, a canopy tree species in the rain forests of Costa Rica, Ph.D thesis, Duke University, Durham, N.C.

Pearcy RW (1983) The light environment and growth of C_3 and C_4 tree species in the understory of a Hawaiian forest, Oecologia, 58, 19-25.

Pearcy RW and Calkin HC (1983) Carbon dioxide exchange of C_3 and C_4 tree species in the understory of a Hawaiian forest. Oecologia, 58, 26-32.

Pearcy RW, Osteryoung K, and Randall D (1983) Carbon dioxide exchange characteristics of C_4 Hawaiian Euphorbia species native to diverse habitats, Oecologia. in press.

Pearcy RW and Robichaux RH (1984) Tropical and subtropical forests. In Chabot B, Mooney HA, eds. Physiological ecology of the North American vegetation, London, Chapman and Hall, in press.

Robichaux RH and RW Pearcy (1980) Environmental characteristics, field water relations and photosynthetic responses of C_4 Hawaiian Euphorbia species from contrasting habitats, Oecologia 47, 99-105.

Robichaux RH and Pearcy RW (1980b) Photosynthetic responses of C_3 and C_4 species from cool shaded habitats in Hawaii, Oecologia 47, 106-109.

Robichaux RH and Pearcy RW (1983) Evolution of C_3 and C_4 plants along an environmental moisture gradient: patterns of photosynthetic differentiation in Hawaiian Scaevola and Euphorbia species, Amer. Jour. Bot. in press.

Sobrado MA, and Medina E (1980) General morphology, anatomical structure, and nutrient content of sclerophyllous leaves of the "bana" vegetation of the Amazonas, Oecologia 45, 341-345.

Thrower, NJW and Bradbury DE (1977) California Chile mediterranean scrub atlas, Dowden, Hutchinson and Ross. Inc., Stroudsburg, Pennsylvania.

MEASURING GAS EXCHANGE OF PLANTS IN THE WET TROPICS

C. FIELD (Dept. of Biology, University of Utah, Salt Lake City, UT 84112 USA)
H. A. MOONEY (Dept. of Biological Sciences, Stanford University, Stanford, CA 94305 USA)

ABSTRACT

Nearly all types of gas-exchange devices produce accurate results given adequate precautions, but devices differ greatly in expense, portability, range of parameters measured, sample handling capacity, and experience required for accurate measurements. No single device represents the best solution for all gas-exchange problems.

Unusual difficulties encountered in studying gas exchange in the vegetation of the wet tropics include transportation, high humidity, unreliable power, and a wide range of leaf sizes. The specific devices which most successfully overcome these difficulties in particular situations may be open or closed systems with or without environmental control. We comment on the advantages and drawbacks of a number of classes of instruments designed to measure the exchange of water vapor or carbon dioxide by plants.

1. INTRODUCTION

Most of the devices designed to measure the exchange of carbon dioxide or water vapor by plants can provide highly accurate data under a variety of circumstances. In discussing those devices and measurement systems best suited to the study of gas exchange in the vegetation of the wet tropics, the important factors are not simply potential accuracy but include a series of considerations which vary in importance depending on the objectives of the research program. Among these considerations are the unit of investigation, the set of parameters measured, sample processing capacity, environmental control capability, portability, ease of use and calibration, sensitivity to environmental factors, and price.

As a result of both technological and biological constraints, gas-exchange measurement systems designed for optimal performance with respect to one set of criteria often perform poorly with respect to others. For example, the power requirements of a device with extensive environmental control capability tend to limit portability, but without environmental control it is very difficult to characterize gas-exchange responses to single environmental factors or even to know whether measured responses represent steady-state behavior. In sum, no single device or system currently available represents the best choice for all gas-exchange research.

Here we consider only field measurements of carbon dioxide or water vapor exchange by single leaves or twigs in the vegetation of the wet tropics. Even within these constraints, several types of devices represent the best available research tools in particular applications. For many objectives, the potential benefits of one type of device can be realized fully only when that device is utilized as part of an integrated program involving more than one type of gas-exchange measurement.

We provide a brief introduction to several considerations in gas-exchange research, as they potentially influence studies in a single habitat type. These comments are directed primarily toward those with limited experience in making gas-exchange measurements. We do not review the large literature on gas-exchange techniques since this has been done by Sesták et al. (1971) and Coombs and Hall (1982). Many of the published studies concerning gas exchange in the vegetation of the wet tropics are cited in this volume in the chapter by Mooney, Field, and Vazquez-Yanes.

Before discussing the drawbacks and advantages of particular types of instruments, it is helpful to outline a few of the unusual problems encountered in the wet tropics.

2. POTENTIAL PROBLEMS

While none of the difficulties associated with gas-exchange studies in the vegetation of the wet tropics is really unique to that habitat, unusual problems may be expected in three major areas: transportation, equipment maintenance, and measurement difficulties.

2.1. Transportation

Most tropical rain forests are remote from research institutions, placing a premium on portable and physically durable equipment. When equipment is shipped as air freight or air baggage, handling roughness frequently increases with the weight of the case, placing a further premium on modularizing larger systems.

Air travel also imposes the difficulty that compressed gases in cylinders, corrosive chemicals (including NaOH-based CO_2 scrubbers), oxidants (including $MgClO_4$-based desiccants), and radioisotopes are considered hazardous materials and require special clearance for air shipment. Most airlines follow the IATA (International Air Transport Association) guidelines concerning handling, packaging, and labeling requirements for restricted articles and can provide specific information and clearance forms. With appropriate clearance, the IATA regulations do allow non-flammable compressed gases, NaOH, and $MgClO_4$ aboard passenger flights, but many airlines impose regulations stricter than the IATA guidelines on passenger carrying planes. Radioisotopes for purposes other than medical diagnosis are never allowed on passenger flights.

The net result of these limitations is that gas-exchange systems that must be flown to tropical sites should be designed, wherever possible, for minimum reliance on restricted articles. Air freight does provide an alternative to checking equipment as baggage, but when international borders are crossed, air freight may lead to a long and sometimes expensive trip through customs.

Problems imposed by the customs requirements of various countries differ greatly but may be severe. While there can be no doubt that experience with a particular country is the best way to speed equipment through customs, a more secure, if more expensive, approach is that offered through the Carnet program of the International Chamber of Commerce. Under the Carnet program, equipment to be used or displayed temporarily in some country is essentially insured against illegal importation.

2.2. Equipment maintenance

The most severe environmental factor in the wet tropics, at least from the viewpoint of electronic equipment, is the relentlessly high humidity. Effective solutions to the problem include sealed enclosures, electrically heated equipment, liberal use of silica gel packets, and waterproofed circuit boards wherever possible.

The remoteness of tropical field sites is likely to create greater maintenance headaches than the harshness of the environment. The disadvantages of remoteness can be at least partially offset by field calibration capability, by redundancy in the equipment and its modes of operation, by well-stocked repair kits, and by detailed instruction manuals.

2.3. Measurements

2.3.1. Electrical environment. Line power in tropical field sites is often unreliable or non-existent. Where possible, battery-operated equipment provides freedom from power lines, but battery power becomes increasingly impractical as power consumption rises to more than a few watts. Equipment with power requirements approaching 100 watts can be run for an entire day from a large marine-type lead-acid battery (e.g. Field et al. 1982), but the availability of several excellent gasoline-powered generators producing on the order of 500 watts at a weight of less than 30 kg and a price of less than $500 (e.g. Honda EM 500, Kawasaki 550, Yamaha EF600), has largely replaced batteries as a first-choice power supply for equipment with higher power requirements. For equipment operated from generators or poorly regulated line power, it is important that the equipment remain unaffected by voltage or frequency variation. For example, variation in frequency seriously degrades the performance of infrared gas analyzers (IRGAs) if the chopper frequency is tied to the power line frequency (e.g. Horiba, Kyoto, Japan). Some manufacturers avoid this problem by driving the chopper from an internal oscillator (e.g. ADC (Analytical Development Co., Hoddesdon, England); Binos (Leybold-Heraeus GMBH, Hanau, W. Germany)).

2.3.2. Biotic environment. The scale of the vegetation in the wet tropics creates a series of problems for gas-exchange research. Canopy-flux studies and eddy-correlation analysis for whole canopies require massive instrument towers or elaborate rope-and-pulley based sampling schemes (e.g. Lemon et al. 1970). Whole-plant measurements for the canopy trees have never been attempted, even though a giant gas-exchange chamber, approximately 18 m in diameter and 20 m high, was constructed and operated around a prism of Puerto Rican rain forest (Odum, Jordan, 1970). For whole-plant studies, the difficulties of isolating a canopy tree may be more severe than the problems of enclosing one.

With smaller sampling units, the major problems presented by the vegetation of the wet tropics are access and leaf size variation. The difficulties of canopy access can be ameliorated with sturdy research towers such as those recently erected in the Queensland rain forests, but towers are very expensive and provide access to highly limited volumes of rain forest canopy (J. J. Landsberg, personal communication). An alternative solution, studying gas exchange characteristics of excised leaves or branches, offers possibilities for some species (Koyama, 1981) but requires much more study before it can be evaluated as a general approach.

The leaf-size variation among plants of the wet tropics presents problems primarily because few gas-exchange systems efficiently handle a wide range of leaf sizes. Chambers affixed to a small portion of a leaf (as in the commercial porometers produced by Delta-T (Cambridge, England) and Li-Cor (Lincoln, Nebraska)) offer a good solution for short-term, single leaf-side measurements. For longer-term, controlled-environment measurements, paired single-sided cuvettes like those described by Sharkey et al. (1982) provide opportunities for studying gas exchange in parts of large leaves but have not been used in the field and should be tested on many species to insure that the leaf segment enclosed by the cuvettes is sufficiently independent of the rest of the leaf. Currently, little is known about the effects on photosynthesis and transpiration of leaf damage, either from herbivores or from trimming leaves to fit in a gas-exchange cuvette. When leaves must be trimmed for measurements, they should be cut well in advance of the measurement and should, wherever possible, be compared with undamaged leaves.

2.3.3. Physical environment. In most gas-exchange devices, measurements of stomatal conductance are increasingly sensitive to errors in humidity measurement as humidity increases. At the very high humidities typically encountered in the wet tropics, even modest calibration errors can produce large errors in values for stomatal conductance (Fig. 1). For open gas-exchange systems, it is also important that humidity effects on IRGA signals and flow rates be incorporated in the calculations for photosynthesis (von Caemmerer, Farquhar, 1981).

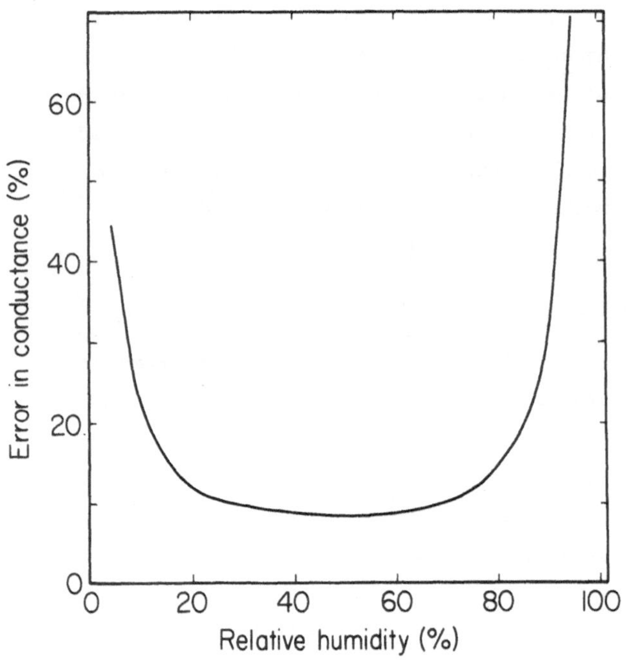

FIGURE 1. Effect of relative humidity in a gas-exchange cuvette on the error in leaf conductance resulting from a 2% calibration error in the humidity sensor of a steady-state porometer. Calculations assume leaf and air temperatures of 25°C and are based on the procedure in Field et al. (1982) and von Caemmerer and Farquhar (1981).

3. SYSTEM PHILOSOPHIES

In principle, all gas-exchange devices operate either as closed systems, open systems, or as some combination of the two. In closed systems, flux of carbon dioxide or water vapor is given by their rate of change in concentration. Open systems measure these fluxes as the product of gas flow rate through the system and the difference in concentration between the incoming and outgoing gas streams. A common variant on the open system philosophy is to measure the rate at which water vapor must be removed (Lange et al. 1969) or carbon dioxide must be added (Koller, Samish, 1964) to maintain the incoming and outgoing gases at similar concentrations of H_2O or CO_2. Canopy-flux and eddy-correlation techniques are variants of the open system approach in which the study vegetation is not enclosed and in which total flow throught the system is calculated from natural wind patterns. Radioisotope-based systems may be open or closed and differ from other systems mainly in that concentrations of carbon dioxide or water vapor are measured in terms of the activity of $^{14}CO_2$ or $^{3}H_2O$. Closed systems are inherently simpler than open systems because the former need to measure only concentration, instead of concentration and flow rate. However, closed systems never provide truly steady-state measurements, because the analysis requires a concentration change as a parameter in the calculation of either carbon dioxide or water vapor flux.

Devices operating on each of the system philosophies - open, closed, or mixed - have been designed to measure either the flux of carbon dioxide or water vapor, or the flux of both. With all but radioisotope techniques, it is generally simpler to measure water vapor exchange than CO_2 exchange, largely because water vapor is typically two orders of magnitude more common than CO_2 in the atmosphere and because the instruments for measuring water vapor concentration, as a function of either humidity or dewpoint, are much simpler and less expensive than the instruments for measuring CO_2 concentration (IRGA's). Partly for this reason, many more studies examine water vapor exchange without information on photosynthesis, than vice versa. Because stomatal conductance can be calculated only from measurements of water vapor exchange, studies of carbon dioxide exchange are of limited value when water vapor exchange is not measured simultaneously. On the other hand, measurements of water vapor exchange are often used to make rough estimates of photosynthesis, an extrapolation that is justified only when the relationships among photosynthesis, stomatal conductance, and environment have already been carefully studied.

4. SPECIFIC DEVICES

Gas-exchange devices designed to operate on each of the major system philosophies have been built at many levels of sophistication, accuracy, portability, and expense. In considering classes of instruments with regard to performance in the vegetation of the wet tropics, we have attempted to focus, within each class, on only the most widely used types of devices. Our comments are not intended as endorsements for any specific products. The tremendous variety of approaches to gas exchange insures that we have omitted some important instruments and that we have over-generalized the advantages and drawbacks of others. Many of the drawbacks which currently appear to reflect intrinsic limitations of a particular system may be eliminated as technology advances.

4.1. Radioisotope methods

Most of the gas-exchange devices designed to measure gross photosynthesis as $^{14}CO_2$ uptake have been highly portable instruments without environmental control (e.g. Strebeyko, 1967; Tieszen et al. 1974). The main application of these devices has been to rapidly sample photosynthetic rates from large numbers of leaves. Instruments can be built at relatively low cost.

Gas-exchange devices based on $^{14}CO_2$ exposure usually process many samples in the field but leave a substantial amount of analysis for later laboratory work. This laboratory analysis typically includes oxidizing the sample, capturing the ^{14}C, and quantifying the ^{14}C with liquid scintillation counting. In addition to the expense of processing samples in the laboratory, systems based on $^{14}CO_2$ uptake allow only one sample per leaf, they provide no immediate feedback to the investigator, and they rarely provide a measure of water vapor exchange. Simultaneous analysis of carbon dioxide exchange via $^{14}CO_2$ uptake and IRGA systems indicates variable performance among species and suggests that, in some cases, $^{14}CO_2$ uptake is a poor index of gross photosynthesis (Karlsson, Sveinbjörnsson, 1981). Radioisotope techniques have been used in detailed studies of photosynthetic response characteristics (Bingham et al. 1980) but yield much higher variances than other techniques.

Modification of the $^{14}CO_2$ technique to incorporate a closed diffusion porometer (Bravdo, 1972) or a steady-state (open system) porometer (Bingham, Coyne, 1977) adds the capability to measure water vapor exchange. The double-isotope ($^{14}CO_2$, ^3H_2O) porometer developed by Johnson et al. (1979) also provides simultaneous measurements of carbon dioxide and water vapor exchange.

Overall, radioisotope techniques are well suited to studies requiring large sample sizes but relatively low precision. The difficulty of traveling with radioisotopes as well as the cost of the instruments for the laboratory analysis make radioisotope techniques a best choice for only a restricted range of applications.

4.2. Closed systems

4.2.1. Water vapor exchange. A number of widely available, commercially produced instruments (Li-Cor LI-700, Delta T Automatic Porometer MK3) and many custom-built instruments function as non-aspirated, closed systems similar in concept to the diffusion porometers described by van Bavel et al. (1965), Kanemasu et al. (1969) and Stiles (1970). The most important recent progress in the development of these closed diffusion porometers includes the incorporation of field calibration plates, fast-response humidity sensors, automatic cycling between drying and measurement, and aspirated or stirred measurement chambers.

Closed diffusion porometers are relatively inexpensive, highly portable, and quite simple to operate. They are, in principle, easy to calibrate, but the temperature sensitivity of the calibrations makes the process operationally quite difficult. Depending on the leaf conductances encountered, they can process up to about 100 samples per hour.

Measurements made with these devices are, however, subject to several sources of substantial error. When conductances are low, the measurements may require time intervals long enough that conductance is influenced by conditions in the porometer cup. In the wet tropics this is especially important, because most porometers yield the greatest accuracy when cycled through humidity ranges which are lower than typical ambient humidities. When leaves are in sunny environments, failure to maintain isothermal conditions in the porometer head and the leaf may result in large errors. Shading the leaf and the porometer head in order to decrease temperature gradients may affect conductance directly. Many instruments are sensitive to errors resulting from water vapor adsorption and desorption (Morrow, Slatyer, 1971; Hack, 1980).

In comparing closed diffusion porometers with open (steady-state) instruments, the major advantage of the former is lower cost. The major disadvantages are the inability to perform steady-state measurements, and the difficulty of eliminating thermal gradients as a source of error. Newer closed porometers which include aspirated chambers and calculations of conductance from transpiration appear to solve many of the problems inherent in the earlier instruments.

4.2.2. Carbon dioxide exchange. Of the many devices designed to measure CO_2 exchange in a closed system, the only ones which are especially attractive for field studies in the vegetation of the wet tropics are highly portable systems intended for rapid sampling. Though closed systems can be used in detailed studies of photosynthetic responses, the inherently steady-state operation of open systems tends to make the latter more desirable for mechanistic studies.

Two types of closed CO_2-exchange systems suited to rapid sample processing are the injection sampling technique of Ehleringer and Cook (1980) and the integral-IRGA system incorporated in a commercially available instrument, the LI-6000 (Li-Cor Inc., Lincoln, Nebraska). The injection sampling technique requires only simple equipment in addition to a suitable IRGA. The LI-6000 couples a battery-powered IRGA via an analog-to-digital converter to a microcomputer, providing an integrated system with rapid user feedback.

The advantages of the injection sampling technique include low cost (once the IRGA is available) and high portability for sampling (but not for analysis). The disadvantages of this technique include the difficulty of avoiding elevated CO_2 due to operator exhalation, the delay in operator feedback, the absence of simultaneous measurements of water vapor exchange, the dependence of the results on accurate timing and handling of the syringes, and the dependence on a relatively fixed IRGA station, equipped with a chart recorder and a CO_2-free gas supply.

The LI-6000 overcomes many of the difficulties with the injection sampling technique. It provides several, nearly immediate outputs of carbon dioxide exchange over the duration of a one to two minute sample. It provides a nearly immediate indication of elevated CO_2 due to exhalation, and it provides simultaneous measurements of CO_2 and water vapor exchange. The LI-6000 is very portable and relatively easy to operate. Except for the expense, most of the drawbacks of the LI-6000 are quite subtle. As with all closed systems, it is difficult to know whether measurements represent responses to current or pre-existing environmental conditions. It is also difficult to eliminate the possibility that changing chamber conditions are influencing leaf responses. The LI-6000 also shares a problem with many porometers that measure humidity with the Vaisala Humicap (6061 HM, Vaisala OY, Helsinki, Finland). At high humidities, values for stomatal conductance are increasingly sensitive to humidity errors, and the Humicap is increasingly subject to non-linearity and drift (Vaisala Ref No. 339/HM).

Overall, closed-system carbon dioxide exchange analysis provides attractive opportunities for survey-type studies requiring large sample sizes and limited mechanistic interpretation. The injection sampling technique is markedly less powerful and more error prone than the LI-6000, but both compare favorably with radioisotope techniques. Except for the humicap problems at high humidities, both types of systems should perform as well in the wet tropics as in other vegetation types.

4.3. Open systems

4.3.1. Water vapor exchange. Most open-system porometers currently in use operate on principles developed by Beardsell et al. (1972). These porometers nearly all include well-stirred cuvettes, sometimes with temperature control (e.g. Bingham, Coyne, 1977), and provide a steady-state humidity by diluting transpiration with a controlled flow of dry air. Versions of steady-state porometers may measure transpiration on part of one side of a leaf (Fanjul et al. 1980) or on entire leaves or small branches (Beardsell et al. 1972).

Steady-state porometers are very portable and provide measurements quickly. Most are designed so that the leaf environment is largely unchanged by the presence of the porometer. Devices which use the Vaisala Humicap for humidity measurement are subject to errors at high humidities. The magnitude of these errors is approximately the same, whether the porometer is an open or closed device. Because steady-state porometers must include components that measure humidity, flow, and temperature, complete calibrations may be complex. However, many porometers contain very stable flow- and temperature-sensing devices which require calibration only rarely.

In comparison with closed-diffusion porometers, steady-state porometers tend to be more expensive and more complex but only slightly less portable. Steady-state porometers may be quite easy to use and can process samples as quickly as closed porometers. While still sensitive to humidity errors, steady-state porometers are relatively immune to the temperature problems that compromise the performance of closed porometers.

4.3.2. Carbon dioxide and water vapor exchange. Among open systems for measuring both carbon dioxide and water vapor exchange, portability varies tremendously. At the most portable end are devices built from modified steady-state porometers. These systems have added CO_2-exchange capability either with the addition of injection sampling at a remote IRGA (Fanjul et al. 1980) or by sequentially measuring conductance and photosynthesis with slightly different techniques (Griffeths, Jarvis, 1981). A similarly portable system (Schulze et al. 1982) avoids the low flow rates required in steady-state porometers by providing the measurement cuvette with moist air. This system utilizes a water vapor IRGA to accurately measure the small differences between the water vapor concentrations in the gas streams entering and leaving the cuvette. These highly portable steady-state systems can be used for limited controlled-environment response studies but are primarily intended for making large samples of photosynthesis and transpiration under ambient conditions, fundamentally the same role as the Li-Cor LI-6000.

Open systems designed to provide more environmental control achieve that goal at the expense of portability. The portable steady-state systems of Fanjul et al. (1980), Griffeths and Jarvis (1981), and Schulze et al. (1982) do not provide temperature control, and provide only limited control of humidity and CO_2 concentration. The devices described by Bingham et al. (1980) and Field et al. (1982) provide extensive temperature, humidity, and CO_2 concentration control, making them suitable for detailed studies of photosynthetic responses. However, these systems cannot be carried by one person in one load, they require on the order of an hour to set up in the field, and they rely on gases in cylinders and 110 volt ac power.

The controlled-environment systems of Bingham et al. (1980) and Field et al. (1982) bring to the field the capabilities of most sophisticated, laboratory-based gas-exchange systems. The strength of these controlled-environment systems is that they can provide a means for evaluating leaf responses to individual environmental variables. This process is useful for developing predictive models but difficult without environmental control, because environmental factors tend to be correlated (Jarvis, 1976). Using statistical models, several studies have nonetheless described response curves on the basis of large numbers of measurements from devices without environmental control (e.g. Whitehead et al. 1981).

The controlled-environment systems yield high accuracy and detailed response surfaces but are expensive, relatively difficult to operate and maintain, and subject to travel restrictions. They are not particularly suited to measuring large numbers of samples.

5. CONCLUSIONS

In the wet tropics as in any other habitat, a poorly calibrated or improperly constructed gas-exchange device can yield incorrect results. With gas-exchange research, the sophistication of the instrument is important, but probably not as important as the experience of the investigator.

The importance of the experience of the

investigator makes it difficult to rank types of devices on an absolute scale. Everything else being equal, however, the following generalizations provide useful guidelines. Malfunctions are easiest to detect in systems that provide immediate feedback to the investigator. Radio-isotope techniques rate poorly in this respect and computerized devices often rate highly. Open systems are typically more expensive than closed systems of comparable capabilities, but the open systems often provide more reliable results. Open systems nearly always require well-stirred cuvettes. The most reliable closed systems also incorporate well-stirred cuvettes. Controlled environment systems provide the most powerful basis for generating predictive models, but it is possible to generate predictions from measurements under ambient conditions.

The environment of the wet tropics demands particular care in measuring transpiration and conductance, largely as a result of the sensitivity of the calculations to humidity errors at high humidities. The most effective strategy for dealing with this problem applies to most of the problems one is likely to encounter in the field. Because any measurement is only as good as the calibration behind it, it is very important that gas-exchange devices used in the field are calibrated carefully and frequently.

ACKNOWLEDGMENTS

Thanks to the participants in the Workshop on Physiological Ecology of the Vegetation of the Wet Tropics for a valuable discussion, and thanks to the personnel of the Instituto de Biologia of the Universidad Autonoma de Mexico for providing a splendid setting for the workshop

REFERENCES

Beardsell MF, Jarvis PG and Davidson B (1972) A null-balance diffusion porometer suitable for use with leaves of many shapes, J. Appl. Ecol. 9, 677-690.

Bingham GE and Coyne PI (1977) A portable, temperature-controlled steady-state porometer for field measurements of transpiration and photosynthesis, Photosynthetica 11, 148-160.

Bingham GE, Coyne PI, Kennedy RB and Jackson WL (1980) Design and fabrication of a portable minicuvette system for measuring leaf photosynthesis and stomatal conductance under controlled conditions, Lawrence Livermore National Laboratory, Livermore CA, UCRL-52895.

Bravdo B-A (1972) Photosynthesis, transpiration, leaf stomatal and mesophyll resistance measurements by the use of a ventilated diffusion porometer, Physiol. Planta. 27, 209-215.

Coombs J and Hall DO, eds. (1982) Techniques in bioproductivity and photosynthesis. Oxford, Pergamon Press.

Ehleringer J and Cook CS (1980) Measurements of photosynthesis in the field: utility of the CO_2 depletion technique, Plant Cell Env. 3, 479-482.

Fanjul L, Jones HG and Treharne KJ (1980) A portable system for simultaneous measurement of transpiration and CO_2 exchange, Photos. Res. 1, 83-92.

Field C, Berry JA and Mooney HA (1982) A portable system for measuring carbon dioxide and water vapour exchange of leaves, Plant Cell Env. 5, 179-186.

Griffeths JH and Jarvis PG (1981) A null balance carbon dioxide and water vapour porometer, J. Exp. Bot. 32, 1157-1168.

Hack HRB (1980) The uptake and release of water vapour by the foam seal of a diffusion porometer as a source of bias, Plant Cell Env. 3, 53-57.

Jarvis PG (1976) The interpretation of the variations in leaf water potential and stomatal conductance found in canopies in the field, Phil. Trans. R. Soc. Lond. B. 273, 593-610.

Johnson HB, Rowlands PG and Ting IP (1979) Tritium and carbon-14 double isotope porometer for simultaneous measurements of transpiration and photosynthesis, Photosynthetica 13, 409-418.

Kanemasu ET, Thurtell GW and Tanner CB (1969) Design, calibration and field use of a stomatal diffusion porometer, Plant Physiol. 44, 881-885.

Karlsson S and Sveinbjörnsson B (1981) Methodological comparison of photosynthetic rates measured by the $^{14}CO_2$ technique and infrared gas analysis, Photosynthetica 15, 447-452.

Koller D and Samish Y (1964) A null-point compensating system for simultaneous and continuous measurement of net photosynthesis and transpiration by controlled gas-stream analysis, Bot. Gaz. 125, 81-88.

Koyama H (1981) Photosynthetic rates in lowland rainforest trees of peninsular Malaysia,

Jap. J. Ecol. 31, 361-369.
Lange OL, Koch W and Schulze E-D (1969) CO_2-Gaswechsel und Wasserhaushalt von Pflanzen in der Negev-Wüste am Ende der Trockenzeit, Ber. Dtsch. Bot. Ges. 82, 39-61.
Lemon E, Allen LH Jr and Müller L (1970) Carbon dioxide exchange of a tropical rain forest. Part II, Bioscience 20, 1054-1059.
Morrow PA and Slatyer RO (1971) Leaf resistance measurements with diffusion porometers: precautions in calibration and use, Agric. Meteorol. 8, 223-233.
Odum HT and Jordan CF (1970) Metabolism and evapotranspiration of the lower forest in a giant plastic cylinder. In Odum HT, ed. A tropical rain forest: A study of irradiation and ecology at El Verde, Puerto Rico, pp. I-165 - I-191. Washington, D.C., U.S. Atomic Energy Commission.
Schulze E-D, Hall AE, Lange OL and Walz H (1982) A portable steady-state porometer for measuring the carbon dioxide and water vapour exchanges of leaves under natural conditions, Oecologia 53, 141-145.
Sesták A, Catský J and Jarvis PG, eds. (1971) Plant photosynthetic production: manual of methods. The Hague, Dr. W. Junk NV.
Sharkey TD, Imai K, Farquhar GD and Cowan IR (1982) A direct confirmation of the standard method of estimating intercellular partial pressures of CO_2, Plant Physiol. 69, 657-659.
Stiles W (1970) A diffusive resistance porometer for field use I. construction, J. Appl. Ecol. 7, 617-638.
Strebeyko P (1967) Rapid method for measuring photosynthetic rate using $^{14}CO_2$, Photosynthetica 1, 45-49.
Tieszen LL, Johnson DA and Caldwell MM (1974) A portable system for the measurement of photosynthesis using 14-carbon dioxide, Photosynthetica 8, 151-160.
van Bavel CHM, Nakayama FS and Ehrler WL (1965) Measuring transpiration resistance of leaves, Plant Physiol. 40, 535-540.
von Caemmerer S and Farquhar GD (1981) Some relationships between the biochemistry of photosynthesis and the gas exchange of leaves, Planta 153, 376-387.
Whitehead D, Okali DUU and Fasehun FE (1981) Stomatal response to environmental variables in two tropical forest species during the dry season in Nigeria, J. Appl. Ecol. 18, 571-587.

NUTRIENT BALANCE AND PHYSIOLOGICAL PROCESSES AT THE LEAF LEVEL.

ERNESTO MEDINA
(Centro de Ecología. IVIC. Aptdo. 1827, Caracas).

ABSTRACT

The leaf mineral nutrient content, its seasonal variation and its eco-physiological significance in tree species of tropical plant communities is reviewed. Leaf span of tropical trees ranges from 6-9 months in deciduous species (rain-green type) up to little more than one year in evergreen species (annual evergreens). Leaves of deciduous species are characterized by lower specific leaf weights and higher nitrogen and phosphorus content per unit dry weight. Differences in leaf nutrient content are minimized when expressed on an area basis, this should be the unit for seasonal comparisons because of its relative constancy after full leaf expansion. Photosynthesis is related to leaf nitrogen content, but future studies should include a fractionation of nitrogen in order to differentiate the accumulation of nitrogenous substances not metabolically related to photosynthesis. The study of nutrient use efficiency in production and growth is emphasized to understand the processes of selection in nutrient rich and nutrient poor environments.

1. INTRODUCTION

The main factors determining the distribution of higher plant species in terrestrial habitats are water availability and absolute values of temperature and its seasonal variations. Locally, variations in soil characteristics such as texture and aeration, pH of soil solution, nutrient availability and presence of toxic elements exert strong influence on species behavior and distribution. Physiologically, the effect of the edaphic conditions is ultimately reflected in the transport of water and nutrients from roots to shoots. Therefore, these effects might be interpreted on the basis of the water and nutrient balance of the plant.

Some effects of soil conditions on plant development derive from the hormonal interaction between root and shoot, which in turn depends on the metabolic activity of the root and foliage systems. The development of the root system requires a continuous supply of carbohydrates from the foliage, at the same time root activity provides the shoot not only with water and mineral nutrients, but also hormones such as cytokinins which regulate leaf surface expansion (Wareing 1977).

The structure of a higher plant consists of two absorption surfaces: the leaf surface which absorbs CO_2 and radiant energy, and the root surface where absorption of water and nutrients takes place. Both the leaf and root surfaces show a high turnover rate during the whole life of the plant, and therefore, the analysis of their structure, composition and function reveal short and long-term growth conditions. Obviously, leaves constitute an easier and more accessible material to study, therefore they are generally utilized to assess the productive capacity and nutrient relationships of higher plants.

The study of nutrient balance at the leaf level has considerable eco-physiological significance because it relates to the productive capacity of the plant as follows:

a) requirement of nutrients in quantity and proportion for leaf growth (nutrient import). The amount of nutrients per unit leaf area constitutes an index of the cost of construction of the photosynthetic surface.

b) duration of nutrient utilization in the leaf tissue for production and translocation of organic matter from the leaf. The amount of energy and carbon fixation during the leaf life span related to the nutrient content in the leaf tissue should be an indication of the efficiency of nutrient utilization.

c) exportation and redistribution of mineral nutrients prior to leaf abscission. The recovery of nutrients from senescent leaves and their distribution to other growing organs or storage compartments in the plant reduces the energy expenditure for nutrient acquisition through the roots. This becomes more important when the availability of a given nutrient in the environment is limited.

The study of nutrient balance in leaves is also important in order to assess the mechanisms operating in higher plants which cope with variations in nutrient availability in different environments. In this respect a number of hypotheses and generalizations have been tested (Chapin 1980; Chabot, Hicks 1982). Some of the relevant aspects can be summarized as follows:

a) low nutrient availability is associated with reduced growth rates. Experimentally it can be shown that any species can make a certain adjustment of its growth rate depending on the availability of nutrients. The range of this adjustment, however, is genetically regulated.

b) plants growing in habitats with restricted nutrient supply, such as bog plants in temperate regions, heath plants in mediterranean climates or plants in highly leached and acid soils in the tropics, have a leaf anatomy characterized by a coriaceous texture, thick cell walls and cuticules, and a compact mesophyll. Their nutrient content per unit dry weight is low due to the accumulation of carbon compounds. Under experimental conditions it can be also shown that nutrient deficiencies induce an increase in leaf thickness and reduction in leaf expansion.

c) under conditions of restricted nutrient supply the efficiency of metabolic utilization of certain nutrients might increase. There is no clear cut evidence for this type of adjustment, however, it is known that the nutrient requirements for growth can vary significantly in species of different environments to such an extent that in oligotrophic habitats the occurrence of nutrient flushes during certain times of the year causes accumulation of nutrients in the plants which are not being directly utilized in metabolism (Mooney, Rundel 1980; Chapin 1980). High efficiency of phosphorus utilization for organic matter production, and P accumulation in leaf tissue when fertilized, have been shown in native grasses of Australian and South-American savannas, environments well known for their limited phosphorus availability (Christie, Moorby 1975; Medina et al. 1980).

2. LEAF NUTRIENT CONTENT AND PHYSIOLOGICAL PROCESSES

The relationships between nutrient supply, leaf nutrient content and physiological performance of the leaf have been intensively studied in cultivated plants (Mengel and Kirkby 1982).

Some of these results can be utilized for understanding the behavior of wild plants under natural conditions. Studies with wild plants have been conducted either under cultivation or in nature, although the number of mineral elements analyzed has been limited. In this review only the most commonly reported minerals will be considered: nitrogen, phosphorus, potassium, calcium and magnesium. Of all the physiological aspects which have been studies here, I emphasize only those which appear most directly related to the production process such as photosynthesis and translocation of organic matter.

2.1 Leaf dimension ratios for ecological characterization of leaf structure.

The amount of organic matter invested in the construction of a leaf can result in a different development of photosynthetic area depending on leaf thickness and degree of compactness of the mesophyll. These characteristics can be expressed quantitatively in a simple way, by calculating the specific leaf area (SLA) or specific leaf weight (SLW) (1/specific leaf area) (Stocker 1931; Evans 1972):

$$SLA = Leaf\ Area\ /\ Leaf\ Weight.$$

The measurement of leaf area is straight forward as long as fresh material is used, whereas certain precautions have to be taken in the measurement of leaf weight in order to make the comparisons between species based on SLA reliable. The size of the petiole, for instance, varies considerably among species, therefore, it should be eliminated before weighing a leaf. Furthermore, among megaphyll species which are common in humid tropical forests, the structural development of the secondary veins is very heavy. The measurements of leaf weight for the calculation of SLA in these cases may be misleading, because then the biggest fraction of leaf weight does not correspond to the photosynthetic area, but to the water conducting tissue. Therefore, it is necessary to measure the leaf area/weight relationship in the interveins to reduce the influence of the main veins.

2.2 Leaf life span of tropical trees.

There is a great variation in the patterns of leaf growth in the humid tropics which range from the typical deciduous behavior where the tree remains leafless for a certain period after leaf fall, to the evergreen behavior with one or two periods of leaf production during the year (Medway 1972; Addicott 1978). In tropical forest communities with a pronounced seasonality, leaf production tends to be confined to a certain period during the year, very often the dry season. Leaf life span of deciduous trees varies from 6-8 months, while the evergreen trees produce leaves which turn over every 12-13 months (Monasterio, Sarmiento 1976; Cuenca 1976; Montes, Medina 1977, Marín, Medina 1981; Medina 1982). There are then two fundamental behavior types in tropical trees, the deciduous (or rain-green type, Walter 1973), and the evergreen (or better the annual evergreens of Chabot, Hicks 1982). The deciduous type tends to predominate in seasonal dry environments while the evergreens dominate the more humid forest types. However, there are evergreen species even in the most seasonal dry communities, the most interesting case in this respect being the predominance of evergreen trees in the South American and Australian savannas (Medina 1982). Leaf renewal, in both deciduous and evergreen types, usually takes place in flushes of 1-2 months duration. The complete expansion of a new leaf often takes place in less than one month.

Seasonal variations in specific leaf weights (in leaves of increasing age) in characteristic

Some of these results can be utilized for understanding the behavior of wild plants under natural conditions. Studies with wild plants have been conducted either under cultivation or in nature, although the number of mineral elements analyzed has been limited. In this review only the most commonly reported minerals will be considered: nitrogen, phosphorus, potassium, calcium and magnesium. Of all the physiological aspects which have been studies here, I emphasize only those which appear most directly related to the production process such as photosynthesis and translocation of organic matter.

2.1 Leaf dimension ratios for ecological characterization of leaf structure.

The amount of organic matter invested in the construction of a leaf can result in a different development of photosynthetic area depending on leaf thickness and degree of compactness of the mesophyll. These characteristics can be expressed quantitatively in a simple way, by calculating the specific leaf area (SLA) or specific leaf weight (SLW) (1/specific leaf area) (Stocker 1931; Evans 1972):

$$SLA = Leaf\ Area\ /\ Leaf\ Weight.$$

The measurement of leaf area is straight forward as long as fresh material is used, whereas certain precautions have to be taken in the measurement of leaf weight in order to make the comparisons between species based on SLA reliable. The size of the petiole, for instance, varies considerably among species, therefore, it should be eliminated before weighing a leaf. Furthermore, among megaphyll species which are common in humid tropical forests, the structural development of the secondary veins is very heavy. The measurements of leaf weight for the calculation of SLA in these cases may be misleading, because then the biggest fraction of leaf weight does not correspond to the photosynthetic area, but to the water conducting tissue. Therefore, it is necessary to measure the leaf area/weight relationship in the interveins to reduce the influence of the main veins.

2.2 Leaf life span of tropical trees.

There is a great variation in the patterns of leaf growth in the humid tropics which range from the typical deciduous behavior where the tree remains leafless for a certain period after leaf fall, to the evergreen behavior with one or two periods of leaf production during the year (Medway 1972; Addicott 1978). In tropical forest communities with a pronounced seasonality, leaf production tends to be confined to a certain period during the year, very often the dry season. Leaf life span of deciduous trees varies from 6-8 months, while the evergreen trees produce leaves which turn over every 12-13 months (Monasterio, Sarmiento 1976; Cuenca 1976; Montes, Medina 1977, Marín, Medina 1981; Medina 1982). There are then two fundamental behavior types in tropical trees, the deciduous (or rain-green type, Walter 1973), and the evergreen (or better the annual evergreens of Chabot, Hicks 1982). The deciduous type tends to predominate in seasonal dry environments while the evergreens dominate the more humid forest types. However, there are evergreen species even in the most seasonal dry communities, the most interesting case in this respect being the predominance of evergreen trees in the South American and Australian savannas (Medina 1982). Leaf renewal, in both deciduous and evergreen types, usually take place in flushes of 1-2 months duration. The complete expansion of a new leaf often takes place in less than one month.

Seasonal variations in specific leaf weights (in leaves of increasing age) in characteristic evergreen and deciduous species of two seasonal tropical forests are depicted in Fig. 1. It is clear that the SLW increases with age. In the

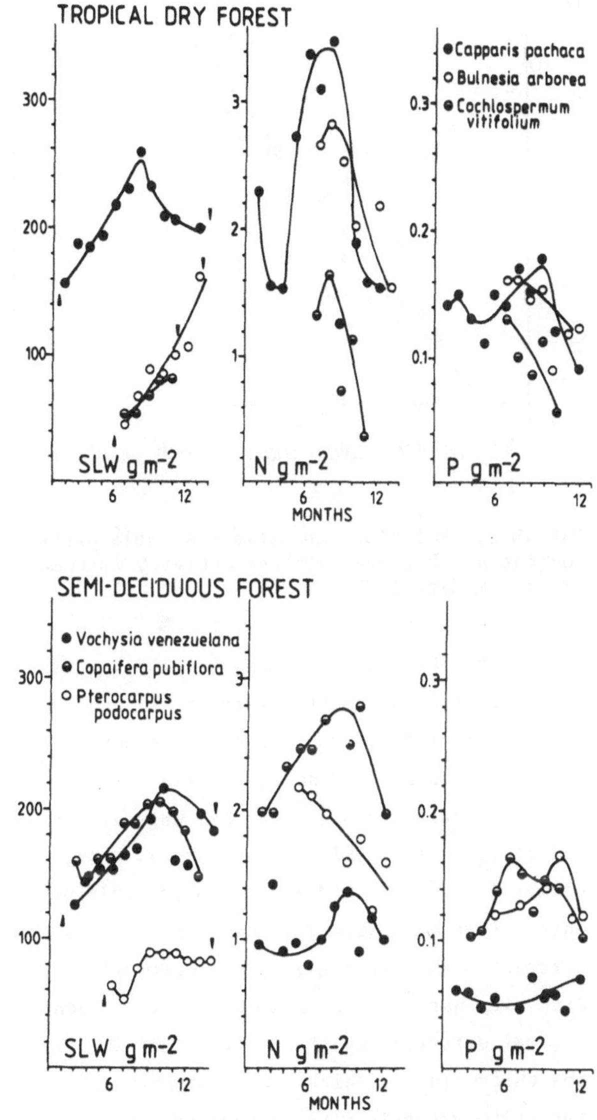

FIGURE 1. Seasonal variations of Specific Leaf Weight (SLW), Nitrogen (N) and Phosphorus (P) contents in evergreen species (C. pachaca, V. venezuelana, C. pubiflora) and deciduous species (B. arborea, C. vitifolium, P. podocarpus) of tropical dry and semi-deciduous (savanna) forests. Upwards pointing arrows indicate leaf production, downward arrows leaf fall.

evergreen species a reversal of this trend can be observed towards the end of the leaf life span. The deciduous species have always lower SLW values, irrespective of leaf age, with no reduction in SLW through time.

The variations in SLW during the year make it valuable to express the seasonal variations in leaf nutrient content on an area basis, because the latter does not change appreciably after full leaf expansion. The accumulation of organic matter in the leaf during its life span and its partial reexportation at the end of the growing period may obscure the true relations of the leaf nutrient content if it is based on leaf dry weight (Woodwell 1974; Ernst 1975; Staaf 1982). In tropical rain forests it is frequent to find leaves in different stages of development simultaneously on the same tree. In these cases it is possible to assess the variations of SLW with leaf age in one collection. The age scale, however, is a relative one (Sobrado and Medina 1980).

2.3 Leaf nitrogen and phosphorus content.

It is well documented that a restriction in nitrogen supply during plant growth causes a strong reduction in the leaf photosynthetic capacity per unit area. Leaf morphology and N content also change correspondingly. With low nitrogen supply the plants tend to produce thicker leaves with higher SLW and lower nitrogen content per unit dry weight. Under experimental conditions it can be shown that there is a highly significant positive correlation between leaf nitrogen content and the photosynthetic rate at saturating light intensities (among others Nevins, Loomis 1970; Medina 1970; Andreeva, Avdeeva 1970; Keller 1972; Motta, Medina 1978; Gulmon, Chu 1981; review by Nátr 1975; Migus, Hunt 1980; Linder et al. 1981; McDonald et al. 1981; Sasahara

1982). This relationship is understandable because photosynthetic rate and activity of RuBP-carboxylase (Fig. 2), soluble protein content and RuBP-carboxylase activity (Fig. 3) and nitrogen content and soluble protein content of leaves (Fig. 4) are strongly correlated, although the correlations are not always linear.

Under natural conditions the nitrogen content of leaves changes markedly with growth conditions and age. Several authors have been able to demonstrate that those changes in nitrogen

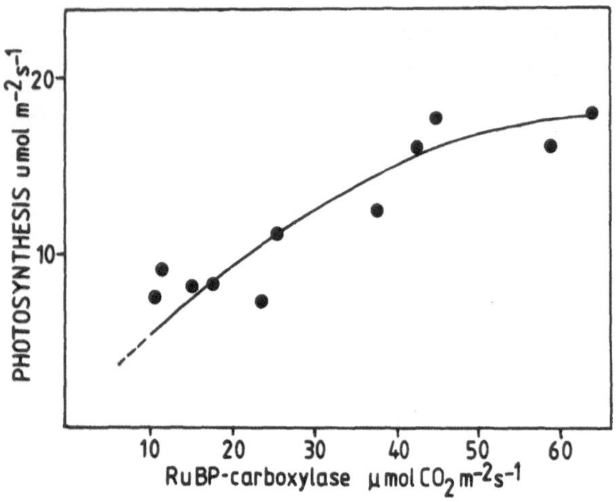

FIGURE 2. Relationship between RuBP-carboxylase activity and photosynthetic rate at saturating light intensity in Atriplex patula (Medina 1970).

content and SLW are also linearly related to photosynthetic rate per unit dry weight (Mooney et al. 1978; Field 1981; Medina 1981). There are, however, some notable exceptions. Small (1972) in a detailed study of the photosynthetic performance and the nitrogen content of bog woody plants did not find any significant correlation between nitrogen content and photosynthetic rate on a per unit weight basis.

Photosynthetic rate and nitrogen content per unit area may be also positively correlated

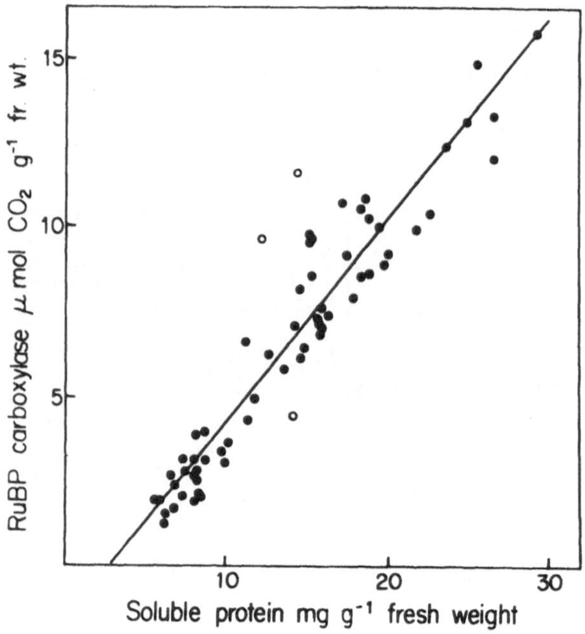

FIGURE 3. Relationship between soluble protein content and RuBP-carboxylase activity in tomato (Motta, Medina 1978).

(review by Nátr 1975; Gulmon, Chu 1981), but variability tends to be higher mainly because changes in nitrogen content are accompanied by variations in SLW. Indeed, SLW and photosynthesis per unit area are negatively correlated (Small 1972; Mooney et al. 1978; Medina 1981) while SLA and photosynthesis per unit weight are positively and often highly correlated (Medina 1981). This seemingly disparity has been attributed to the influence of leaf structure on the diffusion of CO_2 into the chloroplasts, which becomes apparent when the photosynthetic rate is expressed per unit area (Medina 1981).

Deciduous trees frequently produce leaves with higher photosynthetic rates than evergreen trees, even when similar aged leaves are compared (Larcher 1969; Koerner et al. 1979; Mooney, Gulmon 1982). However, there are several reports showing identical photosynthetic

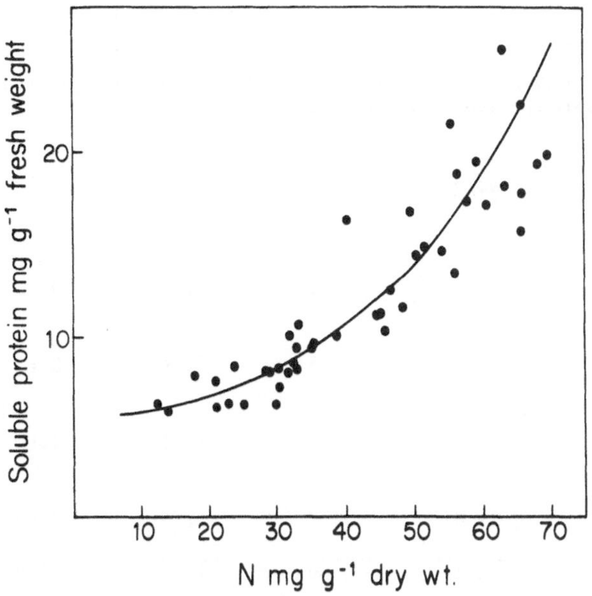

FIGURE 4. Relationship between total organic nitrogen content of tomato leaves and their soluble protein content (recalculated from Motta, Medina 1978).

rates per unit leaf area both in deciduous and evergreen trees coexisting in the same habitat (Saeki, Nomoto 1958; Larcher 1961). In seasonal tropical forests, evergreen and deciduous species co-occur and the deciduous trees produce leaves with considerably higher leaf nitrogen content per unit leaf weight than that of evergreen trees (Loveless 1961; Marin, Medina 1981). This would mean that the photosynthetic rates per unit leaf weight should be considerably higher in leaves of deciduous trees (Medina 1981). However, the nitrogen and phosphorus content of leaves of evergreen trees often have a higher nitrogen and phosphorus concentration per unit leaf area than deciduous leaves due to their markedly higher SLW (Table 1). A similar situation has been described for evergreen and deciduous shrubs in temperate swamps (Schlesinger, Chabott 1977). The photosynthesis-leaf structure-nitrogen content relationships in these species require detailed investigation.

For a deeper understanding of the role of nitrogen in the ecophysiology of higher plants a fractionation of the different nitrogen compounds in leaves should be undertaken, mainly to include protein nitrogen (soluble and insoluble) and soluble non-protein nitrogen (amino acids, amides, alkaloids). Unfortunately, the thorough extraction of the soluble protein fraction from leaves is frequently hampered by the presence of protein precipitating substances, which can be completely avoided by the use of protein protecting agents (Loomis 1974).

Such a fractionation has been reported for shade herbaceous plants (Stoecker, Hacker 1976) and evergreen and deciduous Taiga trees (Chapin, Kedrowski 1983). Stoecker and Hacker (1976) report that plants from oligotrophic sites produce more protein per unit of leaf nitrogen than plants from mesotrophic sites. This may mean that the amount of soluble, non-protein nitrogen is higher in plants with a better nutrient supply, but interactions with other nutrients can not be neglected, such as the potassium-nitrogen relationships which will be discussed below.

Phosphorus contents in leaves of higher plants is normally around one order of magnitude lower than that of nitrogen (Tables 1 and 3). Its deficiency, however, is likely to impair nitrogen utilization, because phosphorus constitutes the universal molecule for energy transfer in organisms. It can also play a very important role as an inorganic ion (ortophosphate) because it intervenes in the energy exportation process from the chloroplast (Mengel, Kirkby, 1982).

Table 1. Specific leaf weight (g/m^2) and nitrogen and phosphorus content per unit leaf dry weight and per unit leaf area of dominant trees of a very dry tropical forest (Marín, Medina 1981).

Species	Leaf life months	SLW	N mg/g	N g/m^2	P mg/g	P g/m^2
Deciduous						
Bulnesia arborea	10-12	91	19.7	1.9	1.36	0.12
Cochlospermum vitifolium	6	54	22.7	1.5	2.14	0.12
Tabebuia billbergiana	6	52	23.1	1.2	1.83	0.10
Pithecellobium carabobense	6	85	22.8	1.7	1.55	0.13
		Averages	22.1	1.6	1.72	0.12
Evergreens						
Capparis linearis	12	485	12.6	5.5	0.69	0.29
C. odoratissima	12	276	11.5	3.2	0.80	0.25
Jacquinia revoluta	12	249	9.7	5.5	0.62	0.20
C. pachaca	12	224	13.5	3.0	0.72	0.16
		Averages	11.8	4.3	0.71	0.23

Under natural conditions phosphate has been hypothesized as the key factor determining distribution of sclerophyll plant communities (Beadle 1966) and is supposed to be responsible for efficiency of nitrogen utilization (Loveless 1961).

2.3.1. The nitrogen-phosphorus relationship. Phosphorus and nitrogen concentrations show similar seasonal variations in leaves. In general, there is an increase in amount per unit leaf area for periods varying from several days to months during leaf growth followed by an almost linear reduction until the leaf is abscised. The increase in concentration can continue well beyond the time of full leaf expansion. Figure 1 shows the seasonal variations of leaf nitrogen and phosphorus concentrations in evergreen and deciduous trees from different plant communities. Observe that the N and P content of the leaves of evergreen and deciduous species can be very similar when expressed on a leaf area basis (compare C. pachaca and B. arborea of the Tropical Dry forest). The seasonal variation of the relative contents of nitrogen and phosphorus can be used to calculate the rates of absorption or retranslocation of these nutrients by means of their ratios. For example, the ratio of N/P tends to increase in evergreens when retranslocation of phosphorus begins earlier or is faster than retranslocation of nitrogen.

Leaf nitrogen and phosphorus contents, at least in adult, fully expanded leaves are strongly correlated (Seth, Bhatnagar 1960; Loveless 1961; Montes, Medina 1977; Sobrado, Medina 1980; Marín, Medina 1981). A quadratic regression between nitrogen and phosphorus was obtained based on a number of studies in tropical forests in Venezuela (Fig. 5). In this curve the deciduous and evergreen trees can be differentiated by their nitrogen and phosphorus content per unit dry weight.

The quadratic relationship implies that at low levels of phosphorus availability (as judged from the leaf content) there is a linear increase of nitrogen content per unit increment of phosphorus content. At high levels of phosphorus, however, the linearity is lost and the curve tends to level off. This means that high N/P ratios should be obtained in plant communities with restricted phosphorus supply. Similar reasoning was applied by Grubb (1977) when analyzing nutrient relationships in tropical montane forests. The values of leaf N and P contents of an array of tree species from four Jamaican montane forests published by Tanner (1977) and from a seasonal sclerophyll forest also in Jamaica (Loveless 1961) can be adjusted to an exponential function further indicating that this kind of relationship may be general.

2.3.2. Specific leaf area and nitrogen content.

In a number of studies in natural plant communities SLA appears to be highly and positively correlated with leaf nitrogen and phosphorus content per unit leaf weight (Cuenca 1976; Montes, Medina 1977; Sobrado, Medina 1980; Linder et al. 1981; Marín, Medina 1981). This kind of relation would be expected from studies under experimental conditions in which leaf expansion is inhibited by deficiencies of N and P (although it is not absolutely clear if it is

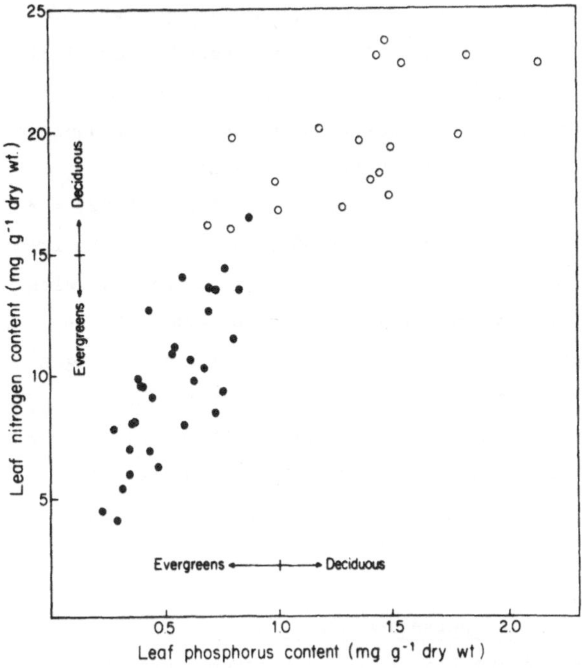

FIGURE 5. Relationship between phosphorus and nitrogen content per unit dry weight in adult leaves of tropical trees from different plant communities (with data from Cuenca 1976; Sobrado, Medina 1980; Marín, Medina 1981). The quadratic equation is: $N = 0.41 + 21.16\ P - 5.22\ P^2$, $r^2 = 0.73$.

a direct effect on leaf metabolism, or an indirect one mediated by inhibition of root growth). The selection of woody plant species with leaves which are thick and low in nitrogen and phosphorus in nutrient poor environments seems then to be related to their metabolic characteristics.

2.3.3. Retranslocation of nitrogen and phosphorus.

The retranslocation of the nitrogen and phosphorus contents of the leaves before abscission constitutes an important process at the ecosystem level because it may amount to a substantial fraction of the nutrient requirement for new growth (Thomas, Grigal 1976; Stoecker, Hacker 1976; Chapin, 1980; Ryan, Bormann 1982; Gray 1983; Chapin, Kedrowski 1983). It appears that approximately 50% of the nitrogen and 60% of the phosphorus is retranslocated before leaf fall (Table 2). The percentage amount of retranslocation seems to be much more pronounced in those species with higher leaf concentrations, as observed by Staaf (1982) and Chapin and Kedrowski (1983) in woody plants from temperate regions. The evergreen species, with a sclerophyll character, do not appear to be more efficient in the conservation of nitrogen and phosphorus as was expected from the observations of Small (1983) with evergreen bog plants, and Gray (1983) in a comparison of evergreen and deciduous shrubs in Southern California.

2.4 Leaf potassium content.

Interpretation of leaf potassium contents is difficult because of the extreme leachability of this element (Tukey 1970). Potasssium is not strongly bound to any structure or macromolecule in the leaf, which partially explains its high mobility. However, it has been shown that potassium concentration is reduced with leaf age, with 10-20% of adult leaf content (depending on stand age) being retranslocated before leaf fall (Ryan, Bormann 1982).
The potassium level in the leaf is important because of its role in the regulation of photosynthate transport through the phloem and in the maintenance of an active level of protein synthesis. An inverse relationship between leaf potassium content and soluble non-protein nitrogen has been observed (Murty et al. 1971; Mengel, Kirkby 1982). Therefore, high potassium is associated with a high proportion of leaf nitrogen in protein. Furthermore, in herbaceous plants grown under conditions of restricted nitrogen supply a tendency to potassium accumulation and an increase in leaf succulence (amount of water per unit leaf area) have been observed (Loetsch 1971; Delgado, Medina 1978). For these reasons, the amount of potassium in leaf tissues has to be investigated whenever the fractionation of protein and non-protein leaf nitrogen is undertaken.

Another aspect of the role of potassium in the functioning of leaves is the regulation of photosynthesis. It is well known that a

Table 2. Apparent retranslocation of nitrogen and phosphorus prior to leaf abscission calculated as the difference between maximum content and content in newly fallen leaves. In parenthesis one standard deviation (Source: 1. Marîn, Medina 1981; 2. Cuenca 1976; 3. Montes, Medina 1977; 4. Sprick 1979).

Species	No Sp	N%	P%	Source
Deciduous				
Tropical dry forest	5	65 (18)	64 (20)	(1)
Semi-deciduous forest (savanna)	6	44 (16)	59 (15)	(2,3)
Evergreens				
Tropical dry forest	5	53 (10)	37 (13)	(1)
Semi-deciduous forest (savanna)	5	43 (10)	62 (15)	(2,3)
Amazon forests	15	48 (14)	66 (15)	(4)
Overall Average		50 (15)	60 (18)	

potassium deficit increases leaf stomatal resistance because of the impairment of the normal opening mechanism of the guard cells (Fischer 1968; Nátr 1972), but it also increases the mesophyll resistance to CO_2 diffusion, probably because of its interactions with proteins (Peoples, Koch 1979). Nitrogen deficiencies, which cause an accumulation of potassium in leaf tissue of Commelinaceae, may be the cause of the increased drought tolerance observed in these plants (Mothes 1932). There is almost no information about these effects in forest trees (Keller 1972; Kramer, Kozlowski 1979).

2.5 Calcium and Magnesium content.

The physiological role of calcium for the maintenance of active transport systems in cellular membranes, and its structural role as component of the middle lamella in the cell wall has been well established (Epstein 1972; Kinzel 1982). The requirements of calcium for normal growth are thought to be genetically controlled, and more or less independent of the amount of calcium in the environment (Loneragan, Snowball 1966). It is noteworthy, for example, that the Graminae, Cyperaceae and Juncaceae tend to exclude calcium from their tissues, while dycotyledons tend to accumulate in their leaves (Kinzel 1982). Foliar calcium is found in ionic form, mainly as salts of malic and citric acids, associated to carboxylic groups of the cell wall (Mengel, Kirkby 1982), as well as in insoluble form as calcium carbonate and crystals of calcium oxalate.

For a long time the existence of calcicole and calcifuge plants in the temperate regions has been recognized (for a thorough review of this theme, see Kinzel 1982). It has been demonstrated that there are plants which maintain high levels of soluble calcium in their tissues, like the Crassulaceae, and are considered calciotrophs, while others are stimulated to produce oxalic acid in the presence of calcium ions, and precipitate them as calcium oxalate, being therefore, considered as calciophobes. Virtually nothing is known regarding the calcium requirements of tropical trees, and no data seem to be available on the calcium fractions. It is known, however, that trees from calcareous soils from dry areas and the sun exposed leaves of the tree canopy contain relatively high amounts of calcium, most of it in insoluble form. Within this context it is noteworthy that trees from tropical rain forests, growing on very acid and leached soils of the upper Río Negro region, also contain considerable amounts of oxalate crystals in their leaf tissues. Figure 6 shows the abundance of calcium oxalate crystals in plants from dry and wet tropical forests.

Calcium is transported to the shoot in the transpiration stream, but is practically immobile in the phloem. Therefore, the amounts of calcium deposited in the leaves have to be circulated externally in the ecosystem through litter fall. An analysis of the seasonal variation of calcium content in leaves shows a steady increase in concentration with age (among others Tamm 1951; Woodwell 1964, Ernst 1975, Montes, Medina 1977; Sobrado, Medina 1980; Marín, Medina 1981).

The role of magnesium in leaf function has been clearly established. It is a component of the chlorophyll molecule and also works as a widespread enzymatic cofactor. Of the enzymes activated by magnesium RuBP-carboxylase occupies an outstanding position. Magnesium deficiences, therefore, are always associated to a reduction in photosynthetic capacity of the leaf, mainly because there is a marked reduction in chlorophyll synthesis.

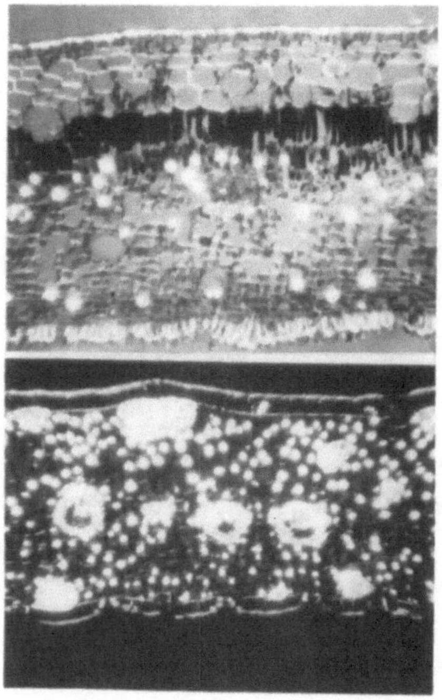

FIGURE 6. Transversal sections of leaves of Clusia sp (above) from the calcium poor upper Río Negro basin, and Jacquinia berterii (below) from Guánica forest on calcareous soils (Puerto Rico) which show calcium oxalate crystals. Pictures taken under polarized light (courtesy Dr. V. García).

Magnesium is ecologically very significant because of its abundance in serpentine soils (Proctor, Woodell 1975), which frequently reaches toxic levels. The molar quotients Ca/Mg in leaf tissues are normally well above one, but can be considerably lower in serpentine soils. Serpentine soils are widespread in the tropics, both in humid and arid areas, but practically nothing is known of the nutrient relationships of the plants growing on them. The simplest question whether there are serpentine specialists in the tropics has not been answered.

Magnesium is more mobile than calcium and its concentration in the leaves shows a more heterogenous behavior during the season. Some species accumulate magnesium in their leaves, while others show a clear reduction towards the end of the leaf life span.

3. NUTRIENT CONTENT AND SPECIFIC LEAF AREA OF TROPICAL FORESTS.

The wide variation in leaf structural characteristics and nutrient content of tree species of a series of tropical forests under different rainfall and temperature regimes is shown in Tables 3 and 4. These tables contain data obtained mainly in our laboratory, but they cover many important tropical forest types, and in addition, they were collected and analyzed using the same techniques, so they are directly comparable. Comparisons between lowland tropical forests have been published among others by Golley et al. (1975); Klinge (1976) and Furch and Klinge (1978); tropical montane forests have been reviewed by Grubb (1977); Tanner (1977) and Grimm and Fassbender (1981).

Only some aspects of Tables 3 and 4 will be emphasized. The deciduous species always have higher values of SLA than the evergreen species of the same or different forests. These higher values of SLA correspond also per higher nitrogen and phosphorus contents per unit leaf dry weight. Among the evergreen communities, those of the elfin forest on the mountain tops of the Caribbean islands and those of the low Caatinga on sandy soils in the upper Río Negro region have very low values of nitrogen and phosphorus, a characteristic clearly associated with the extreme accumulation of structural carbohydrates in the leaves (Grubb 1977; Sobrado, Medina 1980). A similar leaf structure and nutritional status to that of the low Amazon Caatinga has been reported for a Sarawak forest by Peace and MacDonald (1981).

The dry forests represented in Table 4 have higher contents of leaf potassium and calcium than those of the wet forests. This could be due to lesser leaf leaching because of the lower

Table 3. Average values for Specific Leaf Area, and nitrogen and phosphorus concentration for several tropical forest types. In parenthesis one standard deviation.

Forest	No. Species	SLA m^2/g	N mg/g (s.d.)	P
Tropical Dry Forest (1)				
Deciduous	7	127 (45)	21.0 (2.7)	1.58 (0.30)
Evergreens	4	36 (10)	11.8 (1.6)	0.71 (0.07)
Semi-deciduous forest (2)				
Deciduous	4	110 (23)	17.1 (2.2)	1.15 (0.46)
Evergreens	9	73 (16)	12.1 (4.7)	0.57 (0.20)
Montane forests				
Cloud forest (2)	7	68 (26)	11.7 (5.2)	0.82 (0.49)
Elfin forest (3)	7	47 (11)	9.9 (2.8)	0.63 (0.21)
Tabonuco forest (3)	7	83 (33)	13.6 (3.8)	0.54 (0.09)
Amazonian forest (4)				
Mixed forest on oxysol	7	74 (12)	12.7 (2.7)	0.60 (0.08)
Tall Caatinga	6	76 (33)	11.6 (4.6)	0.73 (0.19)
Low Caatinga	14	47 (12)	7.4 (2.4)	0.50 (0.17)

Authors: (1) Marín, Medina 1981; (2) Cuenca 1976; (3) Medina et al. 1981; (4) Sprick 1979.

Table 4. Average values of potassium, calcium and magnesium for several tropical forest types. In parenthesis one standard deviation.

Forest type (1)	No. species	K	Ca mg/g (s.d.)	Mg
Tropical dry (1)	11	17.0 (8.1)	15.8 (8.8)	3.6 (1.8)
Sub-tropical dry (4) (calcareous; Pto. Rico)	22	16.3 (7.0)	20.3 (9.0)	3.1 (1.8)
Semi-deciduous (2)	6	6.5 (2.9)	7.7 (3.6)	2.9 (1.6)
Montane forests				
Cloud forest (2)	7	5.5 (2.1)	8.7 (3.2)	2.6 (1.3)
Elfin forest (3)	7	5.1 (1.6)	6.7 (5.3)	1.6 (1.2)
Tabonuco (3)	7	4.8 (2.2)	6.3 (3.4)	1.7 (0.6)
Serpentine (4) (Maricao, Pto. Rico)	15	7.7 (4.7)	8.3 (8.8)	4.4 (3.7)
Amazonian forests (5)				
Mixed forest on oxysol	6	4.6 (1.2)	1.9 (0.7)	1.0 (0.2)
Tall Caatinga	7	6.2 (1.1)	4.4 (1.6)	1.5 (0.5)
Low Caatinga	11	6.4 (3.0	5.8 (3.4)	1.4 (0.3)

Authors: (1) Marín, Medina 1981; (2) Cuenca 1976; (3) Medina et al. 1981; (4) Medina et al. unpublished; (5) Sprick 1979.

rainfall amounts. In the case of calcium, the reason might also be the lack of leaching from the soil complex. Most arid soils tend to be alkaline often showing a significant accumulation of calcium in the upper layers. This is more pronounced in areas with soil of calcareous origin such as the Guánica forest in southern Puerto Rico. The lower extreme of calcium and magnesium accumulation is observed in the trees of Amazonian forests of the upper Río Negro region, mainly in the mixed forest on oxysols. These forests have been described as extremely impoverished in alkali-earth metals (Furch, Klinge 1978). The humid forests of Maricao, Puerto Rico, growing on serpentine outcrops at about 700 m above sea level show the highest concentrations of magnesium. The nutritional relationships of the plants growing in this humid montane forest deserve detailed investigation, because serpentine mountains are rather frequent in tropical regions.

Values of the mineral nutrient content of leaves and their seasonal variations represent only a first orientation on the diversity of nutrient use strategies in natural plant communities. They have to be complemented with in depth studies on the uptake dynamics, metabolism and distribution in the plant body, and the cycling process. These studies may eventually lead to the understanding of selection processes in nutrient poor habitats.

REFERENCES

Addicott FT (1976) Abscission strategies in the behavior of tropical trees. In Tomlinson PB and Zimmermann MH, eds. Tropical trees as living systems, pp. 381-398. Cambridge, Cambridge University Press.

Andreeva TF and Avdeeva TA (1970) Fraction I protein and the photosynthetic activity of leaves, Fiziol. Razt. 17, 225-233.

Beadle NCW (1966) Soil phosphate and its role in molding segments of the Australian flora and vegetation, with special reference to xeromorphy and sclerophylly, Ecology 47, 992-1007.

Chabot BF and Hicks DJ (1982) The ecology of leaf life spans, Ann. Rev. Ecol. Syst. 13, 229-259.

Chapin FS III (1980) The mineral nutrition of wild plants, Ann. Rev. Plant Physiol. 11, 233-260.

Chapin FS III and Kedrowski RA (1983) Seasonal changes in nitrogen and phosphorus fractions and autumn retranslocation in evergreen and deciduous Taiga trees, Ecology 64, 376-391.

Christie EK and Moorby J (1975) Physiological responses of semiarid grasses I. The influence of phosphorus supply on growth and phosphorus absorption, Austr. J. Agric. Res. 26, 423-436.

Cuenca G (1976) Balance nutricional de algunas leñosas de dos ecosistemas contrastantes: Bosque nublado y Bosque decíduo, Tesis de Licenciatura, Escuela de Biología, UCV, Caracas.

Delgado M and Medina E (1978) Leaf succulence and potassium accumulation. In Singh JS and Gopal B, eds. Glimpses of Ecology, pp. 417-424. Int. Sci. Publ., Jaipur, India

Epstein E (1972) Mineral nutrition of plants: principles and perspectives. John Wiley and Sons, Inc. New York.

Ernst W (1975) Variation in the mineral contents of leaves of trees in Miombo woodland in South Central Africa, J. Ecology 63, 801-807.

Evans GC (1972) The quantitative analysis of plant growth. University of California Press, Berkeley.

Field C (1981) Leaf age effects on the carbon gain of individual leaves in relation to microsite. In Margaris NS and Mooney HA, eds. Components of productivity of Mediterranean regions: basis and applied aspects, pp. 41-50. Dr. W. Junk, The Hague.

Fischer RA (1968) Stomatal opening:role of potassium uptake by guard cells, Science 160, 784-785.

Furch K and Klinge H (1978) Towards a regional characterization of the biogeochemistry of alkali and alkali-earth metals in northern South America, Acta Cient. Ven. 29, 434-444.

Golley F. McGinnis JT, Clements RG, Child GI and Duever MJ (1975) Mineral cycling in a tropical moist forest ecosystem. Athens, University of Georgia Press.

Gray JT (1983) Nutrient use by evergreens and deciduous shrubs in Southern California. I. Community nutrient cycling and nutrient use efficiency, J. Ecology 71, 21-41.

Grimm U and Fassbender HW (1981) Ciclos bioquímicos en un ecosistema forestal de los Andes Occidentales de Venezuela. I. Inventario de las reservas orgánicas y minerales (N, P, K, Ca, Mg, Mn, Fe, Al, Na), Turrialba 31, 27-36.

Grubb PJ (1977) Control of forest growth and distribution on wet tropical mountains with special reference to mineral nutrition, Ann. Rev. Ecol. Syst. 8, 83-107.

Gulmon SL and Chu C (1981) The effects of light and nitrogen on photosynthesis, leaf characteristics, and dry matter allocation in the Chaparral shrub, Diplacus aurantiacus, Oecologia 49, 207-212.

Keller T 1972. Gaseous exchange of forest trees in relation to some edaphic factors, Photosynthetica 6:197-206.

Kinzel H (1982) Pflanzenoekologie and Mineralstoffwechsel. Eugen Ulmer Verlag. Stuttgart.

Klinge H (1976) Bilanzierung von Hauptnaehrstoffen im Oekosystemen tropischer Regenwal, Biogeographica 7, 59-76.

Koerner CH, Scheel JA and Bauer H (1979) Maximum diffusive conductance in vascular plants, Photosynthetica 13, 45-82.

Kramer PJ and Kozlowski TT (1979) Physiology of woody plants. Academic Press New York.

Larcher W (1961) Jahresgang der Assimilations und Respirationsvermoegens von Olea Europaea L. ssp. sativa Hoff., et Link., Quercus ilex L. und Quercus pubescens Willd. aus dem noerdlichen Gardaseegebiet, Planta 56, 575-606.

Larcher W (1969) The effect of environmental and physiological variables on the carbon dioxide exchange of trees, Photosynthetica 3, 167-198.

Linder S, McDonald J and Lohammar T (1981) Effect of nitrogen status and irradiance during cultivation on photosynthesis and respiration in birch seedlings (Betula verrucosa Ehrh.), Technical report No. 12, 19 p. Energy Forestry Project, Swedish Agricultural University.

Loetsch B (1971) Sukkulenz und Kaliumspeicherung von Stickstoffmangel Pflanzen, Z. Pflanzenphysiol. 64, 393-399.

Loneragan JF and Snowball K (1969) Calcium requirements of plants, Austr. J. Agric. Res. 20, 465-478.

Loomis WD (1974) Overcoming problems of phenolics and quinones in the isolation of plant enzymes and organells, Methods in Enzymology 31, 528-544.

Loveless AR (1961) A nutritional interpretation of sclerophylly based on differences in the chemical composition of sclerophyllous and mesophytic leaves, Ann. Bot., ns, 25, 168-183.

Marín D and Medina E (1981) Leaf duration, nutrient content and sclerophylly of very dry tropical forest trees, Acta Cient. Ven. 32, 508-514.

McDonald J, Lohammar T and Linder S (1981) Effect of leaf nitrogen content on CO2 exchange in a number of Salix clones, Technical Report No. 16. Energy Forestry Project. Agricultural University Sweden. 19 p.

Medina E (1970) Relationships between nitrogen level, photosynthetic capacity, and carboxydismutase activity in Atriplex patula leaves, Carnegie Institution Year Book 69, 655-662.

Medina E (1981) Nitrogen content, leaf structure and photosynthesis in higher plants. Report to UNEP Study Group on Photosynthesis and Bioproductivity. London.

Medina E (1982) Physiological ecology of neotropical savanna plants. In Huntley BJ and Walker BH, eds. Ecology of Tropical Savannas, pp. 308-335. Springer-Verlag Berlin.

Medina E, Cuevas E and Weaver P (1981) Composición foliar y transpiración de especies leñosas de Pico del Este, Sierra de Luquillo, Puerto Rico, Acta Cientifica Venezolana 32, 159-165.

Medina E. Mendoza A and Montes R (1978) Nutrient balance and organic matter production in the Trachypogon savannas of Venezuela. Tropical Agric. (Trinidad) 55, 243-253.

Medway Lord (1972) Phenology of a tropical rain forest in Malaya Biol. J. Linn. Soc. 4, 117-146.

Mengel K and Kirkby EA (1982) Principles of Plant nutrition. 3rd Edition. International Potash Institute. Bern, Switzerland.

Migus WN and Hunt LA (1980) Gas exchange rates and nitrogen concentrations in two winter wheat cultivars during the grain-filling period, Canadian J. Bot. 58, 2110-2116.

Monasterio M and Sarmiento G (1976) Phenological strategies of plant species in the tropical savanna and the semi-deciduous forest of the Venezuelan Llanos, J. Biogeography 3, 325-356.

Mooney HA, Ferrar PJ and Slatyer RO (1978) Photosynthetic capacity and carbon allocation patterns in diverse growth forms of Eucalyptus, Oecologia (Berl.) 36, 103-111.

Mooney HA and Gulmon SL (1982) Constraints on leaf structure and function in reference to herbivory, BioScience 32, 198-206.

Mooney H and Rundel P (1979) Nutrient relations of the evergreen shrub, Adenostoma fasciculatum, in the California Chaparral, Bot. Gaz. 140, 109-113.

Montes R and Medina E (1977) Seasonal changes in nutrient content of leaves of savanna trees with different ecological behavior, Geo-Eco-Trop 1, 295-307.

Mothes K (1932) Ernaehrung, Struktur und Transpiration. Ein Beitrag zur Kausalanalyse der Xeromorphosen, Biol. Zb. 52, 93-223.

Motta N and Medina E (1978) Early growth and photosynthesis of tomato (Lycopersicum esculentum L.) under nutritional deficiencies, Turrialba 28, 135-141.

Murty KS, Smith TA and Bould C (1971) The relation between the putrescine content and potassium status of Black Currant leaves, Ann. Bot. 35 687-695.

Nátr L (1972) Influence of mineral nutrients on photosynthesis of higher plants, Photosynthetica 6, 80-99.

Nátr L (1975) Influence of mineral nutrition on photosynthesis and the use of assimilates. In Cooper JP, ed. Photosynthesis and Productivity in different Environments, International Biological Programme 3, pp. 537-555. Cambridge University Press. Cambridge.

Nevins DJ and Loomis RS (1970) Nitrogen nutrition and photosynthesis in sugar beet (Beta vulgaris L.), Crop Science 10, 21-25.

Peace WJ and MacDonald FD (1981) An investigation of the leaf anatomy, foliar mineral levels, and the water relations of trees of a Sarawak forest, Biotropica 13, 100-109.

Peoples RP and Koch DW (1979) Role of potassium in carbon dioxide assimilation in Medicago sativa L, Plant Physiol. 63, 878-881.

Proctor J and Woodwell SRJ (1975) The ecology of serpentine soils, Adv. Ecol. Res. 9, 255-366.

Ryan DF and Bormann FH (1982) Nutrient resorption in northern hardwood forests, BioScience 32: 29-32.

Saeki T and Nomoto N (1958) On the seasonal change of photosynthetic activity of some deciduous and evergreen broadleaf trees, Bot. Mag. (Tokyo) 71, 235-241.

Sasahara T (1982) Changes in size and number of mesophyll cells, nitrogen content and photosynthesis with leaf order in Brassica spp, Ann. Bot. 50, 379-383.

Schlesinger WH and Chabot BF (1977) The use of water and minerals by evergreen and deciduous shrubs in Okefenokee swamp, Bot. Gaz. 138, 490-497.

Seth SK and Bhatnagar HP (1960) Interrelations between mineral constituents of foliage, soil properties, site quality and regeneration status in some Shorea robusta forests, Indian Forester 86, 590-601.

Small E (1972) Photosynthetic rates in relation to nitrogen cycling as an adaptation to nutrient deficiency in peat bog plants. Can. J. Bot. 50, 2227-2233.

Sobrado MA and Medina E (1980) General morphology, anatomical structure and nutrient content of sclerophyllous leaves of the "Bana" vegetation of Amazonas, Oecologia (Berl.) 45, 341-345.

Sprick E (1979) Composición mineral y contenido de fenoles foliares de especies leñosas de tres bosques contrastantes de la región Amazónica. Tesis de Licenciatura, Escuela de Biología, UCV, Caracas.

Staaf H (1982) Plant nutrient changes in beech leaves during senescence as influenced by site characteristics. Acta Oecologia, Oecol. Plant. 3, 161-170.

Stocker O (1931) Transpiration und Wasserhaushalt in verschiedenen Klimazonen. I. Untersuchungen an der arktischen Baumgrenze in Schwedisch Lappland. Jahrb. wiss. Bot. 75, 494-549.

Stoecker G and Hacker E (1976) Untersuchungen ueber Stickstoff-Blattspiegel-Werte einiger Bodenpflanzen naturnaher Berg-Fichtenwaelder. Saisonale Veraenderungen und artspezifischen Relationen, Flora 165: 65-94.

Tamm CO (1951) Seasonal variation in composition of birch leaves, Physiologia Plantarum 4: 461-469.

Tanner EVJ (1977) Four montane rainforests of Jamaica: A quantitative characterization of the floristics, the soils and the foliar mineral levels, and a discussion of the interrelations. J. Ecology 65: 883-918.

Thomas WA and Grigal DF (1976) Phosphorus conservation by evergreenness on mountain laurel. Oikos 27, 19-26.

Tukey HB (1970) The leaching of substances from plants. Ann. Rev. Plant Physiol. 21, 305-324.

Wareing PF (1977) Growth substances and integration in the whole plant. In Jennings DH, ed. Integration of activity in the higher plant, pp. 337-366, Symposium XXXI, Society for Experimental Biology. Cambridge University Press. Cambridge, UK.

Walter H (1973) Die Vegetation der Erde. Die tropischen und subtropischen Zonen, IIIe. Auflage VEB Gustav Fischer Verlag, Jena.

Woodwell GM (1974) Variation of nutrient content of leaves of Quercus alba, Quercus coccina, and Pinus rigida in the Brookhaven forest from bud-break to abscission, Am. J. Brot. 61, 749-753.

EPIPHYTIC VEGETATION: A PROFILE AND SUGGESTIONS FOR FUTURE INQUIRIES

DAVID H. BENZING
Department of Biology, Oberlin College
Oberlin, Ohio 44074 USA

ABSTRACT

Vegetation anchored in forest canopies has been little studied despite its prominence in the moist tropics and many drier sites. Especially in wet montane communities, much of the productive biomass may be generated by plants attached to bark and leaf surfaces. Epiphylls include a broad range of microbes and thallophytes; they may be most noteworthy for their involvement in nitrogen fixation and effects on the longevity and photosynthetic performance of host leaves. Vascular epiphytes are also diverse and employ a variety of adaptive features to secure moisture and mineral nutrients under what are often very stressful conditions. A number of agencies interact, particularly in drier canopies, to suppress the growth of higher plants rooted there. Unusual absorptive tissues and habits, and mutualistic plant/animal associations designed to mitigate constraints on resource procurement, abound in this second assemblage of tree crown inhabitants. Large loads of epiphytes and epiphylls may have a significant influence on the structure, nutritional dynamics and productivity of a forest; they certainly often promote mineral retention and encourage animal activity. Many of the factors involved in the adaptive biology of epiphytic vegetation and the roles these organisms play in host ecosystems have physiological bases. Models and suggestions emphasizing function are offered as possible directions for research on epiphytes and epiphylls.

1. INTRODUCTION

There are quite a few classic studies of epiphytic vegetation (e.g., Schimper, 1888; Went, 1940; Richards, 1952), but the emphasis has always been descriptive rather than experimental. My purpose here is two-fold: to sketch a general profile of epiphytism as an adaptive strategy and then pose questions about how plants growing on forest canopy surfaces may influence associated biota and hence the quality, structure and performance of host ecosystems. Pertinent to this latter subject are those environmental constraints which account for peculiarities of the canopy dweller's form and physiology, and thus indirectly determine the nature of its interactions with other organisms. The term "epiphyte" is restricted here to include only those tracheophytes that, more often than not, spend their entire lives growing on other plants. Also considered are epiphylls (mostly lower plants and microbes) that may or may not be restricted to the phylloplane. Excluded from this discussion, unless specifically mentioned, are stranglers, haustorial parasites and hemiepiphytes (those organisms which begin life as vines, later to lose their stem, but not root, contact with soil).

2. CHARACTERISTICS OF FOREST CANOPIES AND THEIR IMPACT ON RESIDENT VEGETATION

2.1. Humidity. Tropical tree crowns offer a tremendous variety of growing conditions, mostly as functions of exposure and that vital component for epiphytes and epiphylls, humidity (Sugden, Robins, 1979). At one climatic extreme, sites receive daily rain or fog; air saturation may never fall below 80-90%, and sodden bark soon becomes so heavily laden with vegetation and associated litter that attached plants need few, if any, extraordinary adaptations to counter the occasional short-lived dry spells that occur in even the wettest forest canopies. Indeed, many grow just as well, some even more luxuriantly, at ground level. Personal observations suggest that, as humidity decreases and the disparities between the physical nature (particularly the moisture status) of phorophyte (supporting tree) surfaces and local soils intensify, resident epiphyte populations become progressively more specialized and demanding of arborescent support (Fig. 1). A xeric epiphyte which falls on moist

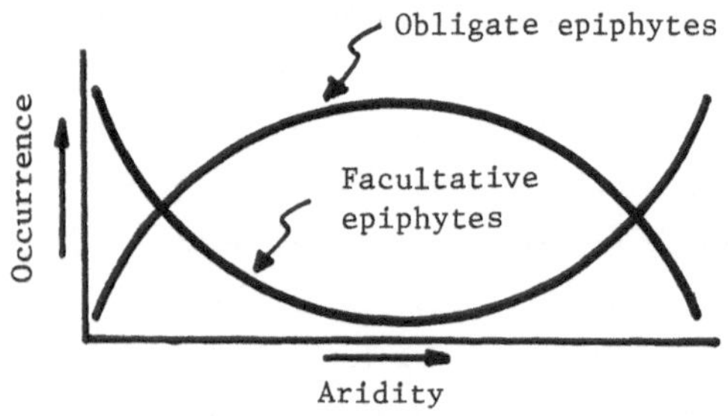

Figure 1. A model illustrating how differences in the water-retaining qualities of bark and soil influence the proportional occurrence of obligate and facultative epiphytes.

soil will usually die (e.g., Benzing, Renfrow, 1971; Benzing, Seemann, 1978). Toward the arid end of the climatic continuum, distinctions between arboreal and terrestrial habitats blur once more and facultative epiphytism again becomes a viable option. Choice of substratum may be further relaxed on drier sites by irradiances that are diminished much less by canopy passage than that penetrating to floors of moister forests. It is possible that some conspecifics segregated by substratum preference at all humidity levels are ecotypically distinct.

2.2. Nature of the substratum. To be successful, seeds of epiphytes must germinate on appropriate bark exposures in suitable tree crowns. Whereas a given microsite may provide acceptable light levels and access to canopy fluids for one species, another may require nurse patches of moisture-retaining moss and lichens in addition. Bark is a temporary, unpredictable medium for all organisms anchored there. Small segments exfoliate, twigs and larger branches fall away, and persistent surfaces often become shaded by expanding crown growth (Benzing, 1979). Ultimately, the phorophyte dies, an event followed shortly by death of most of its attached plant life. In effect, epiphytes and epiphylls must be opportunists; they have to be mobile and fecund enough to deal with the uncertainties engendered by a dispersed, heterogeneous, unstable substratum.

2.3. Constraints on nutrient and moisture procurement. Most forest canopy residents encounter substantial spatial and temporal constraints on moisture and nutrient procurement. Shrouds of clinging vegetation and associated litter hold moisture and essential ions in continuous supply for the cloud and rainforest epiphyte and epiphyll, but inhabitants of more typical canopies subsist on resource pulses from brief, widely spaced rainstorms that leave little residual moisture suspended aboveground. Mineral nutrients coursing down the forest profile in throughfall and stemflow originate from atmospheric rainout, particulates scrubbed from the air and plant surfaces, and leachates from phorophyte, epiphyll and epiphyte tissues higher in the canopy. En route, some mineral ions may be extracted by phorophytes, epiphytes and nutrient-scavenging epiphylls (Jordan et al., 1980), others may be added by cyanophytic lichens and free-living nitrogen fixers (Forman, 1975; Lang et al., 1976; Pike, 1978; Sen Gupta et al., 1982). Total nutrient ion concentrations in these fluids are usually low (only a few ppm;

Benzing, Renfrow, 1980), but fluctuations can be such that a large part of the supply arrives in a relatively small proportion of the total liquid volume (e.g., Kellman et al., 1982). Availability of essential mineral ions to epiphytic and epiphyllous plants will depend on several factors, including climate and the identity and nutritional status of the phorophyte (Tukey, 1970; McColl, 1970; Benzing, Renfrow, 1974; Schlesinger, Chabot, 1977).

2.4. Stress and epiphyte diversity. In view of the demanding physical character of the tropical forest canopy biotope, it is not surprising that most of the more than 60 families contributing to the world's epiphyte floras appear only where bark surfaces are moist much of the time. Climatic constraints on mineral ion procurement and carbon gain in arid locations mandate ever more specialized adaptations. "Extreme" epiphytes that inhabit the harshest sites are so limited in photosynthetic capacity and slow to mature that great resource economy is apparently required to generate a sustaining level of reproductive power (Benzing, 1978; Fig. 2).

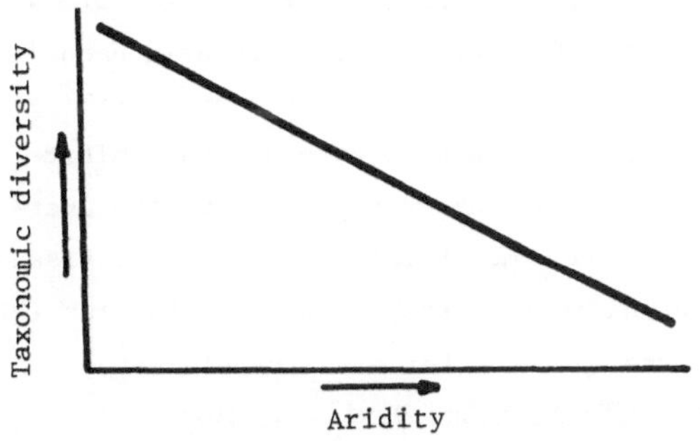

Figure 2. A model illustrating how climatic stress reduces epiphyte diversity by exacerbating the impact of habitat patchiness and disturbance in forest canopies.

Diversity in high and middle taxonomic levels drops off rapidly along the moisture continuum, but the few genera that have become sufficiently stress-tolerant without sacrificing too much regenerative capacity are often quite large (e.g., Dendrobium, Pepermomia, Tillandsia).

3. ADVANTAGES OF CANOPY EXISTENCE

3.1. Resource economy. Access to sunlight is usually offered as the major impetus for epiphytism, sometimes with prose that has a more romantic than scientific flavor. Popular accounts, and even some textbooks, conjure up images of deprived, soil-bound ancestors lured from the darkest forest depths into lofty perches bathed in life-giving light. Indeed, there are many ways to conceptualize the emergence and benefits of epiphytism. Some, like the version cited, wax philosophic, others are much more mechanistic. In epiphytes, proportionately little biomass need be committed to support and transport tissue. Thus, a greater share of the organism's material resources can be employed directly for propagation--a trend which seems to reach its zenith in shootless Orchidaceae and rootless Tillandsia (Benzing, Ott, 1981). But this, like so many other advantages acquired through adaptive change, comes at a price--in this case, lower productivity than that achieved by the more abundantly provisioned phorophyte. Were this not so, forest survival would be jeopardized by a blanket of herbaceous vegetation outstripping all competitors in the quest for light. Other ramifications of this distinction in performance are considered later.

3.2. Escape from herbivory. A second advantage, if not impetus, to epiphytism may have been abatement of a major environmental threat, herbivory. Many plant predators are confined to soil or consume low-growing vegetation. Although severe insect damage is not commonly encountered among epiphytes, herbivores have been reported feeding on them (e.g., Huxley, 1978). Circumstantial evidence of herbivore pressure is mixed. In Bromeliaceae, only the epiphytic bromelioids produce heavy armaments such as those borne by their close terrestrial relatives and so many dry-land, perennial herbs. Spines are substantially reduced or absent among epiphytic cacti. On the other hand, extrafloral nectaries do occur in many canopy-dwelling orchids (Bentley, 1977). Terrestrial and epiphytic orchids alike often synthesize alkaloids (Slaytor, 1977) and harbor raphids, as do aroids and bromeliads. Siliceous epidermal inclusions are widespread among bromeliads of many habitat preferences (Tomlinson, 1969).

Surveys of phytotoxins, mechanical repellents and irritants among terrestrial and arboreal elements in all heavily canopy-adapted families would be instructive. If herbivore pressures are indeed relaxed for the epiphyte compared to its soil-rooted counterpart, then chemical deterrents should be more prevalent in soil-rooted plants vis-à-vis their epiphytic relatives. To be most effective, these surveys should be accompanied by censuses of herbivore diversity and abundance.

Herbivory could, of course, be further reduced by the low apparency of many epiphytes (hyper-dispersions), the poor nutritional quality (nitrogen content in particular) of their tissues, or the generally pronounced toughness of epiphyte leaves. Crypsis or mimicry may afford considerable protection on certain widely exploited supports, as it does for some Australian mistletoe parasites which avoid marsupial predators by simulating their toxic Eucalyptus hosts (Barlow and Wiens, 1977).

4. ADAPTATIONS FOR EPIPHYTISM

Epiphytes, above all those extreme forms from more arid sites, are conspicuously designed for life in tree crowns and other substrata of similar physical quality. They have adopted numerous features, although not many truly novel ones, to counter the constraints prevailing in forest canopy habitats. An outline of the general characteristics, shared by most epiphytes, which enhance such basic functions as water and mineral procurement and utilization, holdfast, and reproduction, is provided in Table 1. Some additional adaptations will also be covered in the following section.

4.1. Mineral nutrient procurement and use. Nowhere is the epiphyte's design for canopy life more apparent than in the way access to mineral ions has been increased to better exploit nutritive solids and solute-charged fluids moving down the canopy profile. Certain Araceae, Bromeliaceae, Polypodiaceae and others produce special interception devices. Platycerium (Polypodiaceae), bird's-nest Anthurium, the trash-basket orchids (Cyrtopodium), Astelia (Liliaceae) and many bromeliads create soil substitutes by impounding litter via appropriately shaped individual shoots, root masses and collections of interconnected ramets. "Impoundment" of another sort leading to resource gain is seen in at least six families which offer accommodations to ant colonies in leaves and stems (Huxley, 1980). Many orchids and tillandsioid bromeliads have fabricated what might be regarded as mini-impoundments: the specialized spongy tissues of the orchid velamen and the indumentum of the bromeliad leaf, to cite two well-known examples (Benzing, Ott, 1981; Benzing et al., 1983; Benzing, Pridgeon, 1983).

Residents of moderately to very arid canopies also possess many features shared by sterile-soil terrestrials. Members of both groups exhibit long life cycles, inherently slow growth, evergreenness, and some at least maintain low tissue levels of nitrogen, phosphorus and potassium (but not necessarily of other nutrient elements; Benzing, 1978; Benzing, Davidson, 1979). Virtually all epiphytes are iteroparous, so some unexpended resources can be recycled between successive sexual efforts.

4.2. Water procurement and conservation. The impoundment techniques just described for mineral ion procurement are equally efficacious for interception and retention of water. Tanks, if sizable, serve to tide arid-canopy epiphytes over rainless periods, but this strategy is ineffective in very dry forests. Several heavily trichomed atmospheric bromeliads without catchments reportedly absorb water from moist air (De Santo et al., 1976), but it is not known whether this capacity is significant under natural conditions. Other not so novel mechanisms are also employed by xeric canopy dwellers. Xeromorphic foliage is exhibited to some degree in all but the exceptional wet-forest and deciduous epiphytes. CAM is common (Kluge, Ting 1978), not an unexpected finding given the versatility of this metabolic process and the rapid shifts in humidity that characterize so many canopy habitats. With the possible exception of some orchids (Avadhani et al., 1982), C4 photosynthesis appears to be absent in canopy dwellers. Some orchids (Benzing et al., 1982), a number of ferns and a few other epiphytes are drought avoiders. Here, short-lived leaves operate via the C3 pathway and expend moisture extravagantly. Succulent stems, which persist several seasons, are much less porous and possess little photosynthetic capacity. High mineral ion costs associated with ephemeral leaves and unpredictable moisture supplies, even during the rainy season, probably preclude the drought-deciduous option on most bark surfaces.

5. FUTURE INQUIRIES ON EPIPHYTE NUTRITION AND WATER BALANCE

Our knowledge of epiphyte nutrition and water balance is scanty. Future inquiries must be much more comprehensive in order to determine whether different patterns of resource procurement and utilization exist in epiphytic versus terrestrial plants. It is unclear at this time just how unique constraints on epiphytes are and how novel the adaptive responses have become.

5.1. A second look at water and carbon balance. While the biochemical bases of carbon assimilation may differ little, if at all, between canopy-dwelling and soil-rooted plants, the ways in which certain drought-adapted epiphytes modulate shifts between C3 and CAM activity and regulate water loss may be unique. We have noted that epiphytes anchored on exposed surfaces often face large fluctuations in evaporative demand and moisture supply; therefore, guard cells fine-tuned for response to short-term changes in atmospheric humidity would presumably offer considerable benefit. Curiously, certain xeric tillandsias bear permanently occluded stomata (Billings, 1904; Tomlinson, 1969; C. Martin, personal communication). Many of these same bromeliads exhibit paradoxical combinations of hygromorphic and xeric shoot features: much attenuated leaves, a foliar epidermis whose thin cell walls bear delicate cuticles (Tomlinson, 1969) and CAM photosynthesis (Medina, 1974; Medina, Troughton, 1974). Conceivably, moisture is held more tenaciously than leaf anatomy would indicate, or atmospheric conditions required to relieve water deficits recur often enough, perhaps nightly, or these plants need not maintain hydratures as high as those required by most homoiohydrous forms.

Unlike those of most terrestrials, roots of many epiphytes are exposed and photosynthetic. The trophic importance of this condition must vary with the taxon. Chlorophyll levels are usually low, but among "shootless" orchids, root systems have replaced stems and leaves as carbon gainers (Benzing, Ott, 1981). How organs which lack stomata but engage in active photosynthesis operate with adequate water-use efficiency in relatively xeric habitats is unclear. Roots of some orchids, including the shootless taxa, possess devices at the velamen/cortex interface that may regulate ventilation (Benzing et al., 1983). Canopy-dwelling aroids and rootless orchids, lacking such contrivances, might deploy their more weakly pigmented green tissues primarily for carbon conservation and simple maintenance in a manner similar to cactus phyllodes that idle while under severe water stress (Kluge, Ting, 1978). As long as photosynthetic capacities are modest and permeabilities continuously low, these organs may operate near optimum water-use efficiencies, and at least those orchid roots equipped for CAM can also recycle dark-respired CO_2 (Benzing, Ott, 1981).

Drought, whether climatic or physiological, promotes the presence of organic osmotica which may, in part, help to stabilize membranes subjected to severely depressed water potentials (Aspinwall, Poleg, 1981; Wyn Jones, Storey, 1981). Many epiphytes must rely on similar mechanisms to cope with moisture stress. Which organic agent is best suited for a specific case may depend on the periodicity and severity of the drought experience and the fertility of the habitat. Rapidly metabolized molecules might be the best choice for the organism whose water potential fluctuates rapidly, while canopy dwellers with less permeable shoot surfaces might be better served by more stable ones. A non-nitrogenous osmoticum could be the better choice for an exceptionally oligotrophic taxon. Efforts to identify and track the rise and ebb of compounds suspected of reducing desiccation injury or aiding in its repair should be examined in epiphytes with different water balance and nutritional profiles.

Finally, there is a paucity of data on a subject not previously mentioned: stress reaction and rebound in epiphytes. Most canopy dwellers should gain from adaptations that allow a rapid return to prestress activity levels once contact with a resource pool is renewed. Sala and Lauenroth (1982) found that some arid-land grasses possess capacities to maximize the impact of those small, scattered rainfall events that deliver most of their moisture. Pressures for similar adaptations exist in most forest canopies and have probably left their mark on the plants that grow there.

5.2. Mineral nutrition revisited. More refined analyses are needed to determine how the many versions of orchid velamina and bromeliad trichomes operate to retard transpiration and promote salt and water balance under various field conditions. Biophysical aspects of these structures merit special attention. Böttger et al., (1980) have demonstrated that velamen tissue from aerial roots of *Vanilla planifolia* has a very high buffering capacity that may be important for the initial mobilization of ions in passing canopy fluids. Broader surveys should also include the velamentous roots of Araceae which, so far, are only presumed to be comparable to those of Orchidaceae. Claims concerning the absorptive qualities of foliar hairs associated with several polypodiaceous ferns (Müller et al., 1981) and *Astelia* (Oliver, 1930) also need experimental verification. Orchid trichomes, at least those borne by members of Pleurothallidinae, are apparently nonabsorptive (Benzing, Pridgeon, 1983). Except for certain bromeliads (Benzing et al., 1976), a few orchids (Benzing, Pridgeon, 1983), poikilohydrous *Polypodium polypodioides* (Stuart, 1969), two additional polypodiums and *Pleopeltis angusta* (Müller, et al., 1981), shoots of epiphytes have not been analyzed to note whether they possess unusual surface refinements for resource procurement. Judging by the high shoot/root ratios of many canopy dwellers, leaves seem likely to play a prominent role in resource acquisition in many additional taxa.

Oligotrophs supposedly lack special mechanisms that enhance their capacity to absorb mineral ions from very dilute sources (Grime, 1977; Chapin, 1980) but more thorough surveys are needed to substantiate this hypothesis. To my knowledge, no epiphyte has been so examined up to now. Potentials for luxury consumption, particularly of phosphorus, do exist in many plants, including canopy-dwelling *Tillandsia circinnata* (Benzing, Renfrow, 1980), but the occurrence and nature of the process has been studied far more in lower plants and microbes. Uptake kinetics under varying conditions of supply should be examined to note how epiphytes are adapted at the subcellular level to cope with the temporal constraints on mineral procurement and the chemical peculiarities of their aerial habitats. Much could be learned from comparisons of the behavior of terrestrials with that of epiphytes suspected of having similar capacities. Do any plants that grow on bark match others indigenous to infertile acid or alkaline soils in their ability to alter rhizosphere pH and deploy special rhizodermal transfer cells for procurement of iron and other scarce ions (Romheld et al., 1982)? Do epiphytes mobilize nutrients by releasing organic chelating agents or can they take advantage of those generated by rhizosphere microbes, as do many terrestrials (Clarkson, Hanson, 1980)? Finally, perhaps the term "air plant" has validity. Even some eutrophic crops may gain a significant fraction of their nitrogen requirement by absorbing atmospheric ammonia under certain conditions (Faller, 1972; Porter et al., 1972; Farquhar et al., 1980). Ammonia compensation points may be lower among oligotrophs owing to their depressed nitrogen status. A greater uptake capacity plus a lower demand could make this gas a major nitrogen source for some epiphytes and epiphylls. Inputs from nitrogen fixers are discussed shortly.

Ectotrophic, and especially vesicular-arbuscular mycorrhizas (VAM), are commonplace in terrestrial habitats. Supposedly, few soil-dependent plants would grow as well or even survive without them (Malloch et al., 1980; Pirozynski, 1981; but see Lamont, 1982). Potentially similar mutualisms are known in at least a few epiphytes, including some ferns, aroids and orchids (Ruinen, 1953; Furman, 1959; Benzing, 1982). So far, no bark inhabitant has been shown to gain nutritional advantage from its mycobionts, nor were the observed fungi comparable to those involved with so many soil-rooted plants. Nevertheless, several observers claim or imply that fungal endophytes substantially influence a bark-dwelling orchid's phorophyte preference (Sanford, 1974; Jonsson, Nylund, 1979), the health of its supports (Ruinen, 1953), and the epiphyte's vegetative habit (Johansson, 1977). None of these intriguing proposals has received the attention it deserves.

Difficulties in maintaining hyphal outgrowths and promoting survival of mutualistic free-living microbes must be especially pronounced toward the arid end of the canopy continuum. Dispersal could be a problem in all cases since propagules of VAM-forming phycomycetes are too massive to insure adequate infection rates (D. Janos, personal communication). Conversely, those of the orchid rhizoctonias are much smaller, a fact that may allow this family's epiphytes to maintain fungal symbioses, at least as juveniles. Of course, there is no reason to expect mycorrhizal associations to be equally important in terrestrial and epiphytic communities. Atmospheric bromeliads with their strictly mechanical root system (and, in a few cases, none at all) prove that mycorrhizas are not always essential for life attached to bark.

Information on all these aspects of epiphyte nutrition and water balance will prove useful, but none considered in isolation will shed much light on a broader question: How do epiphytes affect biogeochemical cycles and energy relationships in supporting ecosystems? Insights on these phenomena will require an integrated knowledge of the epiphyte's problems of supply, its methods of procurement, mineral- and moisture-use patterns, and cost-benefit relationships, as discussed below.

6. INTERACTIONS: CANOPY DWELLERS AND SUPPORTS

One must examine the entire biota associated with epiphytes and epiphylls to appreciate how these organisms influence their ecosystems. Interactions fall into three more or less discrete subject areas: (1) the canopy dweller's choice of, and subsequent effects on, its support; (2) its services to and/or benefits from animals and nearby microbes; (3) its impact on the nutritional status, structure and performance of the community at large.

6.1. Support preferences. Generally, there is a lack of specificity between phorophytes and epiphytes. Trees harboring one kind of resident will usually accommodate many others, a point underscored by cultivated supports that serve alien epiphytes and their normal associates equally well (e.g., Yeaton, Gladstone, 1982). There are exceptional canopy dwellers, however, which are confined to a single species, but even here general cultural requirements rather than precise chemical preferences may explain the specificity. *Ophioglossum palmatum* in Florida grows exclusively on the palm *Sabal palmetto* (Mesler, 1975) where it and a few other ferns are almost alone in exploiting old, persistent leaf bases filled with moist humus. More elaborate causes for specificity have been proposed in some other cases. The presumption that a fungal endophyte can impose its host preference on certain orchid symbionts (Jonsson, Nylund, 1979) rests on the rather dubious rationale of an obligate, specific association between an epiphyte and a particular mycobiont--a point that receives little support in the literature on orchid mycorrhizas (Warcup, 1975; Hadley, 1982). At the opposite extreme, some trees remain totally devoid of vascular crown residents in any location. Considering the ill effects epiphytes may bring to an over-accommodating host, one wonders whether repulsion may not sometimes be promoted through natural selection.

Some supports certainly provide better anchorage than others. Bark texture, porosity and stability are all obvious determinants of host quality. Seeds cannot attach themselves to smooth canopy surfaces, nor can substrata which exfoliate too quickly harbor developing seedlings for very long. Bark pH, ion exchange properties, and other chemical phenomena have considerable influence on the distribution of nonvascular epiphytes (Pike, 1978; Becker, 1980), but their effects on vascular plants remain largely unexplored (except see Frei, Dodson, 1972). Dense-canopied trees intercept too much light and precipitation to provide suitable sites for most epiphytes. Finally, trees that generate nutritive leachates should be good hosts if all other requirements are met. As mentioned earlier, some woody plants are much "leakier" than others, but even the most parsimonious of these could assure more than enough sustenance for a resident oligotroph accustomed to subsistence on atmospheric inputs alone (Billings, 1904; Wherry, Buchanan, 1926).

6.2. Host decline. Many observers equate dense epiphyte loads with what is, at this point, best given the mildly accusatory title "host decline." Indeed, epiphytes are considered parasites in many local cultures and bear colloquial names attesting to that fact. Evidence includes reduced leaf displays, chlorosis, and inordinately large numbers of dead and dying branches and twigs in densely laden tree crowns. The most heavily involved axes often bear disproportionate shares of the infestation, leading several observers to suggest topical rather than systemic causes (Ruinen, 1953; Johansson, 1977). Fungi originating from orchids and ferns allegedly harm certain East Asian trees (Ruinen, 1953). Hyphal connections bridging epiphyte roots and stems and phorophyte vasculature can be extensive, and sometimes trees do generate reaction tissue. But no attempt has yet been made to demonstrate epiparasitism by direct measurement of nutrient exchange in any of these cases. Since many orchid fungi have parasitic or saprophytic qualities in free culture which phytobionts can suppress in their own tissues, epiphytosis (Ruinen, 1953) may represent a relatively virulent pathological condition for the support, one which offers little direct nutritional benefit to the epiphyte. Johansson (1977) believes that some epiphytic orchids may have become so successful as epiparasites that their leaf areas have been much diminished as a result (but see Benzing, 1979; Benzing, Ott, 1981).

Harm may come to phorophytes through simple shading. Cypress trees in the southeastern United States, so burdened with thick shrouds of the bromeliad Tillandsia usneoides that only the most robust branch tips extend into bright light, seem to keep barely ahead of the encroaching festoons (Billings, 1904). Epiphytes may injure their supports by other means as well. Phytotoxins capable of weakening phorophytes may be released by epiphytes, as they are by some tree-inhabiting lichens (Orús et al., 1981) but so far no one has offered evidence for this. Stranglers such as Ficus and Clusia normally become free-standing arborescent forms after their supports die. On a smaller scale, some epiphytic orchids appear to kill supporting branches by girdling (Cook, 1926). Dense mantles of epiphytes and associated debris may encourage other as yet unobserved pathogens to attack underlying phorophyte tissue.

6.3. Nutritional piracy. Canopy residents need not draw sustenance directly from phorophyte tissues to use their supports as nutrient sources. Precipitation and leachates falling through infested forest canopies are exposed to a variety of thallophytic and vascular nutrient scavengers before reaching the roots below. Some portion of these descending mineral ions is pirated en route, and in effect locked up in the epiphyte's relatively long-lived tissues for terms that may measure in years (Benzing, Renfrow, 1971; Benzing, Seemann, 1978; Benzing, Davidson, 1979) before again becoming available for redistribution to other plants. Leaves of many epiphytes persist for several years against an average of about a year for many tropical, wet-forest, broad-leaf trees. If the supporting ecosystem is sufficiently infertile, stresses on adjacent vegetation, particularly that with higher nutrient turn-over rates, should theoretically intensify as epiphyte biomass expands (Fig. 3). Moreover,

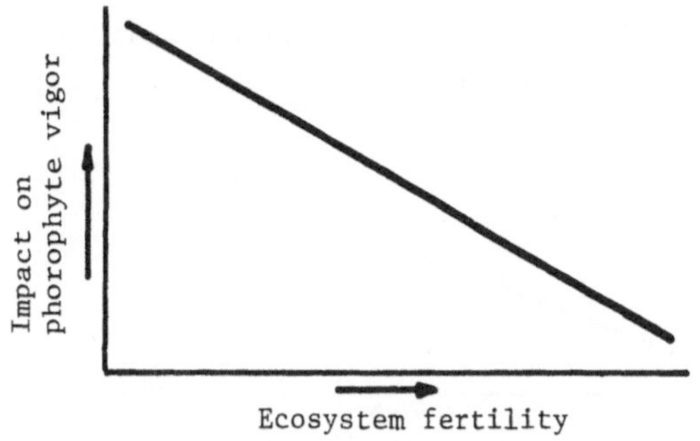

Figure 3. A model illustrating how the adverse effects of habitat infertility on phorophyte vigor may intensify as the pirating activities of associated epiphytes and epiphylls mount.

piracy should self-amplify when conditions on a site permit. Denied enough of some critical element, hosting canopies will brighten as their leaf area indices diminish, thus encouraging ever denser epiphyte loads. As the process continues, even larger percentages of circulating mineral ions can be preempted as more bark surface is exposed for colonization. Viewed from the epiphyte's perspective, this positive feedback is clearly advantageous, especially for the more oligotrophic forms whose vulnerability to shade outweighs that to mineral insufficiency (Benzing, Seemann, 1978). On more fertile sites, even the most sciophytic residents of a canopy will have little impact on the nutritional status of supports. The essential ions they might immobilize will be but small fractions of the larger pools present in those ecosystems.

7. INTERACTIONS: EPIPHYTES AND ANIMALS

Animals are drawn to epiphytes for many reasons. I will not discuss visitations performed to collect floral rewards, disperse seeds or act as herbivores. Of concern here are the varied and sometimes intricate symbioses which help canopy-dwelling plants cope with mineral insufficiency and achieve importance within supporting forest ecosystems. The prime contribution of canopy vegetation to the activity and welfare of animals may be simply in providing safe harbor in a hostile world of climatic extremes and abundant predators (Fig. 4). Tree crowns will have to be stripped of associated vegetation class by class to determine how each population or plant type located there affects other canopy inhabitants. So far, no systematic attempt has been made to accomplish this. Neither have most of the obvious mutualistic relationships involving epiphytes or epiphylls received sufficient study in situ to justify more than speculative comments.

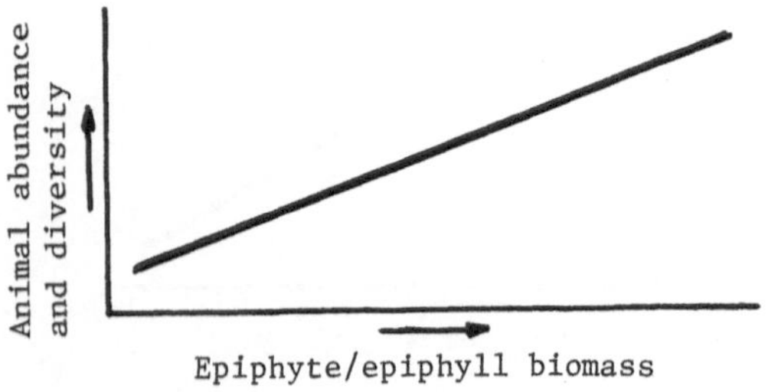

Figure 4. A model illustrating how epiphyll-epiphyte biomass promotes animal abundance and diversity in forest canopies.

Many epiphytic bromeliads provide especially inviting nesting and breeding sites for other organisms, some of which have few or no alternatives. A number of insects with aquatic juveniles oviposit only in bromeliad tanks; among the mosquito inhabitants are several carriers of serious human illness (Pittendrigh 1946). Bromeliad microcosms are probably among the most complex of those created by plants, but virtually unexplored are aerial impoundments associated with the aroid, fern and other debris accumulators mentioned earlier. Phytolema may be diverse even within a single group. Bromeliad tanks support different kinds of communities depending on shoot size, shape and exposure. Those subjected to high light develop food chains with an autotrophic component (Laessle, 1961). Less complex are the detritus-based systems encouraged by heavier litter fall and deep shade. In either case, older leaf bases harbor aquaphobic or amphibious rather than strictly aquatic taxa. Younger axils toward the rosette's center collect and hold fluids fit for exploitation by gill breathers and swimmers.

Checklists of tank residents have been recorded for many years (e.g., Picado, 1913; Laessle, 1961) but little has been noted about community structure and dynamics. The more thoroughly investigated Heliconia inflorescence bract microcosms become fairly elaborate through predictable successional change (Seifert, 1982). Tanks of bromeliads, among other epiphytes, persist much longer and thus provide opportunity for even greater biotic refinements. Since many tank formers, particularly the bromeliads, depend on their impoundments for most mineral nutrient inputs, they must be subject to considerable selective pressure to make those receptacles accommodating to appropriate life forms. Bromeliad leaves apparently do not release oxygen into adjacent fluids (Benzing et al., 1971), but a good argument can be made that leaf shape and pigmentation in many cases have evolved to make the tanks more enticing and secure to cryptic inhabitants (Benzing, Friedman, 1981).

Epiphytes engage in several ant-plant associations, two of which grant nutritional returns to botanical partners (Madison, 1979; Huxley, 1980). Arboreal carton nests create rooting media for so-called "nest garden" plants representing diverse families (e.g., Araceae, Bromeliaceae, Gesneriaceae, Orchidaceae, Piperaceae). How the more obligate members of this group move among germination sites is controversial (Ule, 1901; Madison, 1979; Dressler, 1981). A few seem to be cultivated by ants after the fact, if not actually sown by nest occupants. Enticements may include edible seed appendages (Madison, 1979). Diaspores of some nest-inhabiting Coryanthes contain what appear to be lipid deposits in the normally empty space between testa and rudimentary embryo (C. Dodson, personal communication). Physiological bases for nest preference have not been examined. Ant-nest communities merit much closer scrutiny with an emphasis on succession and causes of aggregation, physiological adaptations of plants to the acidic character and other chemical peculiarities of the ant carton, and for evidence that protection is gained through nest culture.

Quite intriguing is the near total restriction of ant-fed myrmecophytes to the canopy milieu. One or more epiphytic members of at least five families (Asclepiadaceae, Bromeliaceae, Nepenthaceae, Piperaceae and Rubiaceae) and Polypodiaceae harbor ant colonies in their shoots (Huxley, 1980; Thompson, 1981). Orchidaceae will probably be added to this list. No elaborate trophic rewards beyond those designed for pollinators are offered, as one would expect if ants are to provide maximum nutritional inputs. Unlike their canopy-dwelling counterparts, terrestrial myrmecophytes often feed, as well as house, ants in return for protection, a behavior encouraging a more closed trophic cycle. Phytobionts undoubtedly receive some minerals from their mutualists, but these inputs are probably minor compared to those secured from soil.

Juxtaposed to this odd asymmetry in ant exploitation is the rare occurrence of carnivory in epiphytes despite its potential to afford plants direct access to the same rich mineral ion source provided through trophic myrmecophytism. Where botanic carnivores occur, light and moisture are usually plentiful. Thompson (1981) and Givnish (1982) propose that drought and shade in canopy habitats render the cost of traps and the secretions required to operate them prohibitive. Ant nidification, in contrast, is less expensive. No special lures need be fabricated to attract the gravid ant-queen other than a domicile: a swollen, hollow stem, an expanded leaf base, or an invaginated lamina. These can then do double duty, continuing to fulfill the original task of nutrient and moisture storage and/or carbon gain. The same cannot be said for secretions or for deeply cyanic, pigmented, foliar traps which routinely exhibit relatively short lives and seem ill designed for photosynthesis, at least in shade light. Moreover, hollow pseudobulbs of epiphytic orchids, and stems and shoots of Rubiaceae and bromeliads, remain serviceable longer than the individual leaves of any carnivorous plant, allowing longer payback schedules.

While the logic offered to explain the differing animal-mediated nutrient procurement mechanisms used by terrestrial as opposed to canopy-dwelling plants has intuitive appeal, there is no direct evidence to support this hypothesis. Many wet-forest epiphytes probably experience greater mineral nutrient than moisture stress, and no small portion of them are heliophilic. Surveys of these organisms, and of others that maintain water-filled catchments in drier locations, could turn up additional botanic carnivores and other plants that, while they have similar dependencies and equally unconventional nutrient sources, fail to satisfy any existing definition of botanical carnivory. Actually, there are taxa that obtain mineral ions from animals by routes other than those associated with recognized carnivorous and myrmecophytic modes. Both plant carnivory and trophic myrmecophily are too restrictive as currently defined to encompass all the ways plants use animals for significant nutritional gain (Benzing, 1980).

8. EPIPHYTES/EPIPHYLLS: EFFECTS ON ECOSYSTEM FERTILITY AND PRODUCTIVITY

8.1. Epiphytes and epiphylls as storage compartments. The influence epiphytes have on the size of aggregate mineral nutrient pools and on nutrient distribution among compartments in supporting ecosystems must vary with humidity. On stable sites heavily dependent on atmospheric inputs to balance mineral losses, a massive epiphyte/epiphyll presence made possible by wet conditions may actually enhance the forest's nutrient-capturing and -retaining capacities without unduly complicating procurement for any community member (Fig. 5). Here the negative

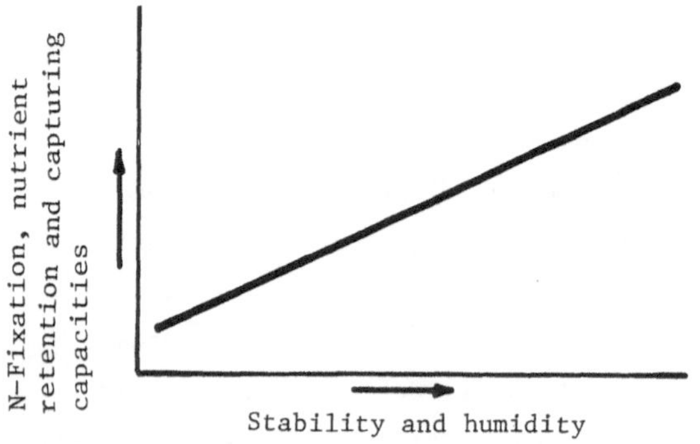

Figure 5. A model illustrating how nitrogen fixation and nutrient storage and retention capacities of a tropical forest ecosystem attributable to its epiphylls and epiphytes are promoted by humidity and stability.

connotation of "piracy" renders that term less appropriate in describing mineral ion fluxes between epiphytes and epiphylls, supports and other plants. Rather than continuously accumulating mineral-containing biomass, as expanding populations do, mature loads of bark- and leaf-associated vegetation probably approach nutritional steady states (Fig. 6); mineral ion influx from rainfall and leachates is apt to be balanced by efflux from spent reproductive tissue and decomposing vegetative organs. Trees equipped with mycorrhizal canopy roots, a phenomenon apparently restricted to very moist forests, may even tap epiphyte debris and associated fluids before they leave the canopy (Nadkarni, 1981). Finally, the mere existence of a substantial epiphyte/epiphyll mass expands nutrient storage capability in any ecosystem. Whether a well-developed canopy flora pays a photosynthetic return commensurate with the host foliage displaced and the resources coopted is an open question. At present, no values seem to be

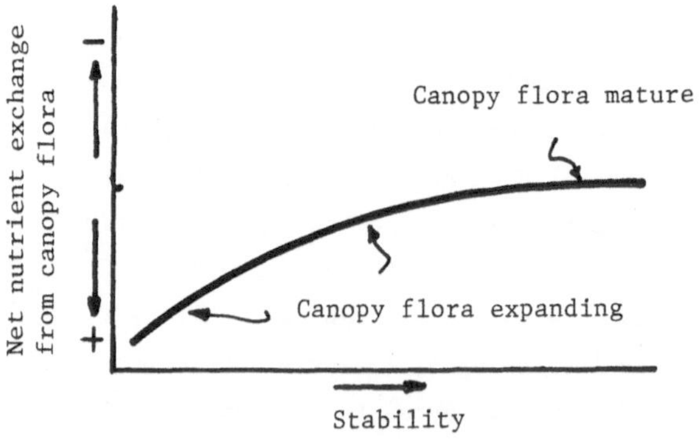

Figure 6. A model illustrating how disturbance influences the role epiphytes/epiphylls play in forest mineral-nutrient dynamics. Canopy floras that are typically expanding on less stable sites are usually co-opting some essential ions at the expense of other vegetation. Canopy floras with more mature profiles approach steady states, releasing as much nutritive matter as they acquire.

available on the magnitudes of epiphyll development in tropical forests. Only a few biomass studies provide data on epiphytes and all of these deal with moderately to very wet systems (Fittgau, Kinge, 1973; Edwards, Grubb, 1977; Golley et al., 1978; Tanner, 1980; Grubb, Edwards, 1982). Most report that epiphytes constitute only a few percent of the total aboveground biomass, a statistic which obscures the fact that they may contribute far more than that to the aggregate leaf surface area. Edwards, Grubb (1977) did calculate epiphyte weight at about 50% of that for tree leaf biomass in a New Guinea lower montane rain forest. Tanner (1977) observed values as great as 35% in Jamaica. No survey published to date provides assessments of epiphyte physiology or nutrient turnover.

8.2. Nitrogen fixation. Some evidence suggests that an extraordinary amount of nitrogenase activity can be associated with epiphytes in wet forests (Sengupta et al., 1981). Nitrogen reduction on leaf surfaces was shown to be especially common among mostly orchidaceous epiphytes. It may be that these plants are particularly accommodating to diazotrophs because of appropriate textures and extended leaf longevities. Epiphylls colonize leaf surfaces slowly but, should substrata persist long enough, fixers among them can build up to levels that create substantial nitrogen inputs (Barkman, 1958; Bentley, Carpenter, 1980). Should colonization of phyllospheres prove to be more actively epiphyte-assisted, an even more intriguing dimension will be added to forest canopy biology. Three terrestrials that routinely form mycorrhizas release sizable quantities of several metabolities from their roots (unlike three others that are nonmycorrhizal), presumably to enhance appropriate fungal infections (Schwab, 1982). Energy expenditures of this sort would also serve the epiphyte well if nitrogen returns were cost-effective, as they appear to be for certain crops. Bacterial isolates from leaf surfaces sprayed on rice shoots induced yields equal to or greater than those of controls provisioned with substantial urea amendments via roots (Sen Gupta et al., 1982).

At the very least, one would expect epiphytes to promote ecosystem fertility if only by encouraging nitrogen fixation through retention of moisture that would otherwise fall through uninfested tree crowns. This humidifying mechanism could vastly expand canopy hospitality for nitrogenase activity (Fig. 5). Moreover, canopy supplements may result more from autotrophic activity than are those elaborated by soil microbes. Particularly in the rhizosphere, fixation is performed largely by heterotrophs and is probably more energy- but less moisture-limited than that occurring on leaf and bark surfaces.

8.3. Effects of epiphytes/epiphylls on community productivity. Personal observations and those few data mentioned earlier (Fittkau, Klinge, 1973; Edwards, Grubb, 1977; Golley et al., 1978; Tanner, 1980; Grubb, Edwards, 1982) indicate that epiphytes/epiphylls generate much, even most, of the photosynthetic tissue present in many wet forest canopies. The consequences of this condition must be substantial because crown residents influence forest productivity in many ways that correlate with their numbers. Beyond contributing photosynthate to the community total, they pirate phorophyte nutrients, shade adjacent foliage and promote fertility (Fig. 5). An epiphyte/epiphyll presence also increases genetic diversity in tree crowns, perhaps to the extent of promoting a more effective partitioning and use of resources there. Under some conditions, forest canopies and their associates might produce larger and/or more competent aggregate biomasses than would be possible were epiphytes and epiphylls absent (Fig. 7). Very

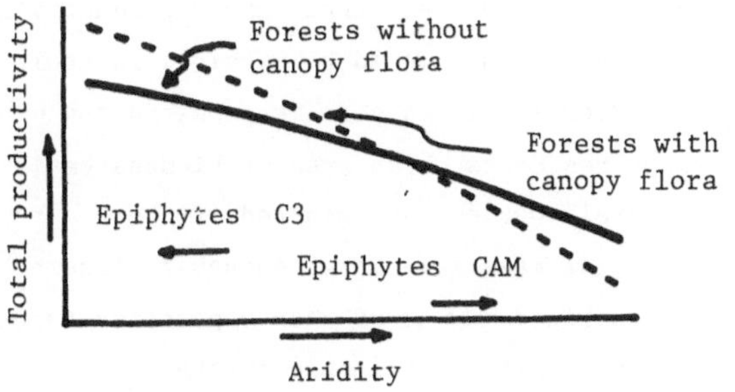

Figure 7. A model illustrating how, in a series of mature forests of varying wetness, the presence of well-developed epiphyte/epiphyll loads may alter ecosystem productivity relative to levels that would prevail in their absence.

likely, the types of canopy residents involved would determine the kind and extent of that impact.

Trees have access to more continuous moisture supplies than do most of the plants growing in their crowns, consequently the material resources they allot to foliage should, in theory, produce relatively high yields. Little is known about epiphylls, but epiphytes are not always vigorous producers. Adapted for drought, usually to a greater degree than their supports, these plants probably achieve higher water-use efficiencies, but their return on nutriental investments (particularly nitrogen and perhaps phosphorus) may not follow suit. Analyses have shown that, while C3 plants often contain only fractionally more foliar nitrogen than do CAM plants, their photosynthetic rates are far greater (Larcher, 1980). This would indicate that nitrogen-use efficiency for carbon gain is probably lower in the CAM forms. However, given their propensity for long-lived foliage, returns could be much greater over the long term. Epiphytes in humid canopies are likely to be C3 types whose production and patterns of mineral use are more comparable with those of their supports.

One could reasonably hypothesize that sizable epiphyll/epiphyte loads in stable, wet forests would substantially increase nutrient storage and retention and promote nitrogen fixation (Fig. 5) and furthermore, that canopy productivity could be greater than would be possible in their absence. Even if carbon gain is slower among the canopy residents on an area or biomass basis, that shortfall would be eliminated if the aggregate leaf mass were large enough to insure that the combined outputs of both vegetative types exceeded the maximum attainable by the trees alone (Fig. 7). Conversely, epiphylls and epiphytes situated in dry forests should have a more negative influence on the community's productivity, especially where fertility is low. Here, scarce mineral ions committed to support the modest carbon gains of CAM epiphytes and poikilohydrous thallophytes would, to some extent, routinely come at the expense of biomass with far greater photosynthetic capacity. Disturbance would have an impact across the entire range of forest types. Communities subjected to fairly frequent losses of nutrient capital could never achieve the mature nutritonal steady state (Fig. 6); their epiphyll/epiphyte loads would be constantly regenerating and accumulating mineral ions at the expense of supports. Broadly based inquiries into the participation of canopy residents in the nutritional dynamics of a variety of tropical forests should be carried out. Without them, knowledge of tropical forest structure and function will remain incomplete.

9. CONCLUDING REMARKS

There are many more questions that can be asked about epiphytes and epiphylls, but most are well beyond the usual purview of the physiological ecologist. Still, knowledge of adaptive physiology would aid in their interpretation considerably. Why, for instance, are the neotropical epiphytes more diverse than their Old World counterparts despite the greater extent and dispersion of paleotropical forests (Madison, 1977)? What phylogenetic constraints have prevented large, seemingly well-disposed taxa from adapting to canopy life (e.g., Asteraceae, Leguminosae)? Conversely, what combinations of features allow other families to be so successful in canopy habitats? Quite apart from their effects on light reception by host leaves, do epiphylls influence leaf longevity? Has their presence served as a selective force in shaping other parameters of leaf biology? Does vegetation anchored in the canopy contribute substantially to gap formation, and hence tropical forest diversity (Strong, 1977)? Far more important than any of these relatively narrow concerns, of course, is the influence of epiphytes and epiphylls on tropical forest structure and dynamics. Only a mechanistic approach will allow us to unravel the complex interactions that determine these relationships.

TABLE 1. Features associated with, and presumably adaptive for, epiphytism.

1. Holdfast, adventitious roots.
2. Small seeds, aerial dispersal, high vagility.
3. Zoophilous pollination.
4. Xeromorphy (including evergreenness), CAM, high water-use efficiency.
5. Iteroparity.
6. High nitrogen- and phosphorus-use efficiency for reproductive purposes.
7. Considerable vegetative reduction.
8. Devices which promote access to unconventional mineral sources (e.g. soil substitutes, ant-provided).

9. Devices for prolonging contact with canopy fluids (e.g. orchid velamen, bromeliad foliar indumentum).

ACKNOWLEDGMENTS

Helpful comments on parts of this manuscript were provided by E. Medina, F. Putz, A. Renfrow and M. Zimmerman.

REFERENCES

Aspinwall D and Paleg LG (1981) Proline accumulation: physiological aspects. In Paleg LG and Aspinwall D, eds. The physiology and biochemistry of drought resistance in plants. Academic Press.

Avadhani PN, Goh CJ, Rao AN and Arditti J (1982). Carbon fixation in orchids. In Arditti J, ed. Orchid biology: reviews and perspectives II. Ithaca, Cornell Univ. Press.

Barkman JJ (1958) Phytosociology and ecology of cryptogamic epiphytes. Van Gorcum & Co., N.V. Assem. Netherlands.

Barlow BA and Wiens D (1977) Host-parasite resemblance in Australian mistletoes: the case for cryptic mimicry. Evolution 31, 69-84.

Becker VE (1980) Nitrogen fixing lichens in forests of the southern Appalachian mountains of North Carolina. The Bryologist 83, 29-39.

Bentley BL (1977) Extrafloral nectaries and protection by pugnacious bodyguards. Ann. Rev. Syst. Ecol. 8, 407-427.

________ and Carpenter EJ (1980) Effects of desiccation and rehydration on nitrogen fixation by epiphylls in a tropical rainforest. Microb. Ecol. 6, 109-114.

Benzing DH (1978) The life history profile of Tillandsia circinnata (Bromeliaceae) and the rarity of extreme epiphytism among the angiosperms. Selbyana 2, 325-337.

________ (1979) Alternative interpretations for the evidence that certain orchids and bromeliads act as shoot parasites. Selbyana 5, 135-144.

________ (1980) The biology of the bromeliads. Eureka, Calif., Mad River Press.

________ (1982) Mycorrhizal infections of epiphytic orchids in southern Florida. Amer. Orchid Soc. Bull. 51, 618-622.

________ Bent A, Moscow D, Peterson G. and Renfrow A (1982) Functional correlates of deciduousness in Catasetum integerrimum (Orchidaceae). Selbyana 7, 1-9.

________ and Davidson E (1979) Oligotrophic Tillandsia circinnata Schlecht. (Bromeliaceae): an assessment of its patterns of mineral allocation and reproduction. Amer. J. Bot. 66, 386-397.

________, Derr J and Titus J (1971) Factors affecting the water chemistry of microcosms associated with the epiphytic bromeliad Aechmea bracteata. Am. Mid. Natur. 87, 60-70.

________ and Friedman WE (1981) Patterns of foliar pigmentation in Bromeliaceae and their adaptive significance. Selbyana 5, 224-240.

________, Friedman WE, Peterson G. and Renfrow A (1983) Shootlessness, velamentous roots, and the pre-eminence of Orchidaceae in the epiphytic biotope. Amer. J. Bot. 70, 121-133.

________, Henderson K, Kessel B and Sulak J (1976) The absorptive capacities of bromeliad trichomes. Amer. J. Bot. 63, 1009-1014.

________ and Ott DW (1981) Vegetative reduction in epiphytic Bromeliaceae and Orchidaceae: its origin and significance. Biotropica 13, 131-140.

________ and Pridgeon AM (1983) Foliar trichomes of Pleurothallidinae (Orchidaceae): functional significance. Amer. J. Bot., in press.

________ and Renfrow A (1971) The biology of the atmospheric bromeliad Tillandsia circinnata Schlecht. I. The nutrient status of populations in South Florida. Amer. J. Bot. 58, 867-873.

________ (1974) The nutritional status of Encyclia tampensis and Tillandsia circinnata on Taxodium ascendens and the availability of nutrients to epiphytes on this host in South Florida. Bull. Torrey Bot. Club 101, 191-197.

________ (1980) The nutritional dynamics of Tillandsia circinnata in southern Florida and the origin of the "air plant" strategy. Bot. Gaz. 141, 165-172.

Benzing DH and Seemann J (1978) Nutritional piracy and host decline: a new perspective on the epiphyte-host relationship. Selbyana 2, 133-148.

Billings FH (1904) A study of Tillandsia usneoides. Bot. Gaz. 38, 99-121.

Böttger M, Soll H and Gasché A (1980) Modification of the external pH by maize coleoptiles and velamen radicum of Vanilla planifolia Andr. Z. Pflanzenphysiol. 99, 89-93.

Chapin FS (1980) The mineral nutrition of wild plants. Ann. Rev. Ecol. Syst. 11, 233-260.

Clarkson DT and Hanson JB (1980) The mineral nutrition of higher plants. Ann. Rev. Plant Physiol. 31, 239-298.

Cook MT (1926) Epiphytic orchids, a serious pest on citrus trees. J. Dept. Agric. Puerto Rico 10, 5-9.

DeSanto AV, Alfani A and DeLuca P (1976) Water vapour uptake from the atmosphere by some Tillandsia species. Annals of Botany 40, 391-394.

Dressler RL (1981) The orchids. London (Eng.), Harvard University Press.

Faller VN (1972) Sulphur dioxide, hydrogen sulfide, nitrous gases and ammonia as sole sources of sulphur and nitrogen for higher plants. J. Plant Nut. Soil Sci. 131, 120-130.

Farquhar GD, Firth PM, Wetselaar R and Weir B (1980) On the gaseous exchange of ammonia between leaves and the environment: determination of the ammonia compensation point. Plant Physiol. 66, 710-714.

Fittkau EJ and Klinge H (1973) On biomass and trophic structure of the central Amazonian rain forest ecosystem. Biotropica 5, 2-14.

Forman R (1975) Canopy lichens with blue-green algae: a nitrogen source in a Colombian rain forest. Ecology 56, 1176-1184.
Frei, Sister JK and Dodson CH (1972) The chemical effect of certain bark substrates on the germination and early growth of epiphytic orchids. Bull. Torrey Bot. Club 99, 301-307.
Furman TE (1959) Structural connections between epiphytes and host plants. Toronto, Proc. 9th Int. Bot. Congr. 2, 127.
Givnish TJ (1982) On the adaptive significance of leaf height in forest herbs. Amer. Natur. 120, 353-381.
Golley FB, Richardson T and Clements RG (1978) Elemental concentrations in tropical forests and soils of northwestern Colombia. Biotropica 10, 144-151.
Grime JP (1977) Evidence for the existence of three primary strategies in plants and its relevance to ecological and evolutionary theory. Amer. Nat. 111, 1169-1194.
Grubb PJ and Edwards PJ (1982) Studies of mineral cycling in a montane rain forest in New Guinea. J. Ecol. 70, 623-648.
Hadley G (1982) Orchid mycorrhiza. In Arditti A, ed. Orchid biology: reviews and perspectives II. Ithaca, Cornell Univ. Press.
Huxley CR (1978) The ant-plants Myrmecodia and Hydnophytum (Rubiaceae) and the relationships between their morphology, ant occupants, physiology and ecology. New Phytol. 80, 231-268.
________ (1980) Symbiosis between ants and epiphytes. Biol. Rev. 55, 321-340.
Johansson DR (1977) Epiphytic orchids as parasites of their host trees. Amer. Orchid Soc. Bull. 46, 703-707.
Jonsson L and Nylund JE (1979) (Favolaschia dybowskyana (Singer) Singer (Aphyllophorales), a new orchid mycorrhizal fungus from tropical Africa. New Phytol. 83, 121-128.
Jordan C, Golley F, Jerry Hall and Jan Hall (1980) Nutrient scavenging of rainfall by the canopy of an Amazonian rain forest. Biotropica 12, 61-66.
Kellman M, Hudson J and Sanmugadas K (1982) Temporal variability in atmospheric nutrient influx to a tropical ecosystem. Biotropica 14, 1-9.
Kluge M and Ting IP (1978) Crassulacean acid metabolism. New York, Springer-Verlag.
Laessle AM (1961) A micro-limnological study of Jamaican bromeliads. Ecology 42, 499-517.
Lamont B (1982) Mechanisms for enhancing nutrient uptake in plants with particular reference to Mediterranean South Africa and western Australia. The Bot. Rev. 48, 597-689.
Lang GE, Reiners WA and Heier RK (1976) Potential alteration of precipitation chemistry by epiphytic lichens. Oecologia (Berl.) 25, 229-241.
Larcher W (1980) Physiological plant ecology, 2nd edn. New York, Springer-Verlag.
Madison M (1977) Vascular epiphytes: their systematic occurrence and salient features. Selbyana 2, 1-13.
________ (1979) Additional observations on ant-gardens in Amazons. Selbyana 5, 107-115.
Malloch DW, Pirozynski KA and Raven PH (1980) Ecological and evolutionary significance of mycorrhizal symbioses in vascular plants. Proc. Natl. Acad. Sci. U.S.A. 77, 2113-2118.
McColl JG (1970) Properties of some natural waters in a tropical wet forest of Costa Rica. BioScience 20, 1096-1100.
Medina E (1974) Dark CO_2 fixation, habitat preference and evolution within the Bromelaiceae. Evolution 28, 677-686.
________ and Troughton JH (1974) Dark CO_2 fixation and the isotope ratio in Bromeliaceae. Plant Science Letters 2, 357-362.
Mesler MR (1975) The gametophytes of Ophioglossum palmatum L. Amer. J. Bot. 62, 982-992.
Müller L, Starnecker G and Winkler S (1981) Zur Ökologie epiphytischer Farne in Sudbrasilien. I. Saugschuppen. Flora 171, 55-63.
Nadkarni N (1981) Canopy roots: convergent evolution in rainforest nutrient cycles. Science 214, 1023-1024.
Oliver WRB (1930) New Zealand epiphytes. J. Ecology 18, 1-50.
Orús MI, Estevez MP and Vicente C (1981) Manganese depletion in chloroplasts of Quercus rotundifolia during chemical simulation of lichen epiphytic states. Physiol. Plant. 52, 263-266.
Picado C (1913) Les Broméliacées epiphytes considérée comme milieu biologique. Bull. Scientif. France et Belgique 47, 215-360.
Pike LH (1978) The importance of epiphytic lichens in mineral cycling. The Bryologist 81, 247-257.
Pirozynski KA (1981) Interactions between fungi and plants through the ages. Can. J. Bot. 59, 1824-1827.
Pittendrigh CS (1946) Bromeliad malaria in Trinidad, W.W. I. Amer. J. Trop. Med. Hyg. 26, 47-66.
Porter LK, Viets FG and Hutchinson GL (1972) Air containing nitrogen-15 ammonia: foliar absorption by corn seedlings. Science 175, 759-761.
Richards PW (1952) The tropical rain forest: an ecological study. Cambridge (Eng.) Univ. Press.
Romheld V, Marschner H and Kramer D (1982) Responses to Fe deficiency in roots of "Fe-efficient" plant species. J. Plant Nutrition 5, 489-498.
Ruinen J (1953) Epiphytosis. A second view on epiphytism. Ann. Bogor. 1, 101-157.
Sala OE and Lauenroth WK (1982) Small rainfall events: an ecological role in semiarid regions. Oecologia (Berl.) 53, 301-304.
Sanford WW (1974) The ecology of orchids. In Withner CL ed. The orchids: scientific studies. New York, John Wiley & Sons.
Schimper, AFW (1888) Die epiphytische vegetation Amerikas. Jena (Gustav Fischer). Bot. Mitt. Tropen. II, 162 pp.
Schlesinger WH and Chabot BF (1977) The use of water and minerals by evergreen and deciduous shrubs in Okefenokee Swamp. Bot. Gaz. 138, 490-497.

Schwab SM (1982) Quantitative and qualitative comparison of root exudates of mycorrhizal and nonmycorrhizal plant species. In Abstracts: Botanical Society of America, Publication #162. Bloomington, Indiana.

Seifert R (1982) Neotropical Heliconia insect communities. Quart. Rev. Biol. 57, 1-28.

Sengupta B, Nandi AS, Samanta RK, Pal D, Sengupta DN and Sen SP (1981) Nitrogen fixation in the phyllosphere of tropical plants: occurrence of phyllosphere nitrogen-fixing microorganisms in eastern India and their utility for the growth and nitrogen nutrition of host plants. Ann. Bot. 48, 705-716.

Sen Gupta B, Nandi AS and Sen SP (1982) Utility of phyllosphere N_2-fixing micro-organisms in the improvement of crop growth. Plant and Soil 68, 55-67.

Slaytor MB (1977) The distribution and chemistry of alkaloids in the Orchidaceae. In Arditti J, ed. Orchid biology--reviews and perspectives I. Ithaca, Cornell Univ. Press.

Strong DR (1977) Epiphyte loads, tree falls, and perennial forest disruption: a mechanism for maintaining higher tree species richness in the tropics without animals. J. Biogeography 4, 215-218.

Stuart TS (1969) The revival of respiration and photosynthesis in dried leaves of Polypodium polypodioides. Planta 83, 185-206.

Sugden AM and Robins RJ (1979) Aspects of the ecology of vascular epiphytes in Colombian cloud forests, I. The distribution of the epiphytic flora. Biotropica 11, 173-188.

Tanner, EVJ (1977) Four montane rain forests of Jamaica: a quantitative chracterization of the floristics, the soils and the foliar mineral levels, and a discussion of the interrelations. J. Ecol. 65: 883- 918.

________ (1980) Studies on the biomass and productivity in a series of montane rain forests in Jamaica. J. Ecol. 68, 573-588.

Thompson JN (1981) Reversed animal-plant interactions: the evolution of insectivorous and ant-fed plants. Biol. J. Linn. Soc. 16, 147-155.

Tomlinson PB (1969) Anatomy of the monocotyledons: III. Commelinales-Zingiberales. London, Oxford Univ. Press.

Tukey HB Jr. (1970) Leaching of metabolites from foliage and its implication in the tropical rain forest. In Odum HT ed. A tropical rain forest, Chap. H, p. 155-161. Washington, D.C., U.S. Atomic Energy Comm.

Ule E (1901) Ameisengarten in Amazonasgebiet. Bot. Jahrb. Syst. 30, 45-52.

Warcup JH (1975) Factors affecting symbiotic germination of orchid seed. In Sanders FE, Mosse B and Tinker PB eds. Endomycorrhizas. New York, Academic Press.

Went FW (1940) Soziologie der Epiphyten eines tropischen Urwaldes. Ann. Jard. Bot. Buitenzorg 50, 1-98.

Wherry ET and Buchanan R (1926) Composition of the ash of Spanish moss. Ecology 7, 303-306.

Wyn Jones RG and Storey R (1981) Betaines. In Paleg LG and Aspinwall D eds. The physiology and biochemistry of drought resistance in plants. Academic Press.

Yeaton RI and Gladstone DE (1982) The pattern of colonization of epiphytes on calabash trees (Crescentia alata HBK.) in Guanacaste Province, Costa Rica. Biotropica 14, 137-140.

METHODS FOR VESICULAR-ARBUSCULAR MYCORRHIZA RESEARCH IN THE LOWLAND WET TROPICS

D.P. JANOS (Department of Biology, University of Miami, Coral Gables, Florida 33124 U.S.A.)

1. INTRODUCTION

Mycorrhizal associations influence plant growth (Janos, 1980a), nutrient cycling (Janos, 1983), succession (Janos, 1980b), and community composition (Janos, 1983) in the lowland wet tropics. Mycorrhizae have pervasive importance primarily because they improve the uptake of phosphorus (Bowen, 1980), a plant macronutrient which is only scantily available in most wet-tropical soils (Sanchez, Buol, 1975). Vesicular-arbuscular (VA) mycorrhizal fungi can be lost from tropical soils, however, as a consequence of land use (Janos, 1975, 1980b); practical techniques for their large-scale reintroduction are not available. Indigenous tropical ectomycorrhizal fungi regularly associate with the economically significant family Dipterocarpaceae (Janos, 1983), but the physiological ecology of these associations is virtually uninvestigated. Although research on tropical mycorrhizae is hampered by a lack of trained investigators, more is needed and is merited (Mikola, 1980).

Many of the methods and principles of mycorrhiza research are presented in a recent, general text (Schenck, 1982). In this paper I describe difficulties in mycorrhiza research peculiar to the lowland wet tropics and summarize methods which I have found useful there. I discuss only VA mycorrhizae because ectomycorrhizal species are relatively uncommon except in low-diversity stands and in dipterocarp forests (Janos, 1983).

2. ROOT SAMPLING

Information about the belowground ecosystem in wet tropical forests is sparse. Root excavation is hard dirty work complicated by the interweavings of roots in dense superficial layers. Nevertheless, speculation abounds about "self-feeding" plants that recapture nutrients from their own stemflow, fruits, and litter, a phenomenon that requires feeder root distributions which have not been demonstrated. Questions of root turnover as a proportion of net annual primary production, rates of root herbivory and parasitism, root chemical defenses, and feeder root distribution have yet to be fully answered.

2.1. Root biomass, distribution, and diversity

The feeder root systems of tropical canopy trees are notably superficial, although the vertical distributions of major roots can be comparable to those of temperate-zone trees (Jenik, 1978). Tropical feeder root systems resemble broad, thin plates which can have radii several times the horizontal projections of branch crowns. Even the stilt-roots of palms resurface to form feeder root networks after deeply penetrating the soil near the bole. Jenik (1978) has described the architecture of tropical root systems.

2.1.1. Root abundance and production. Jenik (1971) outlined a procedure for determining root biomass in soil pits. Blocks of soil of a set volume are cut from the sides of pits at selected depths, and roots are soaked from the blocks for collection on screens. Root extraction from clay soils can sometimes be facilitated by adding sodium hydroxide to water in which the soil blocks are soaked, because sodium tends to disaggregate soil colloids. Several authors have presented root size-class biomass distributions as a function of soil depth (e.g., Jenik, 1971; Klinge, 1973a, 1973b, Huttel, 1975; Stark, Spratt, 1977). These confirm that the major proportion of feeder roots occurs within the top 20 cm of lowland wet tropical soil profiles. Root growth into litter and into soil pits has been used to estimate annual root biomass increment as 33% of above-ground wood accumulation for a forest on an Oxisol at San Carlos, Venezuela (Jordan, Escalante, 1980). These authors noted, however, that this estimate does not incorporate root sloughing during the period allowed for growth, so total root production may have been greater than estimated. Higher concentrations of the same secondary compounds in roots than in shoots (McKey, 1979) imply that significant root mortality may occur because of root herbivory and parasitism.

2.1.2. Feeder root distribution and diversity. Horizontal dispersion patterns of feeder roots influence the sources of mineral nutrients that are exploited by individuals, but data on the root distributions of single trees in tropical forests are scarce (see Jenik, 1971). These patterns can be determined by exposing all lateral roots around the base of a trunk, and then following each by excavating it with small hand tools (e.g., Hannah, 1972). Alternatively, blocks of soil can be collected at regular intervals along several radii from the bole of an individual distant from conspecifics. Roots are washed from the blocks, sorted to species, and those of the selected individual are weighed or their volume is measured by water displacement in a graduated cylinder. The roots of many species can be recognized by gross morphological characteristics such as color of the inner bark, branching pattern, and thickness of ultimate rootlets. They are especially easy to distinguish if they exude latex when cut, as do some moraceous species. Profuse feeder roots of some species are found associated with woody litter although relatively sparse elsewhere; their production may be stimulated by microorganisms (St. John, Machado, 1978). Most mature individuals of canopy species are probably unable to recover nutrients from their own stemflow, because roots branch at acute angles forward. Consequently, few feeder roots remain near the boles of old individuals unless adventitious laterals are produced on trunks and main root axes. The feeder roots of forest trees usually overlap those of other individuals. The number of morphotypes of feeder roots in 86 one-liter cubes cut from the soil surface at random locations in lowland forest at two sites in Corcovado National Park in Costa Rica, ranged from two to ten with a mode of six (Fig. 1).

2.2. Sampling for mycorrhizal infection

2.2.1. Characterization of habitats. Two approaches may be taken to assessing the frequency of VA mycorrhizal infection within a habitat. The average number of mycorrhizal feeder roots can be expressed as a proportion of the total number of feeder roots without regard to species. This is determined for randomly located one liter cubes of soil from successive ten centimeter depths. Sufficient samples should be taken to stabilize a plot of average frequency of infection against cumulative number of samples. Although infection decreases markedly with increased depth in soil (St. John, 1980a), samples should be taken at all depths at which feeder roots occur. Because VA mycorrhizal fungi cannot grow unless associated with hosts (Janos, 1980b, 1983), this measure gives an estimate of their abundance within the habitat that reflects the relative root densities and infection levels of host-species. Alternatively, roots of mature individuals of selected species can be excavated from the base of the bole until corticated rootlets are encountered, and the presence of infection can be recorded. Species should be chosen for excavation on the basis of their abundance in order to characterize a major proportion of the stand (e.g., St. John, 1980a). When weighted by species abundance, such data reflect the importance of mycorrhizae to mature individuals and to nutrient cycling within a stand; unless information on the biomass of feeder roots produced by each species is also gathered, this approach is less likely than the first to reflect the abundance of mycorrhizal fungi within the habitat.

2.2.2. Species records. Fewer than one-thousand tropical tree species have been surveyed for mycorrhizae (Janos, 1983). Although finding mycorrhizae establishes that a species can be mycorrhizal, failure to find mycorrhizae requires that more samples be examined before lack of infection can be considered a regular feature of a species. Species of the Lecythidaceae, Sapotaceae and Proteaceae should be carefully examined in this regard, in addition to members of families commonly regarded as non-mycorrhizal (Janos, 1983). It is easiest to sample saplings that can be entirely uprooted, but they should be sampled in tree-fall gaps or open habitats where they receive ample sunlight. Under the conditions of low intensity light that prevail in forest understory, little infection may be supported (see Gerdemann, 1968). Seedlings and saplings that are growing rapidly are likely to be more dependent on mycorrhizae and to sustain more

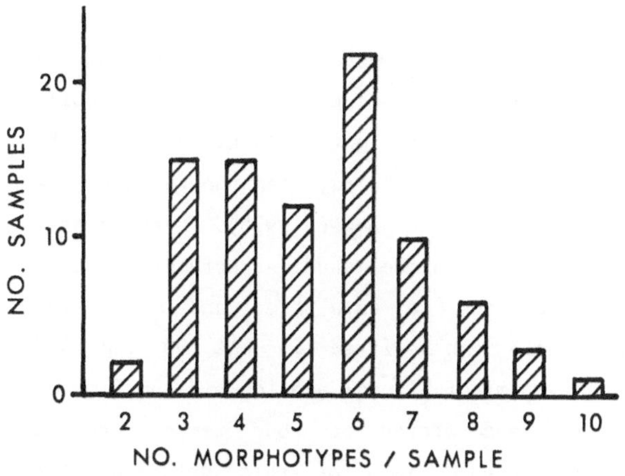

FIGURE 1. Distribution of number of morphotypes of feeder roots in 86 one-liter surface soil cubes from two forested sites in Corcovado National Park, Costa Rica. Distributions for both sites are combined because they do not differ significantly (Kolmogorov-Smirnov two-tailed test; $P > 0.1$).

infection than mature individuals with diminished mineral requirements (Baylis, 1962). Corticated roots should be sampled from the entire root system of an uprooted individual to maximize the chance of detecting patchy infection.

3. INFECTION ASSESSMENT

Samples of young, corticated feeder roots can be cleared and stained to prepare them for examination for mycorrhizal infection. Infection is restricted to root cortices, and is usually prevalent in ultimate-order rootlets, although trees growing under flooded conditions can show greatest infection near main axes of roots (see Keeley, 1980). It is easiest to handle short, narrow, ultimate rootlets attached to penultimate or even lower-order roots for staining. To get crude but quick impressions before staining of whether or not mycorrhizae are present, portions of samples of gently-washed, fresh roots can be examined with a compound microscope for typical VA mycorrhizal fungus hyphae (Fig. 2a) on the root surface.

Such examinations are not definitive, because mycorrhizae can be present when few external hyphae are visible; much external mycelium is lost when roots are extracted from soil. Moreover, abundant external hyphae can be entangled with root hairs, even though roots are uninfected (Bevege, Bowen, 1975). Nevertheless, for non-quantitative studies, screening roots before staining can save time and materials. Root samples can be fixed in FAA (90 ml of 50% ethanol, 5 ml of glacial acetic acid and 5 ml of formalin) and stored at room temperature in tightly-capped vials of this solution for years.

3.1. Root staining

One procedure is widely used with slight modifications for visualizing VA mycorrhizae (Phillips, Hayman, 1970; Kormanik, et al., 1980; Kormanik, McGraw, 1982). It involves i) clearing roots by heating in 10% potassium hydroxide, followed by a water wash ii) bleaching in freshly-mixed, alkaline hydrogen peroxide (3 ml of household ammonia in 30 ml of 10% hydrogen peroxide), followed by a thorough water wash, iii) acidification by soaking for 3 to 4 min in 1% hydrochloric acid, iv) staining by simmering in 0.05% trypan blue in lactophenol (300 g phenol, 250 ml lactic acid, 250 ml glycerin and 300 ml water), and v) destaining in clear lactophenol. Care should be used in heating the strong potassium hydroxide solution which is caustic, and will erode aluminum very quickly. Never use an aluminum pressure-cooker to clear roots with this solution; a stainless-steel pressure-cooker or an autoclave can be used to speed clearing. Unfortunately, the roots of some tropical tree species have a cortical layer of tannin-filled cells that requires such prolonged treatment that the roots disintegrate. Because one must stop short of complete clearing with these roots, I prefer to stain with trypan-blue, coloring fungal structures dark blue instead of the pink to red given by acid fuchsin used by some workers. Use of trypan blue without phenol, however, has not been widely tested. Lactophenol must be used with adequate ventilation, because its fumes are highly toxic. Some investigators have eliminated phenol from the staining solution which reduces its toxicity and allows it to be replenished and reused (Kormanik, et al., 1980). Tropical tree roots, however, often are coarse, and lactophenol may give added clarity necessary for some studies. Sometimes it helps to split coarse roots longitudinally, but sectioning roots is disadvantageous because VA mycorrhizal infection is often sparsely distributed throughout the root cortex.

3.2. Scoring infection

Two general methods are taken to scoring VA mycorrhizal infection in stained root samples (Kormanik, McGraw, 1982); the method of choice depends on the intended use of the data. The approaches differ in accuracy and in the amount of labor involved: non-systematic scanning is based on relatively large samples but only categorizes infection, systematic sampling gives accurate quantitative estimates of the proportion of root length colonized in small samples. Although chemical, colorimetric and auto-flourescent techniques also have been applied, their success has been limited, and they lack general applicability.

3.2.1. Quantification. Non-systematic scanning of feeder roots in petri dishes with a dissecting microscope can be almost as precise as more laborious systematic sampling (Giovannetti, Mosse, 1980), and facilitates rapid survey of large portions of root systems. For each dish of roots spread in a single layer, the proportion of susceptible roots that are mycorrhizal is estimated in broad categories (e.g., 0-5%, 6-25%, 26-50%, 51-75%, 76-100%) as is the intensity of infection (i.e., small colonizations, widely scattered; large colonizations, uniformly distributed through the infected portions of roots; coalescing colonizations). Occasionally, the results of non-systematic scanning should be compared to those of the systematic gridline intersect method which is the most precise (Giovannetti, Mosse, 1980). In the latter method, a square grid is marked on clear acetate and placed under the sample dish; the number of root intersections with gridlines and the infection status at each intersection are recorded for randomly arranged, non-overlapping root pieces. At least 100 intersections should be tallied for estimates of the percentage of root length colonized and the total length of roots in the sample (see Newman, 1966; Giovannetti, Mosse, 1980).

Problems arise in applying these methods to tropical species with many peg-like ultimate rootlets borne on penultimate, and sometimes lower-order roots that are corticated. Unless such roots are cut into small pieces, it is difficult to insure their random arrangement over the grid. Results of the non-systematic procedure are not greatly influenced by this problem, but it is easier to estimate infection categories when roots are in pieces. If roots are not cut for the non-systematic procedure, the reduced field of a dissecting microscope at high magnification can be used for sub-sampling. Infection categories are estimated for haphazardly-selected, non-overlapping microscope fields stratified over several wedge-shaped sectors of a dish. The modal category can be used to represent the whole dish.

In using either procedure described above to quantify infection, it is important to keep in mind that it estimates percent, not total, infection. A plant with a low percent infection of an extensive root system can have more mycorrhizae than one with a small, highly-infected root system (Gerdemann, 1968). The estimate of infected root length given by the systematic procedure may be the most closely correlated with effective mutualistic function. Transport of materials to hosts seems to be rate-limited by the number of hyphal connections to their roots (see Pearson, Tinker, 1975). Great lengths of infection require numerous infection points, because infection does not spread far from its point of entry to the root. Mutualistic function is also influenced by the spread of hyphae through soil, however, and if this is not highly correlated with infected root length, the latter is a poor estimator of mycorrhizal function. For woody species in which accumulated biomass reflects a long history of growth, infection at any one time may be poorly correlated with overall growth. For these reasons, quantifying infection may be most useful in studies of factors that influence infection itself; inference of host response should be cautious.

3.2.2. Recognizing VA mycorrhizal infection. Many different fungi are often visible on root surfaces; VA mycorrhizal fungi must be clearly distinguished from them in scoring infection. Fortunately, the hyphae of VA mycorrhizal fungi are different from those of all other fungi. VA mycorrhizal fungus hyphae are dimorphic: coarse, thick-walled, predominantly aseptate hyphae with unilateral, angular projections form a persistent, major portion of VA mycorrhizal fungus mycelium; emanating from their angular projections these hyphae bear numerous, fine, thin-walled, highly-branched, ephemeral hyphae which are septate at maturity (Fig. 2a).

The anatomical structures from which VA mycorrhizae take their name, vesicles and arbuscules, distinguish infections. After penetrating a root, hyphae grow inter- or intracellularly in the cortex, and in some hosts may form complex intracellular hyphal coils in the outer cortex. Arbuscules, thought to be the major sites of interchange of materials between host and fungus, are formed within cells in the inner cortex on lateral hyphal branches from intercellular hyphae or from hyphae in adjacent cells. These trunk hyphae branch dichotomously to produce ultimate branches barely resolvable by light microscopy (Fig. 2b) which nevertheless are surrounded by and external to the host plasmalemma. With age, the arbuscule disintegrates into a dense, irregular mass of hyphal wall, interfacial material, and entrapped host cytoplasm. Arbuscules are ephemeral, the whole process of formation and degeneration requiring about two weeks. Vesicles (Fig. 2c), filled with lipid droplets at maturity, usually complete the infection cycle when they are formed in the outer cortex. They are inter- or intracellular, intercalary or terminal, thick-walled storage organs. Sometimes they are not preceded by many arbuscules. I have rarely found arbuscules in the roots of many species of tropical trees. Therefore, I have used internal vesicles and associated external typical hyphae to distinguish VA mycorrhizae. It is important that the criteria used to define infection be stated when reporting infection percentages.

3.3. Assessing other feeder root attributes

Knowledge of the dependence of plant species on VA mycorrhizae for mineral uptake, growth and survival (Janos, 1980a) is important for understanding the role of mycorrhizae in succession and competition. Baylis (1975) suggested that dependence on mycorrhizae could be inferred from feeder root morphology. He hypothesized that great dependence on mycorrhizae is correlated with large diameter of ultimate rootlets, and is inversely correlated with the frequency, abundance, and length of root hairs for which VA mycorrhizal fungus hyphae functionally substitute. Infection was significantly related to such aspects of root morphology for 89 tropical plant species (St. John, 1980b), confirming Baylis' hypothesis if it is assumed that lack of infection of a mature individual reflects independence of mycorrhizae. Additional tests of this hypothesis are needed, because if widely applicable it will greatly facilitate determination of plant community composition with respect to dependence on mycorrhizae. At present, dependence must be assessed by time-consuming growth experiments.

4. IDENTIFYING VA MYCORRHIZAL FUNGI

VA mycorrhizal fungus species can only be identified from spores or sporocarps. It may prove possible to identify infections produced by known species, however, because _Gigaspora_ species tend to produce only arbuscular infections, _Acaulospora_ species often produce lobed vesicles, and other anatomical markers in specific hosts may be found. Nevertheless, in order to find such markers, fungus species must be isolated and individuals of a single host species inoculated with each. Because VA mycorrhizal fungi cannot be grown in axenic culture, their isolation is accomplished by collection of spores or sporocarps of single species for inoculation of hosts in sterilized soil.

4.1. Collection

VA mycorrhizal fungi usually are collected by searching for sporocarps, or extracting spores from soil. Sporocarps, fruiting bodies produced at the soil surface or belowground, which commonly range from 1 mm to about 1.5 cm in diameter, can sometimes be found by gently raking away litter and examining the soil surface. In my experience, however, they are rare in lowland tropical wet forest. Moreover, at maturity they can be almost indistinguishable from small clumps of mud. Sieving soil to extract spores and small sporocarps is a more productive way of collecting the fungi.

I have found the following procedure of wet-sieving and decanting (Gerdemann, Nicolson, 1963) to give satisfactory results for several Central American tropical forest soils: i) after scraping litter from the soil surface to minimize organic matter, collect 500 ml of soil (if soils of different bulk-densities are to be compared, dry-mass of an equivalent sample should be determined and spore abundance expressed per gram dry-mass), ii) suspend the sample in approximately 3 l of water by agitating it with a swirling motion after breaking up all clumps of soil by hand, then wait a few seconds for the motion to slow and heavier particles to settle, iii) pour the suspension through a 500 µm sieve, collecting the liquid that passes through; wash the material on the sieve with a gentle stream of water to dislodge spores attached to roots, continuing to collect liquid passing through; examine the material on this sieve for sporocarps, iv) resuspend the material that passed through the sieve, and after allowing a few seconds for settling, pour the suspension through a 250 µm sieve, washing the sieve to clean the retained material of clay particles, and again collecting the liquid passing through; carefully rinse the material from the sieve into a petri dish using a squeeze-bottle, v) repeat the preceding step with sieves of 100 µm and 50 µm if desired. Samples can be separated into more fractions by using sieves of additional pore sizes. Fractions below 100 µm are very difficult to examine because of the abundance of clay particles retained on small sieves.

The material rinsed into petri dishes should be in clear water (if the water is murky the sieve was not washed thoroughly), and can be examined with a dissecting microscope for spores. It consists of bits of vegetable matter, seeds, arthropod remains, live invertebrates, and nematode cysts, and should be sparsely distributed across the bottom of the plate. It is useful to place a grid under the plate to aid examination of the entire dish. Any spores found can be lifted from the plate with a flattened dissecting needle, jewelers' fine forceps, or a small-bore pipette. Spores are placed in a watch-glass or drop of water on a microscope slide for further examination. Although several centrifugation and flotation techniques (see Daniels, Skipper, 1982) are becoming widely used for spore extraction by temperate-zone researchers, I feel the extra handling involves too much chance of loss of the

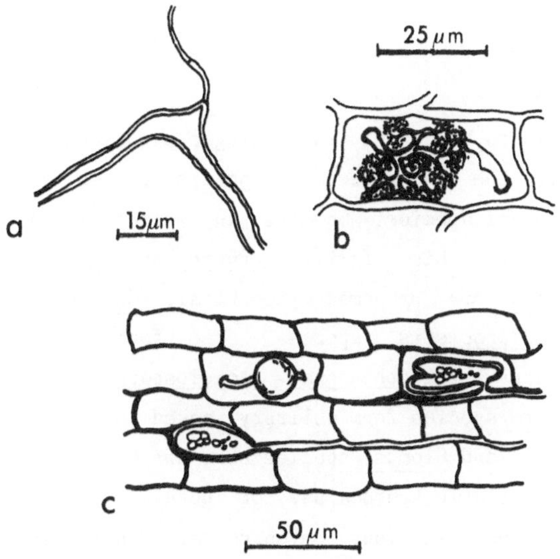

FIGURE 2. a) Typical dimorphic vesicular-arbuscular mycorrhizal fungus hyphae comprising a fine, thin-walled, septate hypha borne on a unilateral angular projection of a coarse, thick-walled, aseptate hypha. b) Arbuscule in the early stages of disintegration within a root cortical cell. c) Vesicles of different shapes between and within cortical cells. Scales are approximate.

very few spores in typical tropical forest soil samples (Janos, 1983).

4.2. Taxonomy

Four genera of fungi in a single zygomycetous family, the Endogonaceae, contain proven VA mycorrhizal species. Species in two additional genera of this family, Entrophospora and Glaziella, are presumed mycorrhizal. Glaziella and Sclerocystis spores are formed in sporocarps, Entrophospora, Acaulospora, and Gigaspora species form solitary, naked spores in the soil, and Glomus species bear spores singly in the soil, in sporocarps, or both. I have encountered all four mycorrhizal genera in Mesoamerican lowland wet forests, but Sclerocystis and Acaulospora species are more common than Glomus species, and Gigaspora species are least common.

Sclerocystis sporocarps are often white when immature, then turn shades of brown or black at maturity. They are usually 1 mm or less in diameter, contain fewer than 100 spores, and may be aggregated in crusts which in some species are covered by a layer of hyphae. The crusts can be formed at the soil surface or slightly above it on a support. The sporocarps, as their generic name implies, are hard, but may be bisected with a razor-blade to reveal a single layer of spores that are radially arranged around a central plexus of hyphae (Fig. 3a).

Acaulospora spores are relatively common in the 250 - 500 µm fraction of particles from wet forest soils. This genus is characterized by lateral formation of each spore on a large-diameter hypha with an inflated end that collapses by the time the spore is mature (Fig. 3b). Unfortunately, spores are seldom encountered still attached to the hyphae that bore them, and are difficult to distinguish among sievings until one has a "search image" for several species gained by examining herbarium specimens.

Glomus and Gigaspora spores are easy to recognize because they almost always retain their attached hypha unless very old and decaying. The way in which this hypha attaches to the spore defines these genera. Glomus spores have straight or funnel-form hyphal attachments (Fig. 3c), Gigaspora hyphal attachments are bulbous swellings (Fig. 3d). Spores of some Glomus species are borne in non-organized sporocarps, sometimes over 1 cm in diameter, which may contain hundreds of spores. With 43 described species (Trappe, 1982), the genus Glomus includes over twice as many species as Gigaspora, and four times as many as the other two VA mycorrhizal genera. Glomus species are difficult to identify because of their simple spore morphologies.

The two genera presumed mycorrhizal each contain a single species at present. Entrophospora has not been reported from the tropics, and I will not describe it. Glaziella aurantiaca (Berk. & Curtis) Cooke has been reported from Carribbean, South Pacific and East Indian beaches. Its sporocarps are large (up to 5 cm dia.), hollow, orange to red spheres with spores scattered in their walls.

Species of all genera can sometimes be identified from a single spore in prime condition, but usually several spores are needed to determine the ranges of diagnostic characteristics. Trappe and Schenck (1982) detailed methodology for identifying species, and described the general features that have taxonomic value. Trappe (1982) presented a synoptic key to the 78 described species in the known mycorrhizal genera.

4.3. Parasitism, predation and dispersal

Spores of VA mycorrhizal fungi are sparse in lowland wet tropical soils. Sporulation may not be induced by environmental conditions that favor the continual presence of young roots which are susceptible to infection, or spores may be destroyed rapidly or removed by parasites, predators or dispersal agents. Fungal hyperparasites are known to attack spores (see Ross, Daniels, 1982), and amoebae and invertebrates may prey upon spores. I found that crickets and cockroaches quickly consume Glomus sporocarps in Panama. Spores are filled with lipid droplets, and when produced in quantity represent a potentially important food resource. Hyphal and vesicle walls contain tannins (Nemec, 1982), however, which may limit predation. Data are accumulating that suggest small mammals may be the principle dispersal agents of at least the sporocarpic species of VA mycorrhizal fungi (Janos, 1983). Soil-borne spores can be dispersed by any agents moving volumes of soil, but in the wet tropics it is likely that they are poorly dispersed. Lack of dispersal is compensated by absence of host-specificity among VA mycorrhizal fungi (Janos, 1980a).

5. EVALUATING EFFECTS ON HOST GROWTH

For many years after their discovery, VA mycorrhizal fungi were thought to be parasites, notwithstanding the lack of resistance to VA mycorrhizal infection of almost all vascular plants. Over the last twenty years they repeatedly have been demonstrated to augment mineral uptake from infertile soils and to increase plant growth. Hence, no question remains of their mutualistic nature. It now seems less important to show that plants grow better with mycorrhizae in sterilized, infertile soils than to assess the degree of dependence of different host species on mycorrhizae, and to determine the effects of single fungus species on host growth.

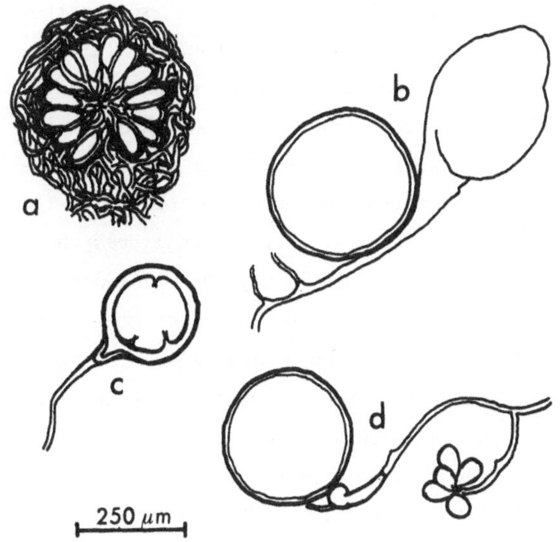

FIGURE 3. Spore features characteristic of vesicular-arbuscular mycorrhizal fungus genera. a) Radially-arranged chlamydospores surrounding a central mass of interwoven hyphae in a bisected Sclerocystis sporocarp. b) Acaulospora azygospore laterally borne on the funnel-shaped stalk of a collapsing hyphal terminus. c) Glomus chlamydospore with a funnel-form, straight hyphal attachment. d) Gigaspora azygospore borne on a bulbous-tipped hypha; a small cluster of auxiliary cells accompanies the spore. All drawings are approximately to the same scale.

5.1. Assessing dependence

Species that are mycotrophic, i.e., capable of benefiting from mycorrhizae, may be facultatively or obligately dependent on mycorrhizae (Janos, 1980a). These categories are the opposite ends of a continuum along which species can be positioned according to their ability to grow with and without mycorrhizae at different levels of fertility. Individuals of an obligately mycotrophic species cannot grow without mycorrhizae at any level of fertility. Although few species are likely to be completely obligate in dependence, most lowland tropical wet forest canopy species may be ecologically obligately mycotrophic, unable to grow without mycorrhizae in the most fertile soil they naturally encounter (Janos, 1980a). Facultatively mycotrophic species can grow as well without mycorrhizae as with them at some levels of fertility; the most facultative species benefit from mycorrhizae only in the most infertile soils. Hypothetical responses illustrating how species may be rank-ordered with respect to dependence on mycorrhizae are shown in Fig. 4. For an ecologically meaningful ranking, one might use all soils of different fertilities on which a group of species to be ranked co-occur, but in practice this can be logistically difficult. Relative dependence can be assessed more conveniently by adding soluble phosphorus to the least fertile of these soils, although care must be taken to insure that differences in phosphorus availability are not quickly eliminated by leaching or immobilization. It is most appropriate to assess dependence relative to phosphorus supply, because this element often limits plant growth, and mycorrhizae have their principal effect on its uptake (Bowen, 1980). Use of several fertility levels will improve resolution of differences among species, but also will increase labor and material requirements.

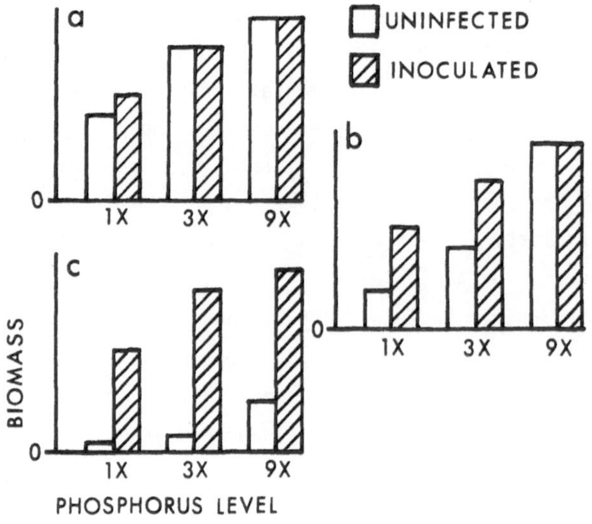

FIGURE 4. Hypothetical responses of three species to phosphorus fertilization and vesicular-arbuscular mycorrhizal inoculation. Species "a" is least dependent on mycorrhizae for growth; species "c" is most dependent.

5.2. Pot-experiment methods

Because VA mycorrhizal fungi cannot be grown in axenic culture, experiments to determine their effects on plant growth necessarily have involved impure inocula. Consequently, control plants must be supplied with all components of the inoculum except mycorrhizal fungi. It is relatively simple to exclude VA mycorrhizal fungi from controls, because the spores of these fungi are larger than organisms that contaminate inocula. Several techniques have been used for soil sterilization to eliminate indigenous mycorrhizal fungi in preparing growth media, and both infected root fragments and spores have been used as inocula. I have had success in the lowland wet tropics with the following methods.

5.2.1. Soil sterilization. I have used methyl-bromide gas with 2% chloropicrin ("Dowfume") at high rates of application (1 1/2 lbs. of gas to approximately 25 one gallon pots of soil) to fumigate heavy tropical clays in the field. Although the gas is lethal to humans and should be used with caution, it is effective for large volumes of soil and is convenient where irradiation or other methods of sterilization are not feasible. Heat and steam sterilization can increase mineral availability in soil; I have observed a fertility increase only once after methyl-bromide and chloropicrin fumigation. In that case, I had fumigated a topsoil that was rich in organic material and, presumably, microbes. Phosphorus availability was doubled, perhaps because fumigation killed all microbes in the soil, liberating phosphorus that had been temporarily immobilized in their biomass.

It is easiest and safest to fumigate soil outdoors by releasing the gas within a fumigation chamber made from an oil drum or large plastic garbage-can to which an airtight cover of slack polyethylene plastic sheeting has been attached. A board with a large sharp nail protruding from it is fixed across the top of the chamber, and the container of gas is gripped through the polyethylene sheet and punctured on the nail. The gas is liquid under pressure, but quickly volatilizes at tropical temperatures. It sinks through the chamber because it is heaver than air which it displaces.

Fumigation is only effective to the extent that the gas can penetrate the full volume of material being treated, so clay soils first should be passed through a 5-8 mm mesh screen to break them apart. Such soils should have a few channels made through them by filling the chamber around pieces of pipe or dowels which are gently removed before sealing the plastic cover. Placing pots of soil in the chamber so that they rest on the rims of the pots below them instead of on the soil in lower pots insures good gas penetration.

The chamber should be kept closed for at least one day, then be opened and allowed to air for several days before the soil is used. Methyl-bromide is colorless, odorless, tasteless, and deadly; chloropicrin is commonly known as "tear gas" and is added to signal leakage because it is easily perceived. Chloropicrin breaks-down more quickly than methyl-bromide, however, so it is advisable to leave promptly after opening the chamber, avoiding breathing fumes from it. Chloropicrin has been suggested to be more effective than methyl-bromide in killing VA mycorrhizal fungi (Linderman, Hendrix, 1982), but the combined gases often are readily available because they are used in agriculture.

5.2.2. Inoculation and control treatments. If the effects of VA mycorrhizal fungi in general are of interest, field-collected feeder roots likely to be infected by several VA mycorrhizal fungus species (Janos, 1980a) are an excellent inoculum, often more infective than spores alone (Powell, 1976). For some determinations, such as the effects of VA mycorrhizal fungi on competition between hosts, a mixed-species inoculum is best. Notwithstanding the lack of host-specificity of VA mycorrhizal fungi, single VA mycorrhizal fungus species might favor different host species. I have used finely cut feeder roots of Theobroma cacao L. and Oenocarpus panamanus Bailey as mixed-species inocula. The major selection criterion for a source of infected roots to use as inoculum is that feeder roots be abundant and easy to collect; cacao forms masses of feeder roots just beneath woody litter in old groves, and O. panamanus, a colonial palm, forms solid mats of feeder roots and older axes intermingled with litter. A second criterion is that the source species be unrelated to the host species under examination in order to minimize the potential for disease transmission. Freshly-collected inoculum should be placed in the root zone of the plants being inoculated so that their roots can contact it as quickly as possible. Root-fragment inocula seem to lose infectiveness rapidly.

Control plants should receive the same quantity of root fragments as inoculated plants, although those for controls are fumigated to kill the mycorrhizal fungi. Other microorganisms contaminating the inoculum can be supplied to control plants by making an infusion from fresh inoculum placed in water overnight, and then filtering this infusion to exclude mycorrhizal fungi. Some investigators filter with Whatman # 1 filter paper, but a 20 µm sieve should be adequate. I have used an 80 µm sieve in the absence of small-spored species of VA mycorrhizal fungi.

Root inoculum containing a single VA mycorrhizal fungus species can be produced by using spores to inoculate a host which produces a large mass of roots in pots of sterilized soil. Spores must be collected from soil using previously described methods and hand sorting. This is laborious where spores are scarce, but the task sometimes can be made easier by transplanting small, infected saplings from the field to pots of fumigated soil. Under these conditions, VA mycorrhizal fungi associated with the transplants occasionally sporulate profusely. Nodulated legumes especially seem to favor sporulation in fumigated soil. Spores can be collected several months after transplanting. Sporocarps are excellent sources of spores for single-species inoculations, but are difficult to find in the field. It may be possible to use rodent feces that contain spores to produce single-fungus root inoculum in those cases in which a single species has been consumed.

Although a host individual is usually infected by more than one VA mycorrhizal fungus species in nature, determination of the effectiveness of single fungus species in increasing host growth may suggest that agricultural practices should be designed to favor certain fungus species. If the effects of single VA mycorrhizal fungus species on host growth are to be compared, effort must be made to eliminate differences in infectiveness of inocula as causes of host growth differences. Equal numbers of spores of different fungi may yield differing amounts of infection, because of dissimilar viability, time required for germination, and production of hyphae. A "most probable number" technique can be used to assess the infectiveness of inocula for later equalization (Daniels, Skipper, 1982). Inoculum is serially ten-fold diluted with sterilized soil to find the dilution at which test plants do not become infected. The number of infective propagules in the undiluted inoculum can be estimated from this.

6. NEEDED RESEARCH

Relatively little mycorrhiza research has been conducted in the tropics (see Janos, 1983), and almost any investigation will contribute significantly to our understanding of mycorrhizae in these habitats. More information is needed concerning root production, levels of mycorrhizal infection in various habitats, the relationship between root morphology and dependence on mycorrhizae, the identities and abundance of tropical mycorrhizal fungi, spore dispersal and predation, and the effects of single species of mycorrhizal fungi on hosts. I suggest that three relatively new lines of research also should be pursued: physiological responses to mycorrhizal infection, conditions under which mycorrhizal fungi are not mutualistic, and effects of mycorrhizae on competition among hosts.

6.1. Physiological responses

Plant responses to mycorrhizal infection have usually been assessed by measuring various morphological features, or biomass. Several other responses, however, merit investigation (Linderman, Hendrix, 1982). Effects of mycorrhizae on plant survival (Janos, 1980a), yield (Sanni, 1976), response to disease (Schenck, Kellam, 1978), and on nodulation of legumes (Mosse, 1977) are being investigated. Changes in anatomy (Daft, Okusanya, 1973) and biochemistry (Krishna, et al., 1981) attributable to VA mycorrhizal infection have been reported. These changes may influence transpiration, and rates of photosynthesis and respiration. Their documentation may suggest the mechanisms by which improved mineral nutrient supply by mycorrhizae contributes to rapid plant growth and maturation, and improves resistance to water stress. Such research should be conducted in ways that enable detection of effects of mycorrhizae in addition to those attributable to improved mineral nutrition. In order to detect consequences of infection beyond those caused by improved phosphorus uptake, control plants can be supplied with soluble phosphorus at a rate that produces the same tissue phosphorus content as that of mycorrhizal plants (Bowen, 1980). Investigation of effects of mycorrhizae on the uptake of elements other than phosphorus requires that control plants be given supplemental phosphorus so that comparisons are not invalidated by large differences in size between infected and uninfected plants.

6.2. Negative effects on hosts

VA mycorrhizal fungi depend on higher plants for energy in the form of carbon compounds (Ho, Trappe, 1973); under some conditions the fungi may be unable to increase mineral uptake sufficiently to amortize their carbon cost to hosts. Although the fungi are usually mutualistic, they can cause both transitory and long-lasting decreases in host growth relative to uninfected controls when phosphorus is readily available (Cooper, 1975), for some host species even in low-phosphate soil (Sparling, Tinker, 1978), at relatively low temperatures (Moawad, 1980), and at relatively low pH (Moawad, 1980). Although decreased host growth is probably directly attributable to the energy cost of mycorrhizae, several different phenomena may account for failure of the fungi to compensate for their cost. Mycorrhizae improve the uptake of minerals that cannot diffuse readily to roots (Bowen, 1980), but when soluble phosphorus is abundant or root densities are very high, improvement of phosphorus uptake by mycorrhizae is not expected. Moreover, factors that inhibit the spread of mycorrhizal fungus hyphae through soil or slow translocation of minerals through hyphae may reduce the effectiveness of mycorrhizae in mineral uptake. An examination of the conditions under which mycorrhizae fail to be mutualistic may give insight to the mechanisms by which they function, as well as having practical implications for maximization of agricultural benefits to be gained from mycorrhizae.

6.3. Effects on competition

Because mycorrhizae can greatly affect the nutrient uptake capabilities of plants, they also may influence the abilities of plants to compete for mineral nutrients. When plants are competing strongly, it is possible that mycorrhizae are important even in the uptake of elements that are not readily locally depleted in soil because of high diffusion rates. Effects of mycorrhizae on competition barely have been investigated (see Bowen, 1980). Their study, however, may reveal general principles of mineral nutrient partitioning related to mycorrhizal interactions that are useful in the design of crop polycultures for the tropics. In addition, the influences of VA mycorrhizae on competition may provide a partial explanation of the coexistence of very many tree species within lowland tropical wet forests (Janos, 1983).

ACKNOWLEDGEMENTS

I thank E. Medina, H. Mooney and C. Vasquez-Yanes for encouraging me to write this paper, and S. Green for reviewing a draft of the manuscript. I am grateful to the Instituto de Biología of the Universidad Nacional Autónoma de México for convening the Symposium and Workshop on Physiological Ecology of Tropical Plants. This paper is contribution No. 102 from the Program in Tropical Biology, Ecology and Behavior of the Department of Biology, University of Miami.

REFERENCES

Baylis GTS (1962) Rhizophagus. The catholic symbiont, Aust. J. Sci. 25, 195-200.

Baylis GTS (1975) The magnolioid mycorrhiza and mycotrophy in root systems derived from it. In Sanders FE, Mosse B, and Tinker PB, eds. Endomycorrhizas, pp. 373-389. London, Academic Press.

Bevege DI and Bowen GD (1975) Endogone strain and host plant differences in development of vesicular-arbuscular mycorrhizas. In Sanders FE, Mosse B, and Tinker PB, eds. Endomycorrhizas, pp. 77-86. London, Academic Press.

Bowen GD (1980) Mycorrhizal roles in tropical plants and ecosystems. In Mikola P, ed. Tropical mycorrhiza research, pp. 165-190. Oxford, Clarendon Press.

Cooper KM (1975) Growth responses to the formation of endotrophic mycorrhizas in Solanum, Leptospermum, and New Zealand ferns. In Sanders FE, Mosse B, and Tinker PB, eds. Endomycorrhizas, pp. 391-407. London, Academic Press.

Daft MJ and Okusanya BO (1973) Effect of Endogone mycorrhiza on plant growth VI. Influence of infection on the anatomy and reproductive development in four hosts, New Phytol. 72, 1333-1339.

Daniels BA and Skipper HD (1982) Methods for the recovery and quantitative estimation of propagules from soil. In Schenck NC, ed. Methods and principles of mycorrhizal research, pp. 29-35. St. Paul, Minn., American Phytopathological Society.

Gerdemann JW (1968) Vesicular-arbuscular mycorrhiza and plant growth, A. Rev. Phytopath. 6, 397-418.

Gerdemann JW and Nicolson TH (1963) Spores of mycorrhizal Endogone extracted from soil by wet sieving and decanting, Trans. Br. Mycol. Soc. 46, 235-244.

Giovannetti M and Mosse B (1980) An evaluation of techniques for measuring vesicular-arbuscular mycorrhizal infection in roots, New Phytol. 84, 489-500.

Hannah PR (1972) Yellow Birch root occupancy related to stump and breast height diameters, Bull. 669. Burlington, Agricultural Experiment Station, Univ. of Vermont.

Ho I and Trappe JM (1973) Translocation of 14C from Festuca plants to their endomycorrhizal fungi, Nature, New Biol. 244, 30-31.

Huttel C (1975) Root distribution and biomass in three Ivory Coast rainforest plots. In Golley FB and Medina E, eds. Tropical ecological systems, pp. 123-130. New York, Springer-Verlag.

Janos DP (1975) Vesicular-arbuscular mycorrhizal fungi and plant growth in a Costa Rican lowland rainforest, Ph.D. dissertation. Ann Arbor, Univ. of Michigan.

Janos DP (1980a) Vesicular-arbuscular mycorrhizae affect lowland tropical rain forest plant growth, Ecology 61, 151-162.

Janos DP (1980b) Mycorrhizae influence tropical succession, Biotropica 12(Suppl.), 56-64.

Janos DP (1983) Tropical mycorrhizas, nutrient cycles and plant growth. In Sutton SL, Whitmore TC, and Chadwick AC, eds. Tropical rain forest: ecology and management, pp. 327-345. Oxford, Blackwell Scientific Publications.

Jenik J (1971) Root structure and underground biomass in equatorial forests. In Duvigneaud P, ed. Productivity of forest ecosystems, pp. 323-330. Paris, Unesco.

Jenik J (1978) Roots and root systems in tropical trees: morphologic and ecologic aspects. In Tomlinson PB and Zimmermann MH, eds. Tropical trees as living systems, pp. 323-349. Cambridge, University Press.

Jordan CF and Escalante G (1980) Root productivity in an Amazonian rain forest, Ecology 61, 14-18.

Keeley JE (1980) Endomycorrhizae influence growth of Blackgum seedlings in flooded soils, Amer. J. Bot. 67, 6-9.

Klinge H (1973a) Root mass estimation in lowland tropical rain forests of central Amazonia, Brazil I, Trop. Ecol. 14, 29-38.

Klinge H (1973b) Root mass estimation in lowland tropical rain forests of central Amazonia, Brazil II, Anais Acad. bras. Ciencias, Rio de J. 45, 595-609.

Kormanik PP and McGraw AC (1982) Quantification of vesicular-arbuscular mycorrhizae in plant roots. In Schenck NC, ed. Methods and principles of mycorrhizal research, pp. 37-45. St. Paul, Minn., American Phytopathological Society.

Kormanik PP, Bryan WC, and Schultz RC (1980) Procedures and equipment for staining large numbers of plant roots for endomycorrhizal assay, Can. J. Microbiol. 26, 536-538.

Krishna KR, Suresh HM, Syamsunder J, and Bagyaraj DJ (1981) Changes in the leaves of Finger Millet due to VA mycorrhizal infection, New Phytol. 87, 717-722.

Linderman RG and Hendrix JW (1982) Evaluation of plant response to colonization by vesicular-arbuscular mycorrhizal fungi A. Host variables. In Schenck NC, ed. Methods and principles of mycorrhizal research, pp. 69-76. St. Paul, Minn., American Phytopathological Society.

McKey D (1979) The distribution of secondary compounds within plants. In Rosenthal GA and Janzen DH, eds. Herbivores: their interaction with secondary plant metabolites, pp. 55-133. New York, Academic Press.

Mikola P (1980) Mycorrhizae across the frontiers. In Mikola P, ed. Tropical mycorrhiza research, pp. 3-10. Oxford, Clarendon Press.

Moawad M (1980) Ecophysiology of vesicular-arbuscular mycorrhiza. In Mikola P, ed. Tropical mycorrhiza research, pp. 203-205. Oxford, Clarendon Press.

Mosse B (1977) The role of mycorrhiza in legume nutrition on marginal soils. In Vincent JM, Whitney AS, and Bose J, eds. Exploiting the legume-Rhizobium symbiosis in tropical agriculture, pp. 275-292. Honolulu, College of Tropical Agriculture, Univ. of Hawaii.

Nemec S (1982) Morphology and histology of vesicular-arbuscular mycorrhizae B. Histology and histochemistry. In Schenck NC, ed. Methods and principles of mycorrhizal research, pp. 23-27. St. Paul, Minn., American Phytopathological Society.

Newman EI (1966) A method of estimating the total length of root in a sample, J. Appl. Ecol. 3, 139-145.

Pearson V and Tinker PB (1975) Measurement of phosphorus fluxes in the external hyphae of endomycorrhizas. In Sanders FE, Mosse B, and Tinker PB, eds. Endomycorrhizas, pp. 277-287. London, Academic Press.

Phillips JM and Hayman DS (1970) Improved procedures for clearing and staining parasitic and vesicular-arbuscular mycorrhizal fungi for rapid assessment of infection, Trans. Br. Mycol. Soc. 55, 158-161.

Powell CL (1976) Development of mycorrhizal infections from Endogone spores and infected root segments, Trans. Br. Mycol. Soc. 66, 439-445.

Ross JP and Daniels BA (1982) Production of endomycorrhizal inoculum B. Hyperparasitism of endomycorrhizal fungi. In Schenck NC, ed. Methods and principles of mycorrhizal research, pp. 55-58. St. Paul, Minn., American Phytopathological Society.

Sanchez PA and Buol SW (1975) Soils of the tropics and the world food crisis, Science 188, 598-603.

Sanni SO (1976) Vesicular-arbuscular mycorrhiza in some Nigerian soils: the effect of Gigaspora gigantea on the growth of rice, New Phytol. 77, 673-674.

Schenck NC (1982) Methods and principles of mycorrhizal research, St. Paul, Minn., American Phytopathological Society.

Schenck NC and Kellam MK (1978) The influence of vesicular-arbuscular mycorrhizae on disease development, Bull. 798. Gainesville, Agricultural Experiment Stations, Univ. of Florida.

Sparling GP and Tinker PB (1978) Mycorrhizal infection in Pennine grassland II. Effects of mycorrhizal infection on the growth of some upland grasses on gamma-irradiated soils, J. Appl. Ecol. 15, 951-958.

St. John TV (1980a) A survey of micorrhizal infection in an amazonian rain forest, Acta Amazonica 10, 527-533.

St. John TV (1980b) Root size, root hairs and mycorrhizal infection: a re-examination of Baylis's hypothesis with tropical trees, New Phytol. 84, 483-487.

St. John T and Machado AD (1978) Evidência da ação de microorganismos na ramificação de raízes, Acta Amazonica 8, 9-11.

Stark N and Spratt M (1977) Root biomass and nutrient storage in rain forest Oxisols near San Carlos de Rio Negro, Trop. Ecol. 18, 1-9.

Trappe JM (1982) Synoptic keys to the genera and species of zygomycetous mycorrhizal fungi, Phytopathology 72, 1102-1108.

Trappe JM and Schenck NC (1982) Taxonomy of the fungi forming endomycorrhizae A. Vesicular-arbuscular mycorrhizal fungi (Endogonales). In Schenck NC, ed. Methods and principles of mycorrhizal research, pp. 1-9. St. Paul, Minn., American Phytopathological Society.

THE ROLES OF PLANT SECONDARY CHEMICALS IN WET TROPICAL ECOSYSTEMS

JEAN H. LANGENHEIM
Department of Biology
University of California
Santa Cruz, CA 95064 U.S.A.

ABSTRACT

Structurally diverse so called "secondary compounds" probably play a multiplicity of roles in the survival of plants within ecosystems. The occurrence, diversity and concentration of some of these chemicals appear to be greater in tropical than temperate ecosystems, and in wet than dry ecosystems. This correlates with the assumption that herbivore and pathogen selection pressures are also greater in these ecosystems. Alkaloids, non-protein amino acids, terpenes and phenolic compounds have received the most attention in the relatively few ecological investigations of these chemicals in the wet tropics. Studies are discussed of the following: 1) floral fragrances as pollination attractants, 2) legume seed toxins against insects, 3) chemical variation on contrasting soil types relative to mammalian and insect herbivory, 4) chemical defense relative to successional status, 5) allelopathy, and 6) environmental constraints on production and variation of terpenoid resins in legumes, and the role of the variability in defense against insects and fungi. Future investigations, taking the perspective of plant physiological ecology, in which the costs of these chemicals in terms of the overall economy of the plant's fitness are determined, would probably necessitate interdisciplinary collaboration due to the inherent complexity of wet tropical ecosystems.

INTRODUCTION

Since plant secondary compounds are generally defined as those for which physiological functions are rarely known, we shall look briefly at their occurrence, known functions and relations to primary metabolism in order to obtain a better perspective for studies in plant physiological ecology. They are a large and structurally diverse group of chemicals--with over 10,000 low molecular weight compounds having been reported, and it has been estimated that only one-fourth of them have been described (Swain, 1977). These compounds have long been the touchstone of structural and synthetic organic chemists and are continuing to be described at a rapid rate. However, in any one plant species, generally only the major compounds of a structural class have been analyzed, with no consideration having been given to the variety of other kinds that might be present.

In this review we shall be concerned only with the groups of secondary compounds in which there has been some tropical research. Among the nitrogen-containing compounds, the alkaloids are the largest and most heterogeneous group, and members have several biosynthetic origins. The non-protein amino acids are a much smaller and generally lesser known group of nitrogen-containing compounds, but have been of considerable interest in ecological studies in the tropics. Phenolic compounds probably have the most widespread occurrence of secondary chemicals, and have received the most attention in various kinds of studies. Terpenes, which comprise an additionally large, structurally diverse group, are increasingly being studied in the tropics for the variety of roles they may play.

Despite their structural diversity, secondary compounds arise from surprisingly few starting materials, i.e., primarily a few common amino acids and acetate. Accumulating evidence also suggests that many secondary compounds are in a state of dynamic equilibrium. When radioactive labeled precursors of these compounds are incorporated into secondary metabolites, the biological half-life may vary from a few minutes to a few days, and the label may be subsequently incorporated into primary or other secondary metabolites. In addition to the possibility of being recycled for use in physiological functions, some serve initially in primary roles, such as growth regulators and coenzymes, transport facilitators, shields against excessive radiation, etc.

However, it is the incredible structural diversity, irregular distribution among plants, and specific location in the plant in space and time that have fascinated chemists, systematists and ecologists, because these characteristics provide the basis for evolutionary and ecological roles in defense against herbivores, pathogens and competitors, as well as in attraction for pollinators and dispersal agents. In fact, this fascination has led some, such as Ehrlich and Raven (1964), Janzen (1969), Whittaker (1970), and Whittaker and Feeny (1971) to take the stand that evolution of these substances is not comprehensible except in an ecological context which includes organisms other than the plants producing them. Swain (1977) has further suggested that the plant's ability to vary the chemical signals sent into the ecosystem could serve as a counterpart of behavior in animals. Analysis of adaptation of heterotrophs to plants has frequently shown their dependence on the chemical profile of the plant (e.g., Atabekov, 1975; Ikeda et al. 1977), but the plant's response is less well understood, i.e., it is not as clear how the chemical variation in the plant relates to herbivore and pathogen attack. One of the most difficult aspects of interpreting the plant's response is that it is subject to multiple selection pressures--often simultaneously on the same organs--from diverse herbivores and pathogens. Another problem in interpreting chemical variation in the plant is that it may not just react passively through its constitutive chemistry, but may respond chemically to attack relatively quickly (Cruickshank, 1963; Stoessel, 1970; Ryan, 1979; Schultz et al., 1982; Schultz, 1983) and possibly even communicate an alarm to other associated plants (Baldwin and Schultz, 1983; Rhoades, 1983). However, Dethier (1959, 1970) has emphasized that secondary compounds may significantly influence the direction and intensity of natural selection in plant populations by mediating the specificity of these many plant-heterotroph interactions. In the physiological context, Mooney et al. (1983) reiterate the further challenge that little is known about the costs of these defensive chemicals. Their costs are not only a function of biosynthetic maintenance and turnover expenditures, but of the value of the elements in their construction, which in turn may be determined by their importance to other activities such as photosynthesis, reproduction, nutrient uptake, etc. This view then emphasizes the interrelationships of secondary compounds to primary functions, while not denying their possible roles in defense (as well as in other ecosystem interactions). Siegler and Price (1976) and Siegler (1977) also have stressed that secondary compounds probably have multiple functions, and Jones (1979) has questioned whether or not our research outlook could be sharpened by ridding the literature of the term "secondary" and relegating those compounds with unknown functions as yet to an "uninvestigated" category. In this manner, a more unified perspective may be gained of the multiplicity of the roles these chemicals may play in terms of the plant's survival and reproduction within the ecosystem.

SECONDARY CHEMICAL OCCURRENCE IN DIFFERENT ECOSYSTEMS

On the assumption that these compounds may be serving "chemical behavioral" roles for plants in ecosystemic interactions, we could anticipate that their occurrence, diversity and concentration would be greater in tropical than temperate ecosystems and in moist than dry tropical ecosystems, because herbivore and pathogen pressures have generally been assumed to be greater under these conditions (Janzen, 1973a,b; Levin, 1976a; Scriber, 1973; Holliday, 1971; Wellman, 1972). For example, the proportion of net productivity consumed by herbivores in tropical forests appears about twice that of temperate forests (3.5-4% vs 7-8%, respectively) and the absolute amount consumed is about 3.5 times greater (Golley 1972; Whittaker 1975). Among tropical ecosystems, higher moisture levels have been shown to be significantly correlated with greater insect abundance (Janzen, Schoener, 1968; Janzen, Pond 1975; Stanton 1975; Bigger, 1976) and with higher diversity of some tropical insects (Strong, 1977).

Numerous analyses have been made of tropical plants for secondary compounds, particularly for commercial usage (e.g., Gottlieb, Mors, 1978, 1980; Mabry, Ulubelen, 1980); in fact, the pharmaceutical value of alkaloids has led to extensive screening surveys. Levin (1976b) utilized a series of world surveys of alkaloids from leaves by Farnsworth and associates (Smolenski et al. 1972, 1973, 1974a, 1974b; Fong et al. 1972) and from seeds by Earle and Jones (1962) to conclude that the incidence of alkaloid-bearing plants is nearly twofold greater in tropical than in temperate floras, and that a latitudinal cline is evident. Using a toxicity index developed by toxicologists for mammals, Levin and York (1978) showed that alkaloids tend to be more toxic in tropical dicot families than in temperate or cosmopolitan families, and in woody than in herbaceous families. Also in New Guinea the percentage of alkaloid-bearing plants, alkaloid concentration and their toxicity index vary among community types (Hartley et al., 1973; Levin, York, 1978). The lowland rainforest is richest in alkaloid-bearing species and these have both the highest concentrations and toxicity indices. Monsoon, foothill and montane rainforests have somewhat lower values; in the higher elevation cloud forests they are even lower and alkaloids are not recorded for the subalpine forest and scrub communities. Levin and York (1978) concluded that the differences in toxicity of these alkaloid profiles seem most likely to be evolutionary by-products of differences in and intensity of both mammalian and insect herbivory (Freeland, Janzen, 1974). Although their toxicity indices were based on mammals, alkaloids are known to defend successfully against insects (e.g., Dolinger et al., 1973; McKey 1974; Janzen et al., 1977; Robinson 1979). A defensive role for alkaloids would explain not only the ecogeographic gradient, but why the incidence of alkaloid-bearing species is greater in families either restricted to or located primarily in the tropics than in temperate regions or with cosmopolitan distributions. The greater toxicity of alkaloid profiles in woody plants than on herbs may be a consequence of greater herbivore pressures on woody plants, which are a more persistent resource and have a greater diversity of within plant niches than herbs. In temperate floras, where more data are available, it has been shown that fewer insects and pathogens occur on herbs than in woody plants (Strong, Levin, 1979).

If alkaloids are playing an important defensive role in tropical communities, changes in alkaloid profiles to cope with different pests from one region to another, or increasing concentrations of these compounds could be anticipated. Geographic variation is evident, e.g., Cinchona in South America shows a reduction in total crystallizable alkaloids southward from the equator (Camp, 1949; Schramm, Scharting, 1961a), and chemical races with ecogeographic correlations have been described (e.g., Strophanthus sarmentosus in Reichstein 1963; Solanum dulcamara in Boll, Anderson, 1962; Sanders, 1963). Also considerable variation occurs within populations upon which selection may be acting. For example, the alkaloid content in the bark of individuals of Cinchona pubescens varied tenfold and those in C. officinalis thirteenfold (Camp, 1949).

A survey of fossil resin (amber) through geologic time indicated the importance of tropical environments in the evolution of terpenoid resin-producing plants (Langenheim 1969, 1975, 1981). Among tree taxa producing copious amounts of resin two-thirds are tropical or subtropical with numerous angiosperm families represented (e.g., Leguminosae, Dipterocarpaceae and Burseraceae). Even among the two copious-producing conifer families (Araucariaceae and Pinaceae), the largest quantities of resin (and hence commercial usage) occur under tropical or subtropical conditions (Langenheim 1969, 1983). We also have suggested that both copious production and variation in resin composition in these tropical legumes may be related to defense against herbivores and pathogens which vary among tropical ecosystems (Langenheim 1969, 1975; Langenheim et al. 1977, 1978, 1981, 1983; Stubblebine, Langenheim 1977, 1980).

EXEMPLARY ECOLOGICAL STUDIES OF SECONDARY CHEMICALS IN THE TROPICS

Among the various potential plant chemical signals that could have been investigated in tropical rainforests, the relatively few studies (often of a pioneering nature) have tended to focus on defensive roles against insects and mammals. A smaller number of studies has been directed toward their acting as attractants for pollination and their having allelopathic effects on competitors.

Floral fragrances as pollination attractants

The scents of flowers often play major roles as attractants to pollinating insects. Odor is of particular importance to nocturnal insects and animals where visual stimuli are essentially absent. Bees in general appear sensitive to volatile chemicals. In Neotropical forests members of the Orchidaceae, Araceae, Gesneriaceae and other plant families have evolved different floral fragrances to attract different euglossine bee pollinators (Dodson et al. 1969; Dodson, 1970, 1975). Williams and Dodson (1972) reported approximately 60 distinct floral fragrance compounds for orchids. Each orchid species has a characteristic combination of fragrant terpene and phenolic compounds which apparently "screen" the kinds of bees attracted. Although several species (up to 40 in one location in Panama) occur in a particular site, frequently only one bee species is attracted to a plant species with a particular fragrance.

The bees are dependent on easily collectible fragrance compounds in order to attract other male bees of their species and form leks where mating takes place. Relatively minor changes in chemical composition of floral fragrances apparently could upset the attraction system, since field experimental evidence indicates that the bees are sensitive to different fragrance composition from population to population. For example, when both compounds are offered Euglossa viridissima is much more frequently attracted to methyl cinnamate in northeastern Mexico, whereas the same species is more strongly attracted to eugenol in western Mexico.

In other studies it has been shown that one component may reinforce the effectiveness of the second or third in producing a characteristic odor (Harborne, 1982). The dynamics controlling this synergism have received little study. The turnover of chemicals in flower color attractants (mainly flavonoids and carotenoids) is about seven to ten days (Harborne, 1982). However, the maximum scent production for pollination may be coordinated with the ripening of pollen, and diurnal variations and production may be coordinated such that odor is timed for the pollinator. These kinds of variation are of considerable consequence in tropical ecosystems, and the remarkable timing and rapid turnover offer fascinating challenges for studies in plant physiological ecology.

Legume seed toxins

Since bruchid beetles and some lepidopterans generally attack seeds of members of the Leguminosae, a dominant element in amphi Atlantic tropical forests, there has been special interest in the role that secondary compounds play in those legumes that conspicuously escape attack as well as in insects which become host specific on them. Toxicity of seed powders and of pure compounds extracted from seed was tested by incorporating them into an artificial diet for rearing larvae of the southern armyworm (Prodenia eridania) (Rehr et al., 1973), which has a highly effective system for detoxifying insecticides and secondary chemicals encountered in a wide array of food plants (Krieger et al. 1971; Gordon, 1961). In these experiments non-protein amino acids (e.g., L-canavanine and β-hydroxy-y-methyl-glutamic acid) were repellent whereas equivalent concentrations of normal amino acids analogs were acceptable. Rehr et al. (1973) also demonstrated that L-dopa (3,4-dihydroxphenylalanine) found in high concentrations in Macuna seeds (generally immune to insect attack) increased mortality, and pupal and adult development in the southern armyworm. In subsequent experiments Janzen et al. (1978) studied dose-dependent effects of a variety of alkaloids, non-protein and protein amino acids on larvae of a bruchid crop pest, Callosobruchus maculatus. Alkaloids were the most toxic of these chemicals, generally having lethal effects at 0.1%. The authors suggested that, since alkaloids generally occur below 1-2% in seeds, natural selection does not have to "drive" alkaloid concentration in seeds to high levels to result in the seed being toxic or inedible to an animal which does not normally prey on the seed. Nevertheless, insect specialists have become adapted to feeding on alkaloid-containing seeds. A greater range in toxicity was exhibited by non-protein amino acids with generally higher concentrations needed to be lethal; however, non-protein amino acids range from 3-10% in many seeds. Certain combinations of non-protein amino acids also had more significant toxic effects than single ones, and even some protein amino acids were toxic at 5% concentration. A variety of other secondary compounds in seeds had detrimental effects on the production of adult beetles at levels representative of those found in nature. Thus the authors further concluded that many of the secondary compounds found in seeds are likely to be toxic to some animal, and also are responsible, at least in part, for the extreme host-specificity of many seed-eating insects.

Analysis of the proportion of total and free amino acid nitrogen of seven L-canavanine-synthesizing leguminous seeds revealed a disproportional allocation of total seed nitrogen to L-canavanine (Rosenthal, 1977). In two of the species over one half of the nitrogen pool was sequestered in L-canavanine and in the other five species one-third of all seed nitrogen occurred in this compound. In all cases L-canavanine represented 80% of the total free amino acids. L-arginine, the principal storage metabolite in many higher plants, also occurs in these seeds. This disproportionate placement of seed nitrogen into canavanine becomes more understandable from studies of Canavalia ensiformis showing that arginase and urease are abundant in canavanine-containing seeds (Rosenthal, 1974, 1983). In fact, the seed urease content correlates well with the level of stored L-canavanine. These enzymes rapidly mobilize the nitrogen of canavanine and facilitate its movement into primary metabolic pathways. Also the arginyl tRNA synthetase in such plants discriminates between L-arginine and L-canavinine so that they avoid production of dysfunctional proteins. Therefore, the plant may "gain" by storing the nitrogen in some seeds as canavanine, rather than as arginine, since canavanine may confer an additional defensive advantage. In fact, selection pressures of herbivores may be responsible for the extremely large quantities of such non-protein amino acids in some seeds. Also a host specific bruchid, such as Caryedes brasiliensis, has several distinctive but interrelated biochemical adaptations similar to that of the plant, Dioclea megacarpa, which the bruchid utilizes as food (Rosenthal, 1977, 1983; Rosenthal et al., 1976, 1977, 1982). It has been proposed that among the first adaptations by Caryedes was the development of arginyl-tRNA synthetase that discriminates between L-arginine and L-canavanine such that it avoids the synthesis of canavanyl proteins. The early bruchids that invaded this legume may have simply excreted canavanine--the other nitrogen-rich metabolites having satisfied their nitrogen needs. Through time, however, C. brasiensis may have become progressively better equipped to cope with and ultimately utilize the nitrogen resource of canavanine by formation of urea from it and then producing ammonia, as does the plant. In this way C. brasiliensis has relative freedom from competition for seeds, and does not have to maintain detoxifying capabilities for a wide range of toxins.

The mobility of nitrogen within the plant has been demonstrated in numerous other instances. For example, in forest vegetation 20-40% of the nitrogen in senescing leaves may be resorbed (Ryan, Borman, 1982) or moved to developing fruits (Derman et al., 1978). Thus, for tropical plants it would appear of considerable interest to continue studies, such as those of Canavalia and Dioclea for non-protein amino acids as well as others for alkaloids, to assess how the nitrogen invested in chemical defense could be translocated throughout the plant to protect other vulnerable organs or to be reused for growth and development. This is an area pinpointed by Mooney et al. (1983) as being generally significant in terms of understanding physiological constraints on plant chemical defenses.

Variation in secondary chemicals on contrasting soil types

In some equatorial areas nutrient-poor white sand soils are associated with blackwater rivers, which are rich in humic acids and low in nutrients. Janzen (1974) attributes some of these conditions to the vegetation growing on these soils being exceptionally rich in phenolic compounds (as well as other secondary chemicals) which flow into the rivers. He further postulates that the total allelochemic investment of plants should be greater on nutrient-poor soils, because the cost to the plant in reduction of fitness due to herbivory would be greater than on nutrient-rich soils.

Several studies have now compared phenolic and other secondary compounds in plants on nutrient-poor and nutrient-richer soils in tropical rainforest sites. In two African rainforests, one of which occurred on very acid, sandy soil (Douala-Edea Reserve, Cameroon) and the other on more nutrient-rich, less acid sandy loam (Kibale Forest, Uganda), the secondary chemical profiles differed greatly (McKey et al. 1978; Gartlan et al., 1980). Mature leaves of the most abundant species examined showed mean concentration of total phenolics and condensed tannins to be significantly greater (± twofold) on the nutrient poor sands than on the more nutrient-rich sandy loam.

Phenolic compounds exist in plants in enormous variety, and ecological significance has been attributed to most of the structural classes (Levin 1971; Swain 1974, 1979). However, condensed tannins have been of particular interest with regard to herbivory, because they have the capacity to form complexes with proteins that are stable to conditions encountered in the gut of some animals and are resistant to proteolysis. Ingestion of tannins may result in greatly increased excretion of nitrogen in the feces, leading to malnutrition or death. Thus the advantage for the plant's producing condensed tannin is that it would be more difficult for the herbivore to overcome and that feeding on tanniferous plants imposes a cost to the herbivores that increases with increasing tannin concentration (Feeny, 1975, 1976; Rhoades, Cates 1976; Rhoades, 1979).

Because this African study was a part of an investigation of anthropoid primates, studies of herbivory of Colobine monkeys was studied. Unlike the two species of _Colobus_ monkeys found in Kibale forest (with less condensed tannin), _Colobus satanas_ in the Douala-Edea Reserve (with high tannins) avoid the leaves of almost all of these abundant tree species, selectively feeding on leaves of the less common deciduous trees, secondary growth vines and seeds with lower phenolic concentrations (McKey et al, 1978). These authors, however, cannot discount the possibility that avoidance of tannin-rich leaves by _C. satanas_ is due to significantly lower contents of nitrogen and minerals, or their toughness, about which they have no information.

Another difference in the chemical profile is that within the two African sites the mature leaves of the great majority of species yielded either tannins or alkaloids; rarely was there detectable quantities of both. Nitrogen content was greater both in the leaves, and in the soil of the sandy loam compared to the sand, and thus may be related to the greater percentage of trees which produced alkaloids on the sandy loam soil.

In other studies with Colobine monkeys, leaf food choice has been found to be complex and influenced by numerous interacting factors. Oates et al. (1977) found a strong negative correlation between condensed tannin concentration and food selection of _Colobus guereza_ in Uganda. Less correlation occurred with alkaloid content. However, in a subsequent study of the south Indian colobine, _Presbytis johnii_, low fiber content was the most reliable predictor of foliage choice, although the stable part of the leaf diet was also characterized by very low condensed tannin content (Oates et al., 1980). Apparently the forestomach microbiota have the ability to detoxify at least some alkaloids in this species.

In the central Amazonian rainforest, there were no statistically significant differences in concentration of total phenolics, condensed tannins and sesquiterpenes in the leguminous species _Copaifera multijuga_ on nutrient-poor sandy and nutrient-richer oxisol soils (Nascimento 1980). Although there were significant differences in nitrogen, potassium, phosphorous, calcium and magnesium available from the two soils, the differences in these nutrients were insignificant in the leaf. Thus, the nutrient availability for the herbivore, maintenance of plant function and production of secondary chemicals is not significantly different between the trees on the two soils, despite the nutrient differences in the soil per se. The level of insect herbivory was low (between 1-2%) and herbivory did not differ significantly between trees on these soils. Although the differences were insignificant in leaf phenolic content on the two soils, significant variation occurred within the upper and lower canopy levels, and between the seedlings and the upper canopy of the mature tree. Significant intrapopulational variation occurred in the leaf sesquiterpenes, however, this variation was insignificant between the two soils. Hence, in this particular species significant variation occurred in the phenolic compounds and terpenes, but they apparently were not related to differences in soil conditions.

Several studies have shown that concentrations of certain phenolic compounds vary with stress and nutrient conditions of the plant (Wender 1970; del Moral 1972). del Moral (1972) proposes that chlorogenic acid (CGA) is representative of a group of phenolics which have originated as regulators of various metabolic systems under stress and have also subsequently acquired allelochemic properties against pathogens, herbivores and competitors. Physiological stress, metabolic regulation, pathogens, herbivores and competitors may interact in phenolic acid production. Physiological stress selects for increased CGA because it can ameliorate the effects of this stress. Energy demands of the plant suggests a plastic response is more adaptive than an ecotypic fixed response and, therefore, ability to produce large concentrations of CGA rapidly is incorporated into the genotype. Higher concentrations and the ability to produce high concentrations rapidly lead to greater defense capacities against herbivores and pathogens; hence a positive feedback loop is established where pathogens select for disease resistance, and both unpalatability and higher concentration can influence herbivores. These studies thus point to the complex interactions between abiotic and biotic factors with regard to the plant's synthesis and utilization of these compounds.

<u>Chemical defense relative to successional status</u>

In studying insect herbivory and defensive characteristics in saplings of 46 canopy trees species in a Panamanian rainforest, Coley (1983) has hypothesized that there is a trade-off between investments in growth and defense between pioneer and persistent species. The levels of herbivory on mature leaves of pioneer species is six times those on persistent species, yet they grow two and one-half times as fast, have lower fiber and phenol content, are less tough and have shorter leaf lifetimes.

Although rates of leaf removal are comparable for species growing in both gap and understory sites, the impact of herbivory is greater in the understory. Thus Coley (1983) proposes that habitat quality is more a major selective force behind evolution of different defensive systems than apparency. High quality habitats are defined as those in which rapid growth is possible, as contrasted with low quality ones in which growth is limited by any abiotic factor, such as light or nutrients. Thus for a given level of herbivore pressure, the advantage of defense should increase as the potential maximum growth rate declines. Because of the apparent trade-off between growth and defense, the poorly defended species should be favored when habitat quality and growth potential are high relative to herbivory. Under these conditions, it would be possible for undefended species to "tolerate" herbivory, if the reduction in productivity due to losses was less than the alternate costs of defense.

Neither phenolic content nor phenolic:protein ratios were significantly correlated with herbivory in this Panamanian rainforest. In 70%

of the species, young leaves (which had to three times the phenolic content) suffered higher damage levels than mature leaves. Although this higher phenolic content in young leaves may be significant in some way to herbivory, Coley suggests that from her evidence the general role of phenolics in insect defense may have been overemphasized in some current theories (e.g., Feeney, 1976; Rhoades and Cates, 1976). However, a community study of this type must be statistically based, and means may obscure patterns of chemical variation (Shultz, 1983), and/or the presence of particular phenolic compounds which could be significant with regard to specific insects. In addition, chemical defense may reside in other kinds of chemicals than phenolics.

Allelopathy

It is often assumed that allelopathy [defined by Rice (1974) as "any direct or indirect harmful effect by one plant or microorganism on another through production of chemicals which escape into the environment"] will not be a significant ecological factor in areas of high precipitation, because toxins are leached from soils too rapidly to have a significant effect. However, Gliessman (1976, 1978) studied two cases that had significant impact upon succession through continued input of phytotoxins, which compensated for rapid decomposition and excessive leaching in the warm moist climate. For example, once dominance of bracken fern (Pteridium aquilinum) has been established, following forest disturbance in Costa Rica, continual production from green fronds of phenolic toxins, leachable by rainfall, provides an effective mechanism for maintaining the fern's dominance. Also during rainless periods an accumulation of toxins builds up and, when rain resumes, input of toxins is even greater. Also Quercus eugeniaefolia in a Costa Rican upland forest has the potential for allelopathic dominance over both understory plants and associated tree species by the continual input of toxins (presumably phenolic compounds) from green leaves, fresh dead leaves, and possibly roots. An effective concentration of toxins is also held within the soil and thus reinforces both those produced by living leaves and leached from litter. In the humid climate of Taiwan, Wang et al. (1967) isolated and identified large quantities of phytotoxins from tropical clays, which have been demonstrated to hold inhibitory substances against leaching by heavy precipitation much better than sandy ones (Jameson, 1968). The same toxin could also effectively inhibit microbial activity in the soil and hence decomposition, which further enhances the opportunity for concentration (Rice, 1974).

Other studies in Mexico have shown significant allelopathic mechanisms by tropical plants. Terpenes and terpene alcohols from leaves and fruits of Schinus molle gave strong inhibitory effects in standard tests on germination of wheat and cucumber (Anaya, Gomez-Pompa, 1971). Anaya and Amo (1978) demonstrated growth inhibitors of Ambrosia cumanensis, both from the living plant and decomposition of organic residues, which allow Ambrosia to compete advantageously against other ruderal second growth species such as Mimosa pudica and M. somnicus, Bidens pilosus, Cassia jalapensis, Achyranthes aspera, Crusia calocephala and Crotolaria sagittalis. The specificity of liberated compounds was demonstrated to depend upon the presence or absence of soil microorganisms. Some of the action of A. cumanensis on these ruderals was due to sesquiterpene lactones, which have specific biological action that depends on concentration and the species over which it acts (Amo, Anya, 1978).

In addition to the effects on seed germination in successional processes, allelopathy can be involved in seed dormancy, prevention of seed decay prior to germination and inhibition of nitrification (Rice, 1977). Allelopathic interactions occur in a wide range of crop-weed combinations which can play a beneficial role in tropical multiple cropping systems and crop rotations (Gleissman, 1983). Also, many of the toxins that have been implicated in allelopathy may be important in protecting the plant against diseases, thus emphasizing again that secondary compounds may play multiple roles.

Long-term study of leguminous resin-producing plants

Detailed, continuing ecological studies of secondary chemicals, either in particular plant taxa or in communities, are rare. Our current long-term evolutionary study of leguminous resin-producing genera has been directed toward understanding the genetic and environmental constraints (both abiotic and biotic) on production and variation in terpenes (Langenheim 1969, 1975). The amphi-Atlantic Caesalpinioid genera *Hymenaea* and *Copaifera* were selected as model systems representative of copious resin-producing tropical angiosperms. Their center of diversity is in rainforest, but in the New World sympatric species of both genera have radiated into all major lowland ecosystem types (Langenheim, 1973; Lee, Langenheim, 1975). Morphologically the two genera are distinct and there are striking similarities as well as differences in their resin chemistry and secretory structures (Langenheim, 1973; Langenheim et al., 1978; Langenheim, 1981). Resin composition is different in the three organs studied (leaf, stem and fruit), which in turn are under different biotic selection pressures. The seeds escape bruchid beetle attack, despite containing neither alkaloids nor non-protein amino acids. In this case, the resins (comprised of both sesquiterpenes and diterpenoids) in the pods appear to deter bruchids (Rehr et al. 1973; Janzen 1973), although rhinochenid weevils are often successful seed predators. The trunk resins also are comprised of sesquiterpenes and diterpenoids but their ratio is strikingly different in the two genera in the New World (Langenheim, 1981); bark beetles apparently are the most common predator in the trunk and are inhibited physically by the resin (S. Wood, pers. comm.). Chemical effects on the bark beetles have not as yet been studied.

The leaf chemistry appears remarkably similar in the two genera, and the leaves have been studied in most detail because: 1) experimental studies could be done under controlled environmental conditions on leaves of seedlings, 2) selection pressures were anticipated to be heavy on seedlings, with leaves especially vulnerable to attack, 3) leaves of seedlings, saplings and adults could be compared. Leaf resins generally are comprised of the same suite of sesquiterpene hydrocarbons, with oxygenated sesquiterpenes present in some resins (Langenheim 1981; Arrhenius et al., 1983). However, quantitative compositional differences are superimposed upon this general qualitative similarity; quantitative variation patterns occur intergenerically, interspecifically and intraspecifically in local populations within widespread species (Langenheim et al. 1978; Langenheim 1981). Variation within populations was relatively low along a geographic gradient of the widespread species, *Hymenaea courbaril*, in Mexico and Central America, but variation among these populations was much greater (Martin et al. 1974). This is in contrast to most South American *H. courbaril* populations and other *Hymenaea* species where variation within a local population is greater than among local populations (Martin et al. 1976; Langenheim et al. 1978). In fact, the greatest intrapopulational differences have been observed in rainforest and related gallery forest ecosystems, where relatively large discrete quantitative differences occur between parent tree and progeny (Langenheim et al., 1977; Stubblebine, Langenheim, 1980; Langenheim, Stubblebine, 1983). Quantitative variation has also been analyzed between tissues, e.g., the young stem and petiole are similar but differ from the leaf blade (Langenheim et al. 1978). Intraleaf variation occurs both spatially in terms of distribution and yield of the leaf secretory pockets (Langenheim et al. 1982) and during development of the leaf (Crankshaw, Langenheim, 1981; Langenheim, Hall 1983). In order to facilitate communication regarding leaf resin quantitative compositional patterns, Compositional Types, based upon major compounds (i.e., those making up at least 10% of the total) were recognized (Langenheim et al., 1980).

Little phenotypic plasticity in these quantitative resin compositional patterns with regard to physical environmental factors, such as photoperiod, irradiance, temperature and moisture status, have been clearly demonstrated in a series of experimental studies of Hymenaea and Copaifera seedlings (Stubblebine et al., 1978; Langenheim et al., 1979, 1981). Tight genetic control of resin composition has also been shown for conifers (Hanover, 1971; Squillace, Dorman, 1972). Yield (i.e., total amount of resin) in Hymenaea leaf resins in seedlings showed little phenotypic plasticity between extremes in moisture status, which is in contrast to monoterpenoids in conifer trees (Von Rudloff, 1975; Hodges, Lorio, 1975) and the perennial labiate Satureja (Gershenzon et al., 1978). With respect to differences in irradiance, however, the yield of resin varied greatly among individuals in both Hymenaea and Copaifera seedlings, and increased significantly from low to high light treatment conditions ($20-200\ \mu E\ m^{-2}\ s^{-1}$) in the control chamber (Langenheim et al. 1982). This great individual variability in leaf resin yield expressed in phenotypic response to irradiance is also shown among seedlings of both Hymenaea and Copaifera in Amazonia rainforest (Nascimento 1980; Langenheim et al. 1983). Since experimental data suggest that inhibitory effects on herbivores are dose-dependent, it would seem that higher resin-yielding individuals would have an advantage under these biotic pressures.

If biosynthesis of the lower terpenoids is directly dependent upon availability of photosynthate (i.e., abundant photosynthate promoting terpene synthesis and/or reducing turnover) as suggested by Loomis and Croteau (1973, 1980), then irradiance on the rainforest floor could be a significant limiting factor for photosynthesis and thus possibly influence terpene accumulation. Greenhouse grown seedlings of both Hymenaea and Copaifera rainforest species showed early light saturation of assimilation rates. In plants grown under 6% (ca. $90\ \mu E\ m^{-2}\ s^{-1}$) and full sunlight (ca. $1500\ \mu E\ m^{-2}\ s^{-1}$), assimilation was usually half saturated between 50 and $100\ \mu E\ m^{-2}\ s^{-1}$ (Langenheim et al. 1983a). In growth chamber-grown seedlings leaf resin yield increased significantly from $20-200\ \mu E\ m^{-2}\ s^{-1}$, whereas it increased little from $90-1500\ \mu E\ m^{-2}\ s^{-1}$ in greenhouse-grown seedlings, with assimilation rates having been light saturated at about the level of the higher light in the growth chamber. However, shaded portions of canopies of saplings of Copaifera in the field had about one half the yield of those in full sunlight, whereas a more complicated relationship occurred within the canopy of emergent trees in the rainforest (Nascimento 1980). These results emphasize the complexities involved in attempting to understand changes in yield of leaf terpenes, and the need for carefully designed experiments to test further the relationship between photosynthetic activity and terpene yield in leaves during different stages of ontogeny of the tree.

One of the most interesting patterns of terpene compositional variation, possibly related to herbivory, occurs between leaves of Hymenaea parent trees and their progeny in rainforest and related gallery forest ecosystems, but not in drier forest ecosystems (Langenheim et al. 1977; Stubblebine, Langenheim 1980; Langenheim, Stubblebine 1983; Nascimento, 1980). Both Janzen (1970) and Connell (1970), in fact, predicted that seedlings would be unsuccessful under the parent tree in rainforests due to predators. However, seedlings often occur in abundance under both Hymenaea and Copaifera parent trees, and an hypothesis has been proposed that terpene variation may play a role in defense against lepidopteran herbivores, and thereby in seedling establishment under or near these rainforest trees. The suggestion, however, does not deny the contribution of various other factors in determining the survival of seedlings. Although various insects have been observed on Hymenaea and Copaifera leaves, lepidopterans (including noctuids, tortricids, geometrids, noctuids, and particularly ecophorids) have been the most conspicuous herbivores (other than leaf-cutting ants) throughout the New World distribution of these legumes. Herbivore damage on all species of Hymenaea and Copaifera has generally been recorded between 1-5%, except for the occasional outbreak years when defoliation of some trees occurs (Nascimento, 1980; Langenheim, Convis, unpublished data).

Due to difficulty in obtaining native herbivore for experiments, as is often the case in the tropics, the beet armyworm (Spodoptera exigua) was used to test the hypothesis that Hymenaea leaf resin could function defensively against a polyphagus lepidopteran with a mixed function oxidase detoxification system of lipophilic substances. In leaf disc palatability tests both unconditioned and conditioned larvae were deterred (Stubblebine, Langenheim 1977). In feeding experiments, in which the armyworm was reared on a series of concentrations of the most common resin Compositional Type incorporated into an agar diet, there were dose-dependent effects on mortality, increased time to pupation and reduction in pupal weight. A subsequent feeding study demonstrated that resins of three quantitative Compositional Types produced differential dose-dependent effects on growth rates (lower larval weights and increased time to pupation) and on mortality of Spodoptera exigua, thus implicating herbivore selection pressures as a factor in determination of quantitative variation patterns (Langenheim et al. 1980).

Although an adult tree is persistent and apparent to insects, rainforest trees may be sufficiently dispersed that a new flush of soft and succulent leaves might escape herbivorous attack due to relatively low apparency; however, those trees attacked could be essentially defoliated, if tree defenses were inadequate and herbivore populations high. This type of "explosive" lepidopteran larval attack has been observed in Amazonian rainforest with scattered individual trees being attacked and nearest neighbors not attacked. In Costa Rican forests there also is a wide gradient of herbivory by lepidopterans, such as noctuids, geometrids and pyralids, from individual trees with no damage to those defoliated (Janzen, 1981).

If certain trees are heavily attacked, exhaustion of food supply in the crown could give rise to migration of late instar larvae to the forest floor directly under trees. Many of these larvae are sufficiently mature to pupate, whereas other are still actively searching for food. If the variation in resin composition has differential effects on these native lepidopterans (as shown for Spodoptera exigua), then the late instar larvae migrating from the crown to the forest floor are likely to have adapted to the resin composition of the adult tree either through developmental habituation or from drastic selection in the early larval stages, both well known phenomena in insects (Hagen 1976; Harcourt 1969). They are sufficiently mobile as late as instar larvae to search for preferred chemical compositions on the forest floor, which greatly increases the likelihood that resin composition of many of the surviving seedlings would be unlike the parent. Since it has been demonstrated experimentally that leaf resin composition is under tight genetic control, and changes in leaf resin with ontogeny of the tree appear minimal, this selection by herbivores could then maintain and possibly increase diversity of leaf terpene composition among adult trees.

In central and southern Brazil, cerrado is a vegetation type of considerable floristic diversity, which supposedly is similar to vegetation which occurred in Amazonia during dry intervals of the Pleistocene. Our field studies in cerrado have shown that the ecophorid lepidopteran, Stenoma ferrocanella, attacked Hymenaea stigonocarpa after the time of peak concentration of leaf resins early in leaf development (Langenheim, Hall 1983). This corresponds to the time of peak water and nitrogen content, and least leaf toughness. Phenolic compounds also have the highest astringency levels at this early stage of Hymenaea leaf development in seedlings (Crankshaw, Langenheim 1981). Thus it would appear that the plant has chemically defended the leaf during its most photosynthetically active period (Langenheim et al., 1983; McKey, 1974, 1979; Mooney, Gulman 1982). This stage in leaf development is also the most nutritious in terms of nitrogen, most succulent and the least tough tissue for the larvae. In fact, the larvae attack in mid-leaf development, before the nitrogen and water content decrease and toughness increases too greatly. Moreover, absence of Stenoma ferrocanella attack on H. stigonocarpa was correlated with significantly higher concentrations of caryophyllene in one population and γ-muurolene in another (Langenheim, Hall, 1983). The deterrent and possibly toxic role of caryophyllene is interesting in that there is a clear dose-dependent effect of caryophyllene on mortality in Spodoptera exigua in lab experiments and a threshold concentration related to damage by Stenoma ferrocanella on H. stigonocarpa in the field. Also when caryophyllene concentration is very high, the total yield of the leaf resins is low. Caryophyllene is a commonly occurring compound in plants, but has been reported to attain relatively high concentrations in only a few. It appears that caryophyllene in Hymenaea and Copaifera could function with some leaf-tying ecophorids similar to limonene, a single monoterpene in xylem resin of Pinus ponderosa, which significantly inhibits some populations of western bark beetles (Dendroctonus) (Smith, 1972). Sturgeon (1979) has further discovered directional selection for limonene in P. ponderosa and that high amounts of this toxin are superimposed upon highly variable combinations of four other monoterpenes.

In another cerrado site dominated by Copaifera langsdorfii, in which defoliation of numerous trees by Stenoma sp. (near assignata) and Chlamydastis chloroloba was observed, high selinenes predominated (Langenheim et al., 1983b). When selinenes are the major compounds, the total resin yield also tends to be high. Despite high levels of damage, there was a threshold concentration of selinenes above which damage levels dropped. Also both caryophyllene and cyperene concentrations were significantly higher, along with the selinenes, in less damaged or undamaged trees. These results suggest that the ecophorids could be influencing quantitative variation directionally in sympatric species of the two genera. Furthermore, McKey (1979) has indicated that this kind of variation in secondary chemicals could protect against a broad spectrum of herbivores and pathogens.

In addition to lepidopteran herbivory on Hymenaea and Copaifera, effects of a leaf-spotting fungus have been studied. Although a number of leaf-spotting fungi have been reported from these genera (Heringer 1971), Pestalotia (Pestalotiopsis) was selected because of its occurrence in a diversity of tropical lowland ecosystems, its ease in culturing for experimental studies, and because infection experiments in the greenhouse showed it to be potentially pathogenic. Because Pestalotia grows within healthy and often young tissues (as is true with many tropical leaf colonists) chemical mechanisms may well be important in restricting this fungal growth in live tissue. Mycelial growth inhibition was correlated with the concentration of caryophyllene oxide (Arrhenius, Langenheim 1983). As precursor to caryophyllene oxide, variation in caryophyllene may reflect fungal selection pressures. However, the possibility of fungal induction of caryophyllene oxides also exists.

Until recently, it was generally assumed that defense mechanisms against insects and pathogens evolved only under their own specific selection pressures, and thus there has been little effort to determine commonalities which might exist between different systems. Apparently an exciting possibility exists that pathogen-induced secondary chemicals can also have a definite effect on insect herbivores (Kogan, Paxton, 1983).

In this study of two important tropical resin-producing genera, it has become clear that variability in the terpenes (and probably phenolic compounds as well) may contribute significantly to their effectiveness in defense. Variability in general has been reported to be higher in trees than in other life forms (Hamrick et al., 1979). At first sight quantitative terpene variation may appear minor and only "noise" resulting from genetic variability. However, it may be a "flexible," perhaps economic "manipulation" of established biochemical pathways. Many terpenes, such as those found in resins, are under relatively simple genetic control (Hanover, 1971; Squillace, Dorman, 1961) and compositional variation could be responsive to different biotic pressures, which are expressed both spatially and temporally among populations, within populations, within the canopy, between seedlings, saplings and adult trees and within the leaf itself.

Additionally, the total amount of resin may vary plastically in response to physical environmental conditions. Although there is little evidence as yet for these trees, changes in both yield and composition of resins may be induced by insect and pathogen attack. In studies of temperate deciduous trees, it has been suggested that chemical variability probably places insects in compromise situations which increase their exposure and susceptibility to natural enemies (Schultz, 1983). Additionally, the presence of complex mixtures of both terpenes and phenolic compounds in Hymenaea and Copaifera may increase the time it takes for insects to adapt, as shown by mixtures of other compounds (Pimentel, Belloti, 1976). Thus even sublethal doses of toxic compounds may remain effective over long periods of time. Tree defense variability may be as significant, or even more so, than uniform resistance. Although there are many ways in which secondary compounds can influence the effectiveness of their enemies, they probably can remain effective through evolutionary time only if variability (both genotypic and phenotypic) is a part of the scheme. Chemical variability as a mechanism for defense appears to be important in both temperate and tropical trees, but sufficient data are currently unavailable to compare the levels of variability. However, our studies at least indicate greater variability in terpenes in rainforest and related gallery forest ecosystems, than in the drier tropical ecosystems.

FUTURE STUDIES

If indeed the complexity of biotic interactions reaches a peak in wet tropical ecosystems, we could anticipate plant production of secondary chemicals to be significant in mediating aspects of pollination, dispersal, defense and competition. As is evident from this review, investigation of the roles of these compounds in plants from wet tropical ecosystems is wide open--with most studies to date constituting pioneering efforts. Undoubtedly, hundreds of new secondary compounds will be described from tropical plants by chemists seeking useful products for medicine, industry, etc., and it is assumed that tropical rainforests are the "place" where rich rewards may be found. However, understanding the variation and the roles these chemicals may play within the plant, and as a part of an ecosystem, is beset with more than the usual conceptual and methodological pitfalls. Under all circumstances, these kinds of studies benefit from the collaboration of chemists and ecologists. In wet tropical ecosystems we can expect increased complexity in every aspect of the work, and thus collaboration of chemists, entomologists, mycologists and ecologists with different specialities is particularly desirable. Researchers from extra-tropical regions also need collaborators living in the tropics who can maintain year-round observations. Furthermore, any depth of understanding involving secondary chemicals will require sustained projects (a problem not unique to the tropics).

In recognition of some of these problems, a multinational group, the International Center for Insect Physiology and Ecology (ICIPE), was established in Nairobi, Kenya to carry out interdisciplinary tropical studies, which include secondary chemicals (Meinwald et al., 1977). Since research at this center emphasizes the effect of plant chemicals upon the physiology and ecology of insects, it would indeed be advantageous to have another tropical center which focuses on the plant physiology and ecology side of the plant/chemical/insect interaction. However, even at ICIPE, the lack of knowledge regarding most native insects has led to the standard use of a monophagus and polyphagus species of Spodoptera to determine antifeedant activities of tropical plant secondary chemicals. Thus natural history studies of herbivores and pathogens are a necessity in order to work out diets for experiments to isolate the effects of particular compounds on the relevant natural populations of insects or fungi. Monitoring changes in secondary chemistry following herbivore or pathogen attack to assess possible induction of chemicals, beyond the first-line constitutive defense, is of considerable interest, but may be particularly difficult to document under tropical field conditions.

The ultimate goal from the perspective of physiological ecology would be to determine the role secondary chemicals play relative to the overall metabolism and fitness (productivity/biomass) and/or to the reproductive potential of the plant (Chew, Rodman, 1979). This goal demands information regarding biosynthesis, turnover rates and the assessment of the balance between utilization of secondary chemicals for defense and other biotic interactions, and reallocating them to other plant physiological functions. In order to move into this exciting realm of determining the costs of the chemicals in the plant's overall economy in tropical wet ecosystems, it seems most feasible at this juncture to select plants carefully for which there is some established biochemical data base and that are amenable to both field and laboratory experimentation.

ACKNOWLEDGEMENTS

This paper grew out of research supported by NSF grants BMS 73-22331 and DEB 76-13607. I also wish to thank K. L. Stopol, G. Fail, G. D. Hall, C. Macedo, R. Gianno, S. P. Arrhenius and L. P. McCloskey for comments on the manuscript.

REFERENCES

Amo S del and Anaya AL (1978) Effect of some sesquiterpenic lactones on the growth of certain secondary tropical species, J. Chem. Ecol. 4, 305-315.

Anaya AL and Gomez-Pompa A (1971) Inhibicion del crescimiento producida por el "Piru" (Schinus molle L). Revista de la Sociedad Mexicana de Historia Natural 32, 99-109.

Anaya AL and Amo S del (1978) Allelopathic potential of Ambrosia cumanensis HBK (Compositae) in a tropical zone of Mexico, J. Chem. Ecol. 4, 289-304.

Arrhenius SP, Foster CE, Edmonds CG and Langenheim JH (1983) Sesquiterpenes in leaf pocket resin of Copaifera (Leguminosae, Caesalpinioideae), Phytochemistry 22, 471-472.

Arrhenius SP and Langenheim JH (1983) Inhibitory effects of Hymenaea and Copaifera leaf resins on an associated leaf fungus, Pestalotia subcuticularis, Biochem. Syst. Ecol. 11, in press.

Atabekov JG (1975) Host specificity of plant viruses, Annu. Rev. Phytopathol. 13, 124-146.

Baldwin IT and Schultz JC (1983) Rapid changes in tree chemistry induced by damage: evidence for communication between plants, Science 221, 277-278.

Bigger M (1976) Oscillations of tropical insect populations, Nature 259, 207-209.

Boll PM and Anderson B (1962) Alkaloid glycosides from Solanum dulcamarum III Differentiation of geographical strains by means of thin-layer chromatography, Planta Mec. 10, 421-432.

Camp WH (1949) Cinchona at high altitudes in Ecuador, Brittonia 6, 394-430.

Chew FS and Rodman JE (1979) Plant resources for chemical defense. In Rosenthal GH and Janzen DH, eds. Herbivores: their interaction with secondary plant metabolites, pp 271-300. New York, Academic Press

Coley PD (1983) Herbivory and defensive characteristics of tree species in a lowland tropical forest, Ecol. Monog. 53, 209-233.

Connell JH (1970) On the role of natural enemies in preventing competitive exclusion in some marine animals and in rain forest trees. Proc. Adv. Study Inst. Dynamics Numbers Popl. (Oosterbeck), 298-312.

Crankshaw DR and Langenheim JH (1981) Variation in terpenes and phenolics through leaf development in Hymenaea and its possible significance to herbivory, Biochem. Syst. Ecol. 9, 116-124.

Cruickshank IAM (1963) Phytoalexins, Ann. Rev. Phytopathol. 1, 351-374.

del Moral R (1972) On the variability of chlorogenic acid concentration, Oecologia 9, 289-300.

Derman BD, Rupp DC and Nooden LD (1978) Mineral distribution in relation to fruit development and monocarpic senescence in Anoka soybeans, Amer. J. Bot. 65, 205-213.

Dethier VG (1959) Food-plant distribution and density and larval dispersals as factors affecting insect populations, Can. Entomol. 91, 581-596.

Dethier VG (1970) Chemical interactions between plants and insects. In Sondheimer E and Simeone JB, eds. Chemical ecology, pp. 83-120. New York, Academic Press.

Dodson CH (1970) The role of chemical attractants in orchid pollination, Biochemical coevolution, Oregon State University Press, 83-107.

Dodson CH (1975) Coevolution of orchids and bees. In Gilbert LE and Raven PH, eds. Coevolution of animals and plants, pp 91-99. Austin, Tx., University of Texas Press.

Dodson CH, Dressler RL, Hollis HG and Adams RM (1969) Biologically active compounds in orchid fragrances, Science 164, 1243-1249.

Dolinger PM, Ehrlich PR, Fitch WL and Breedlove DE (1973) Alkaloid and predation patterns in Colorado lupine populations, Oecologia 13, 191-204.

Earle FR and Jones Q (1962) Analysis of seed samples from 113 plant families, Econ. Bot. 16, 221-250.

Ehrlich PR and Raven PH (1964) Butterflies and plants: a study in coevolution, Evolution 18, 586-608.

Feeny P (1975) Biochemical coevolution between plants and their insect herbivores. In Gilbert LE and Raven PH eds. Coevolution of animals and plants, pp 3-19. Austin, Tx., University of Texas Press.

Feeny PP (1976) Plant apparency and chemical defense. In Wallace JM and Mansell RJ, eds. Biochemical interaction between plants and insects, Recent Adv. Phytochem. 10, 1-40.

Fong HHS, Trojankova M, Trojanet JT and Farnsworth NR (1972) Alkaloid screening II Lloydia 35, 117-149.

Freeland WJ and Janzen DH (1974) Strategies in herbivory by mammals: the role of plant secondary compounds, Amer. Nat. 108, 269-289.

Gartlan JS, McKay DB, Waterman PG, Mbi CN and Struhsaker TT (1980) A comparative study of the phytochemistry of two African rainforests, Biochem. Syst. Ecol. 8, 401-422.

Gershenzon J, Lincoln DE and Langenheim JH (1978) The effect of moisture stress on monoterpenoid yield and composition in Satureja douglasii, Biochem. Syst. Ecol. 6, 33-43.

Gliessman SR (1976) Allelopathy in a broad spectrum of environments is illustrated by bracken, Bot. Jour. Linnean Soc. 73, 95-104.

Gliessman SR (1978) Allelopathy as a potential mechanism of dominance in the humid tropics, Trop. Ecology 19, 200-208.

Gliessman SR (1983) Allelopathic interactions in crop/weed mixtures: applications for weed management, J. Chem. Ecol. 9, in press.

Golley FB (1972) Energy flux in ecosystems. In Wiens JA, ed. Ecosystem structure and function. Oregon State Univ. Ann. Biol. Colloq. 31, 69-90.

Gordon HT (1961) Nutritional factors in insect resistance to chemicals, Ann. Rev. Entomol. 6, 27-54.

Gottlieb OR and Mors WB (1978) Fitoquimica Amazonica: um apreciacão em perspective, Interciencia 3, 252-263

Gottlieb OR and Mors WB (1980) Potential utilization of Brazilian wood extractives, J. Agric. Food Chem. 28, 196-215.

Hagen KS (1976) Role of nutrition in insect management, Proc. Tall Timbers Conf. on Ecol. Animal Control by Habitat Management 6, 221-261.

Hamrick JH, Linhart YB and Mitton JB (1979) Relationships between life history characteristics and electrophoretically detectable genetic variation in plants, Ann. Rev. Ecol. Syst. 10, 173-200.

Hanover JW (1971) Genetics of terpenes: II. Genetic variances and interrelationships of monoterpenoid concentrations in Pinus monticola, Heredity 27, 237-245.

Harborne JB (1982) Introduction to ecological biochemistry, London, Academic Press.

Harcourt DC (1969) The development and rise of life tables in the study of natural insect populations, Annu. Rev. Entomol. 14, 175-196.

Hartley TG, Dunstona EA, Fitzgerald JS, Johns SR and Lamberton JA (1973) A survey of New Guinea plants for alkaloids, Lloydia 36, 217-319

Heringer EP (1971) Arvores uteis da regiao geoeconomica do distrato federal, Cerrado 7, 27-32.

Hodges JD and Lorio PL (1975) Moisture stress and composition of xylem oleoresin in loblolly pine, Forest Science 22, 283-292.

Holliday P (1971) Some tropical plant pathogenic fungi of limited distribution, Rev. Plant. Pathol. 50, 337-348.

Ikeda T, Matsumura F and Benjamin DM (1977) Chemical basis for feeding adaptation of pine sawflies neodiprior rugifrons and neodiprion-swainei, Science 197, 497-499.

Jameson DA (1971) Degradation and accumulation of inhibitory substances from Juniperus osteosperma (Torr.) Litle. In Biochemical Interactions Among Plants, pp. 121-127 Washington, D.C., U.S. Nat'l Acad. Sci.

Janzen DH (1969) Seed-eaters versus seed size, number, toxicity and dispersal, Evolution 23, 1-27.

Janzen DH (1970) Herbivores and the number of tree species in tropical forests, Amer. Naturalist 104, 501-528.

Janzen DH (1973a) Community structure of secondary compounds in plants. Pure & Applied Chemistry 34, 529-538.

Janzen DH (1973b) Sweep samples of tropical foliage insects: effects of seasons, vegetation types, elevation, time of day and insularity, Ecology 54, 687-708.

Janzen DH (1973c) Sweep samples of tropical foliage insects: description of study sites, with data on species abundance and size distribution, Ecology 54, 659-686.

Janzen DH (1974) Tropical blackwater rivers, animals and mast fruiting by the Dipterocarpaceae, Biotropica 6, 69-103

Janzen DH (1979) New horizons in the biology of plant defense. In Rosenthal GH and Janzen DH, eds. Herbivores: their interactions with secondary metabolites, pp 331-348. New York, N.Y., Academic Press.

Janzen DH (1981) Patterns of herbivory in a tropical deciduous forest, Biotropica 13, 271-282

Janzen DH and Schoener TN (1968) Difference in insect abundance and diversity between wetter and drier sites during a tropical dry season, Ecology 49, 96-110.

Janzen DH and Pond CN (1975) A comparison by sweep sampling of the arthropod fauna of secondary vegetation in Michigan, England and Costa Rica, Trans. R. Entomol. Soc. London 127, 33-50.

Janzen DH, Juster HB and Ball EA (1977) Toxicity of secondary compounds to the seed-eating larvae of the bruchid beetle Callosobruchus maculatus, Phytochemistry 16, 223-227.

Jones DA (1979) Chemical defense: primary or secondary metabolism? Am Nat 113, 445-451.

Kogan M and Paxton J (1983) Natural inducers of plant resistance to insects. In Hedin PA, ed. Plant resistance to insects, pp. 153-171. Washington, D.C., ACS Symposium Series 208.

Krieger RI, Feeny PP and Wilkinson CF (1971) Detoxification enzymes in the guts of caterpillars: An evolutionary answer to plant defenses? Science 172, 579-581.

Langenheim JH (1969) Amber: a botanical inquiry, Science 168, 1157-1169.

Langenheim JH (1973) Leguminous resin-producing trees in South America and Africa. In Meggers BJ, Ayensu ES and Duckworth WD, eds. Tropical forest ecosystems in Africa and South America: A comparative review, pp 89-104. Washington, DC, Smithsonian Press.

Langenheim JH (1975) Role of the tropics in evolution of resin-producing trees. Proc. VII Internatl. Bot. Congress, Leningrad, 116.

Langenheim JH (1981) Terpenoids in the Leguminosae. In Polhill RH and Raven PH, eds. Advances in legume systematics, pp. 627-655. Kew, UK, Proc. int. Legume Congress, Royal Bot. Gard.

Langenheim JH (1983) Utilization of resins from tropical trees: past, present and future potential, Memoirs of New York Bot. Garden, in press.

Langenheim JH, Lee YT and Martin SS (1973) An evolutionary and ecological perspective of the Amazonian hylaea species of Hymenaea (Leguminosae, Caesalpinioideae), Acta Amazonica 3, 5-38.

Langenheim JH, Stubblebine WH, Foster CE and Nascimento JC (1977) Estudos comparativos da variabilidada na composicão da resina da folha entre arvore parental e progenie de especies selectionadas de Hymenaea. I. Comparacao de populacoes Amazonica e Venezuelanas, Acta Amazonica 7, 335-354.

Langenheim JH, Foster CE, Lincoln DE and Stubblebine WH (1978) Implications of variation in resin composition among organs, tissues and populations in the tropical legume Hymenaea, Biochem. Syst. Ecol. 6, 299-213.

Langenheim JH, Stubblebine WH and Foster LE (1979) Effect of moisture stress on leaf resin composition and yield in Hymenaea courbaril, Biochem. Syst. Ecol. 7, 21-28.

Langenheim JH, Foster CE and McGinley RM (1980) Inhibitory effects of different quantitative compositions of Hymenaea leaf resins on a generalist herbivore Spodoptera exigua, Biochem. Syst. Ecol. 8, 385-396.

Langenheim JH, Arrhenius SP and Nascimento JC (1981) Relationship of light intensity to leaf resin composition and yield in the leguminous genera Hymenaea and Copaifera, Biochem. Syst. Ecol. 9, 27-37.

Langenheim JH, Lincoln DE, Stubblebine WH and Gabrielli AL (1982) Evolutionary implications of leaf resin pocket patterns in the tropical tree Hymenaea (Caesalpinioideae: Leguminosae), Amer. J. Bot. 69, 595-607.

Langenheim JH and Hall GD (1983) Sesquiterpene deterrence of a leaf-tying lepidopteran Stenoma ferrocanella on Hymenaea stigonocarpa in Central Brazil, Biochem. Syst. Ecol. 11, 29-36.

Langenheim JH and Stubblebine WH (1983) Variation in resin composition between parent tree and progeny in Hymenaea: Implications for herbivory in the humid tropics, Biochem. Syst. Ecol., 11, 97-106.

Langenheim JH, Osmond CB, Brooks A and Ferrar PJ (1983a) Photosynthetic responses to light of Amazonian and Australian rainforest seedlings, Oecologia, in press.

Langenheim JH, Convis CL, Stopol KL, and Stubblebine WH (1983b) Chemical leaf defense of Copaifera and Hymenaea against lepidopteran insects in São Paulo State, Jour. S.P. Forestry, in press.

Lee YT and Langenheim JH (1975) Systematics of the genus Hymenaea (Leguminosae, Caesalpinioideae, Detarieae), Univ. of Calif. Public. in Botany no 69, 109 p, Berkeley, Calif., Univ. of Calif. Press.

Levin DA (1971) Plant phenolics: an ecological perspective, Amer. Nat. 105, 157-181.

Levin DA (1975) Pest pressures and recombination systems in plants, Amer. Nat. 109, 437-451.

Levin DA (1976a) The chemical defense of plants to pathogens and herbivores, Annu. Rev. Ecol. Syst. 7, 121-160.

Levin DA (1976b) Alkaloid-bearing plants: an ecogeographic perspective, Amer. Nat. 110, 261-284.

Levin DA and York BM (1978) The toxicity of plant alkaloids: an ecogeographic perspective, Biochem. Syst. Ecol. 6, 61-76.

Loomis WD and Croteau R (1973) Biochemistry and physiology of lower terpenoids, Recent Adv. in Phytochem. 6, 147-185.

Loomis WD and Croteau R (1980) Biochemistry of terpenoids. In Stumpf PK and Conn E, eds. The biochemistry of plants 4, pp. 363-418, New York, Academic Press.

Mabry TJ and Ulubelen H (1980) Chemistry and utilization of phenylpropanoids including flavonoids, coumarins and lignans, J. Agric. Food Chem. 28, 188-196.

Martin SS, Langenheim JH, Zavarin E (1974) Quantitative variation in leaf pocket resin composition of Hymenaea courbaril, Biochem. Syst. Ecol. 3, 760-787.

Martin SS, Langenheim JH and Zavarin E (1976) Quantitative variation of leaf resin composition in Hymenaea, Biochem. Syst. Ecol. 4, 181-191.

McKey DB (1974) Adaptive patterns in alkaloid physiology, Amer. Nat. 108, 305-320.

McKey D (1979) The distribution of secondary compounds within plants. In Rosenthal GH and Janzen DH, eds. Herbivores: their interaction with secondary plant metabolites, pp. 55-133. New York, Academic Press.

McKey DB, Waterman PG, Msi CH, Gartlan JS and Struhsaker TT (1978) Phenolic content of vegetation in two African rain forests: ecological implications, Science 202, 61-63.

Meinwald J, Prestwich CD, Nakanishi K and Kubo I (1978) Chemical ecology: studies from East Africa, Science 199, 1167-1173.

Mooney HA and Gulman SL (1982) Constraints on leaf structure and function in reference to herbivory, Bioscience 32, 198-206

Mooney HA, Gulman SL and Johnson ND (1983) Physiological constraints on plant chemical defenses. In Hedin PA, ed. Plant resistance to insects, pp.21-36. Washington, D.C., ACS Symposium Series 208.

Nascimento JC (1980) Ecological studies of sesquiterpenes and phenolic compounds in leaves of Copaifera multijuga Hayne (Leguminosae) in a Central Amazonian rainforest, Ph.D. dissertation, Univ. of California, Santa Cruz.

Oates JF, Swain T and Zantovska J (1977) Secondary compounds and food selection by colobus monkeys, Biochem. Syst. Ecol. 5, 317-321.

Oates JF, Waterman PG and Choo GM (1980) Food selection by the South Indian leaf monkey, Presbytis johnii, in relation to leaf chemistry, Oecologia 45, 45-56.

Reichstein T (1963) Chemische Rassen von Strophanthus sarmentosus, Planta Med. 11, 293-302.

Rehr SS, Ball EA, Janzen DH and Feeny PP (1973) Insecticidal amino acids in legume seeds. Biochem. Syst. 1, 63-67.

Rehr SS, Janzen DH and Feeny PP (1973) L-dopa in legume seeds: A chemical barrier to insect attack, Science 181, 81-82.

Rhoades DE (1979) Evolution of plant chemical defense against herbivores. In Rosenthal GH and Janzen DH, eds, Herbivores: their interactions with secondary plant metabolites, pp. 4-48. New York, Academic Press.

Rhoades DF (1983) Responses of alder and willow to attack by tent caterpillars and webworms: evidence for pheromonal sensitivity of willows. In Heden PA, ed. Plant resistance to insects, pp. 55-68. Washington, D.C., ACS Symposium Series 208.

Rhoades DF and Cates RG (1976) Toward a general theory of plant antiherbivore chemistry. In Wallace JM and Mansell RJ, eds. Biochemical interaction between plants and insects, Recent Adv. Phytochem. 10, 168-213.

Rice EL (1974) Allelopathy. New York, Academic Press.

Rice EL (1977) Some roles of allelopathic compounds in plant communities, Biochem. Syst. Ecol. 5, 200-206.

Robinson T (1979) The evolutionary ecology of alkaloids. In Rosenthal GA and Janzen DH, eds. Herbivores their interaction with secondary plant metabolites, pp. 413-448. New York, N.Y., Academic Press.

Rosenthal GA (1983) Biochemical adaptations of the bruchid beetle, Caryedes brasiliensis to L-canavanine, a higher plant allelochemical, J. Chem. Ecol. 9, in press.

Rosenthal GA (1974) The interrelationship of canavanine and urease in seeds of the Lotoideae, J. Exp. Bot. 25, 609-613.

Rosenthal GA (1977) Nitrogen allocation for L-canavanine synthesis and its relationship to chemical defense of the seed, Biochem. Syst. Eco. 5, 219-220.

Rosenthal GA, Dahlman DL and Janzen DH (1976) A novel means for dealing with L-canavanine, a toxic metabolite, Science 192, 256-258.

Rosenthal GA, Janzen DH and Dahlman DL (1977) Degradation and detoxification of canavanine by a specialized seed predator, Science 196, 658-660.

Rosenthal GA and Ball EA(1979) Naturally occurring, toxic nonprotein amino acids. In Rosenthal GA and Janzen DH, eds. Herbivores: their interaction with secondary plant metabolites, pp. 353-385. New York, N.Y., Academic Press.

Rosenthal GA, Hughes C and Janzen DH (1982) L-canavanine, a dietary source for the seed predator Caryedes brasiliensis (Bruchidae), Science 217, 353-355.

Ryan CA (1979) Proteinase inhibitors. In Rosenthal GA and Janzen DH, eds. Herbivores: their interaction with secondary plant metabolites, pp. 599-617. New York, Academic Press.

Ryan DF and Bormann FH (1982) Nutrient resorption in northern hardwood forests, Bioscience 32, 29-32.

Sanders H (1963) Chemische Differenzierung innerhalb der Art Solanum dulcamara L, Plant Med 11, 287-298.

Scriber JM (1973) Latitudinal gradients in larval feeding specialization of the world Papilionidae (Lepidoptera), Psyche 80, 355-373.

Schramm LC and Scharting AE (1961) Alkaloid distribution in Colombian cinchonas, Lloydia 24, 1-26.

Schultz JC (1983) Impact of variable plant defensive chemistry on susceptibility of insects to natural enemies. In Hedin PA, ed. Plant resistance to insects, pp. 39-54. Washington, D.C., ACS Symposium Series 208.

Schultz JC and Baldwin IT (1982) Oak leaf quality declines in response to defoliation by gypsy moth larvae, Science 217, 149-151.

Schultz JC, Nothnagle PJ and Baldwin IT (1982) Seasonal and individual variation in leaf quality of two northern hardwoods trees species, Amer. J. Bot. 69, 753-759.

Siegler DS and Price PW (1976) Secondary compounds in plants: primary functions, Am. Nat. 110, 101-105.

Siegler DS (1977). Primary roles for secondary compounds, Biochem. Syst. Ecol. 5, 195-199.

Smith RH (1972) Xylem resin in the resistance of the Pinaceae to bark beetles, U.S. Dep. Agric. For. Serv. Gen. Tech. Rep. PSW-1 Pac. Southwest Forest and Exp. Stat., Berkeley, Calif.

Smolenski SJ, Silinis H and Farnsworth NR (1972) Alkaloid Screening III. Lloydia 36, 359-387.

Smolenski SJ, Silinis H and Farnsworth NR (1973) Alkaloid Screening I. Lloydia 35, 1-34.

Smolenski SJ, Silinis H and Farnsworth NR (1974a) Alkaloid Screening IV. Lloydia 37, 30-61.

Smolenski SJ, Silinis H and Farnsworth NR (1974b) Alkaloid Screening V. Lloydia 37, 506-536.

Squillace AE and Dorman KW (1961) Selective breeding of slash pine for high oleoresin yield and other characters, Recent Advances in Botany 2, pp. 1616-1621, Univ. of Toronto Press.

Stanton N (1975) Herbivore pressure on two types of tropical forests, Biotropica 7, 8-11.

Stoessel A (1970) Antifungal compounds produced by higher plants. In Stulink C and Runeckles VC, eds. Recent Adv. Phytochem. 3, 143-180.

Strong DR (1977) Insect species richness: Hispine beetles of Heliconia latispatha, Ecology 58, 573-582.

Strong DR and Levin DA (1979) Species richness of plant parasites and growth form of their hosts, Amer. Nat. 114, 1-22.

Stubblebine WH, Lincoln DE and Langenheim JH (1975) Vegetative photoperiodic response and resin composition in Hymenaea courbaril, Biochem. Syst. Ecol. 3, 219-228.

Stubblebine WH, Lincoln DE and Langenheim JH (1978) Vegetative responses to photoperiod in the tropical leguminous tree Hymenaea courbaril, Biotropica 10, 18-29.

Stubblebine WH and Langenheim JH (1978) Effects of Hymenaea courbaril leaf resin on the generalist herbivore Spodoptera exigua (beet armyworm), J. Chem. Ecol. 3, 633-647.

Stubblebine WH and Langenheim JH (1980) Estudos comparativos da variabilidade na composicao da resina da fola entre arvore parental e progenie de especies selectionadas de Hymenaea. II Comparacao de populacoes adicionas Amazonicas e do sul do Brasil, Acta Amazonica 10, 293-307.

Sturgeon KB (1979) Monoterpene variation in ponderosa pine xylem resin related to western pine beetle predation, Evolution 33, 803-814.

Swain T (1977) Secondary compounds as protective agents, Ann Rev Plant Physiol 28, 479-501.

Swain T (1979) Tannins and lignins. In Rosenthal GH and Janzen DH, eds. Herbivores: their interactions with secondary plant metabolites, pp. 657-681. New York, N.Y. Academic Press.

Von Rudloff E (1975) Volatile leaf oil analysis in chemosystematic studies in North American conifers, Biochem. Syst. Ecol. 2, 131-167.

Wang TSC, Cheng SY and Tung H (1967) Extraction and analysis of soil organic acids, Soil Sci. 103, 360-366.

Wellman FL (1972) Tropical American plant disease, Metachen, N.J., Scarecrow Press.

Wender SH (1970) Effects of some environmental stress factors on certain phenolic compounds in tobacco, Recent Adv. Phytochem. 8, 1-29.

Whittaker RH (1970) The biochemical ecology of higher plants. In Sondheimer E and Simeone H, eds. Chemical ecology, pp. 43-70. New York, Academic Press.

Whittaker RH (1975) Communities and ecosystems, 2nd edn. New York, N.Y. Macmillan Publ. Co.

Whittaker RH and Feeny P (1971) Allelochemics: chemical interactions between species, Science 71, 757-770.

Williams NH and Dodson CH (1972) Selective attraction of male euglossine bees to orchid fragrances and its importance in long distance pollen flow, Evolution 26, 84-95.

INSECT-PLANT INTERACTIONS: SOME ECOPHYSIOLOGICAL CONSEQUENCES OF HERBIVORY

RODOLFO DIRZO
(Laboratorio de Ecología, Instituto de Biología, UNAM, Apartado Postal 70-233, 04510 México, D.F.)

1. INTRODUCTION

If an intelligent inhabitant of another planet (IAP) wanted to get an idea of what life on earth is like, and was provided only with literature on tropical biology, the sort of conclusions that IAP would arrive at, would be quite misleading. For example, IAP would be led to conclude that, since the proportion of tropical species of plants and animals is so large, this sort of ecosystem should occupy a very large proportion of the earth's crust; also, and quite disappointingly, IAP would have to conclude that man's knowledge of biology in general, is extremely poor. The significance of this metaphor is quite obvious: tropical rain forests are the most complex assemblages of organisms on this planet and contain a disproportionately large number of species of plants and animals; also, in comparison with other types of ecosystems, tropical ones are very poorly understood. Happily however, books like the present one will assist in closing this gap in knowledge. This is a fundamental necessity because a correct management of the vast array of potentially useful resources available in this sort of system cannot be made in the absence of a sound scientific knowledge in tropical biology. Moreover, the tremendous rate of destruction of these reservoirs of life (see Raven 1977) demands that scientists must get together to work seriously in tropical systems before it is too late; this should be done for reasons of intelligent management of natural resources, but also because this knowledge is essential if we are properly to understand life on this planet.

One of the most striking features of tropical systems is the great number of biotic interactions between organisms at the same, or at different trophic levels. Outstanding among the latter are the spectacular interactions between animals and plants. These include most remarkable cases of mutualistic relationships such as the pollination of plants by a wide variety of animals (bats, birds, and many different kinds of insects); seed dispersal by a host of different animals (including birds and several mammals) and the mutualistic interaction in which ants act as protective agents of plants against herbivores and competing plants (Janzen 1966, 1969a, 1974).

In contrast to these mutualistic interactions, another plant-animal interface well represented in the tropics is that in which the animal acts as a predator consuming either whole plants (e.g., seeds) or, more commonly, consuming plant parts (e.g., leaves). Since in this kind of interaction the prey is killed only under special circumstances the plant's fitness being reduced *via* reduction of growth, competitive and reproductive activities (or a combination thereof), this interaction is particularly suitable for study from a phytocentric viewpoint and with a strong ecophysiological emphasis. Such an ecophysiological approach is fundamental to the understanding of the performance of plants in their natural environment. Since herbivory is presumed to be a major environmental factor of tropical plants, it follows that an ecophysiological approach to the study of herbivory in the tropics should be particularly rewarding. Unfortunately, such an approach has not been seriously attempted for tropical plants. With few exceptions, most studies of herbivory in the tropics have been made either with a strong zoocentric emphasis (see Gilbert 1977), or else have emphasized the possible evolutionary consequences of the interaction (e.g., coevolution) (see Derr 1980; Gilbert, Raven 1975). The aim of this paper is to present some ecological aspects of the interaction between insects and tropical plants to expose some areas of this interface in which the participation of plant physiologists might be necessary to gain a more adequate understanding of this topic. I will concentrate on the interaction with insects since this is the group I am most familiar with and because, at least for some tropical systems, herbivory by insects appears to be of greater impact than that of other animals. The whole of this paper deals with above-ground herbivory only; the study of herbivory on roots presents so many technical problems that there is virtually no information available except for some non-tropical plants and under controlled conditions (see Detling et al. 1980 and Ridsdill Smith 1977).

2. THE EXTENT OF INTERACTION OF PHYTOPHAGOUS INSECTS WITH TROPICAL PLANTS.

2.1. Patterns of folivory on mature plants. An analysis of the foliage of fifty-six different species was carried out at the tropical rain forest of the Los Tuxtlas research station (in South-East Mexico) (R. Dirzo et al. unpublished) with the aim of exploring the possible occurrence of patterns of folivory in the community. The data illustrate the extent of participation of insects in the consumption of leaf tissue in the forest. Table 1 shows the frequency of occurrence of different types of damage on the leaves of the sampled plants. With the exception of a minor proportion of samples (3.5%) whose damage could not be assigned to any herbivore, it is quite clear that insects play the major role of folivory in this community. Only 4% of the sampled leaves bore no evidence of insect damage. Damage by pathogens and vertebrates had a much lower frequency, and when present, it was almost always accompanied by insect damage as well. I was unable to find comparable, quantitative data for other tropical systems although some authors indicate that these trends occur elsewhere. For example, Johnston's (1981) description (and photographs) of leaf consumption in mixed mangrove stands in New Guinea strongly suggest that insects are, by far, the most important herbivores of this system. Likewise, Janzen (1981), based on available data and his long experience with tropical deciduous forests in Costa Rica, reports that "herbivory by large herbivores is probably trivial when compared to that of insects." Whatever the reasons for this pattern, it is a fact that at least for some (quite contrasting) systems, it is against the activities of leaf-eating insects that plants must direct their ecophysiological responses. The data on levels of herbivory by insects show (Fig. 1) the frequency distribution of mean leaf

TABLE 1. The distribution of different types of damage on the foliage of fifty-six species at Los Tuxtlas, Mexico.

Type of damage	Frequency (%)
Insect	96.4
Pathogen	19.6
Mammal	10.7
Unknown	3.5

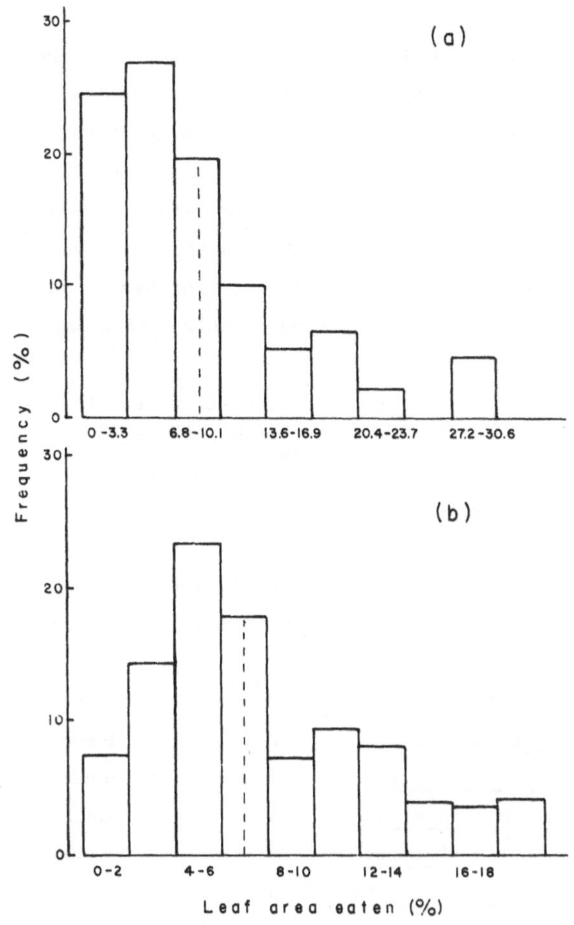

FIGURE 1. The frequency distribution of the mean leaf area eaten (%) for sixty-one species at the tropical rain forest of Los Tuxtlas, Mexico (a) (from R. Dirzo et al., unpublished) and twenty-three species from mangrove forests at Port Morsby, New Guinea (b) (from Johnstone, 1981). The broken line represents the mean for each site.

area eaten (%) for sixty-one species at the tropical rain forest of Los Tuxtlas and twenty-three species of the extremely diverse mangroves at Port Morsby New Guinea (Johnstone 1981). Although the mean and maximum levels of herbivory are higher for Los Tuxtlas than for Port Morsby, the patterns for both forests are strikingly similar. There appears to be three modal groups of damage; low, intermediate and high; likewise the mean and mode are in both cases, lower than 10% and samples with levels of damage of 10% or less, include approximately 70% of the total. Thus, although a few species may show high levels of damage (up to 29% and 18% in Los Tuxtlas and Port Morsby, respectively) the majority of species lose, at the most, 10% of their leaf area to insects.

Janzen (1981) suggests a rather different image for the strongly seasonal deciduous forest of Santa Rosa, Costa Rica. He did not attempt to extract general patterns like the ones shown in Fig. 1; instead he drew attention to the great heterogeneity of leaf consumption by insects, and mentioned the common occurrence of massive defoliations for a large number of tree species. He listed twenty-five species of trees in quite different families that were 90% defoliated by insects other than leaf-cutting ants. He concluded that severe defoliation is a fairly common event in the Santa Rosa forest-although with great variation among individuals of a given species both within and between years. However, apparently Janzen's observations were directed to the most conspicuous episodes of massive defoliation, and I suspect that if a systematic survey is made including defoliation values for a large sample of trees, a different picture could emerge. It is interesting that some preliminary data from an ongoing survey in a seasonal deciduous forest at Chamela (very similar to Janzen's forest), indicates that for the species studied so far, herbivory levels range from 2 to 12% with a mean of 7% leaf area eaten (V. Philip, personal communication); these figures are very close to he ones shown in Fig. 1.

In summary, then, it appears that although cases of massive defoliation occur with some frequency and great variation, a great majority of species and individuals support herbivory levels of 10% or less. More data are needed to evaluate the generality of these trends.

2.2 Folivory on seedlings.

Leaf consumption at the seedling level has been poorly documented so patterns are not readily obtainable. A rough indication of the levels of damage is hinted from an extensive survey of the scores of damage on the leaves from 321 seedlings at Los Tuxtlas (Fig. 2). Four

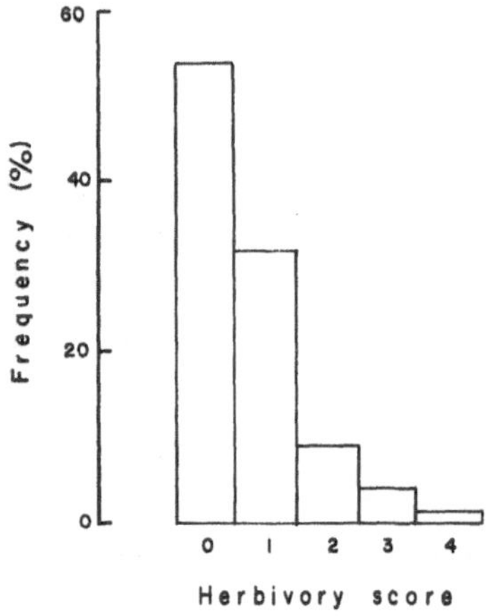

FIGURE 2. The frequency of scores of damage on 895 leaves of 321 seedlings sampled at Los Tuxtlas, Mexico (from R. Dirzo, in preparation). The scores are; 1, 1-25% of the leaf eaten; 2, 25-50%, 3, 50-75%; 4, 75-100%.

categories of damage were defined on the basis of the proportions of leaf area eaten. Although these categories are very gross, it is clear that the vast majority of leaves are undamaged or very slightly damaged (i.e., no more than 25% of the leaf area is eaten) and only a very small proportion of the sample (1%) shows extensive damage (i.e., score 4). The three most abundant species in the sample (Psychotria chiapensis, Nectandra ambigens and Faramea occidentalis) had become established just before the survey was made and might have been exposed to herbivory only for a short period, so it is possible that the levels of herbivory are underestimated. Also, I was unable to record how much of the herbivory was due to insects or to other animals. Much more precise data are available for six primary species whose seedlings have been watched in permanent sites for over 2 years (R. Dirzo, in preparation). Table 2 shows an

TABLE 2. The number of different types of damage present on the leaves of six plant species at Los Tuxtlas, Mexico and the herbivores responsible (from R. Dirzo, in preparation).

Species	No. of different types of damage	Herbivores responsible
Nectandra ambigens	7	Insect
Omphalea oleifera	7	Insect
Pterocarpus hayesii	3	Insect
Brosimum alicastrum	5	Insect
Psychotria chiapensis	6	Insect
Faramea occidentalis	5	Insect

analysis of the different types of damage for the 6 species. It can be seen that although there are several types of damage for a given species, in all cases all types of damage were made by insects. This pattern has been very consistent from month to month and between years. Moreover, a more detailed analysis of the types of damage (Fig. 3) shows that for each plant species there is a predominant type of damage, which does not coincide between species. Thus, for these forest species there appears to be a species-specific spectrum of damage and this might be important when considering the ecophysiological responses of plants to herbivory. Recently obtained data by J. Nuñez and R. Dirzo (unpublished) with two pioneer species at Los Tuxtlas show that seedlings are again attacked by insects, but in contrast, they do not show the same specificity. Now we know that the biochemical profiles of primary and secondary species are different and we expect that other ecophysiological differences might also be involved.

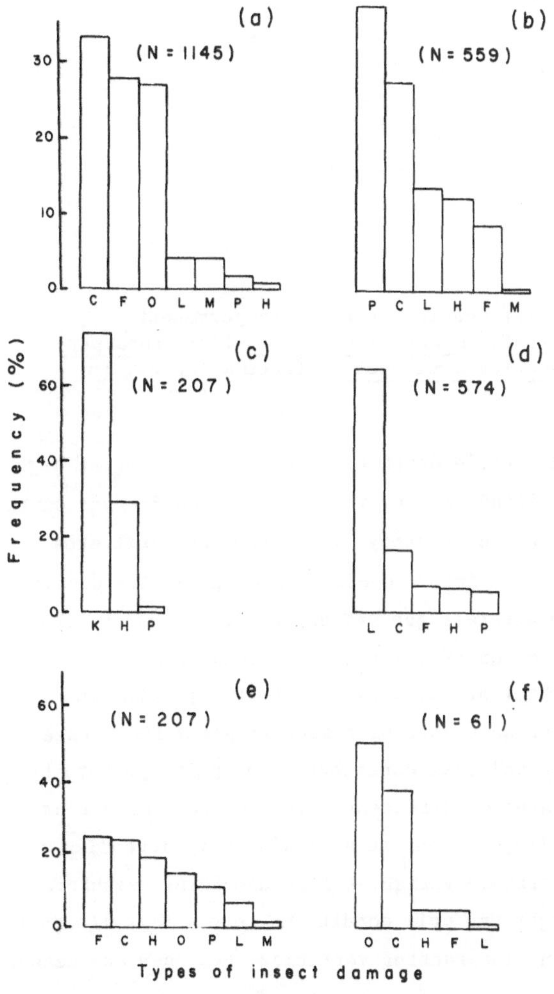

FIGURE 3. The frequency of occurrence of different types of insect damage (coded by each different letter) on the leaves of seedlings of six species at Los Tuxtlas, Mexico. (a) *Nectandra ambigens*; (b) *Psychotria chiapensis*; (c) *Pterocarpus hayesii*; (d) *Faramea occidentalis*; (e) *Omphalea oleifera*; (f) *Brosimum alicastrum* (from R. Dirzo. in preparation).

With regards to the levels of damage on seedlings, an analysis of individual plants and leaves in the permanent quadrats has been made by means of an Index of Herbivory (IH) per plant calculated on the basis of four categories of leaf area damaged (IH=Σ leaves in each category/total no. of leaves). The data for the six species during a complete year of observation are shown in Fig. 4. The salient features of this are (i) the great similarity among species, (ii) the constancy through time, and finally, (iii) that the levels of herbivory remain at a relatively constant value of 1.0. The question that emerges from these results is what is the meaning of these apparently trivial levels of damage in terms of plant performance and fitness? This question is analyzed in Section 3.

2.3. Damage on seeds.

In contrast to foliage, damage to seeds appears to be much more fairly shared by insects and other seed predators mainly mammals in the case of post-dispersal seed predation and mammals and birds in predispersal predation. In Los Tuxtlas, the data gathered during 2 years for a few species indicates a great deal of variation and, so far, no consistent patterns have become apparent. For example, some species that drop their seeds in a rather synchronous manner tend to be extensively damaged (e.g., *Nectandra ambigens*, which is attacked by at least two different species of beetle), yet other synchronous species (e.g., *Brosimum alicastrum*) show no, or very little insect-damage. In contrast, for some species (e.g., *Omphalea oleifera*) that produce seed rather continuously, we have found very low values of attacked seeds per crop. In summary, there seems to be no apparent pattern although, very likely, longer-term observations might detect them.

As far as damage per individual seed is

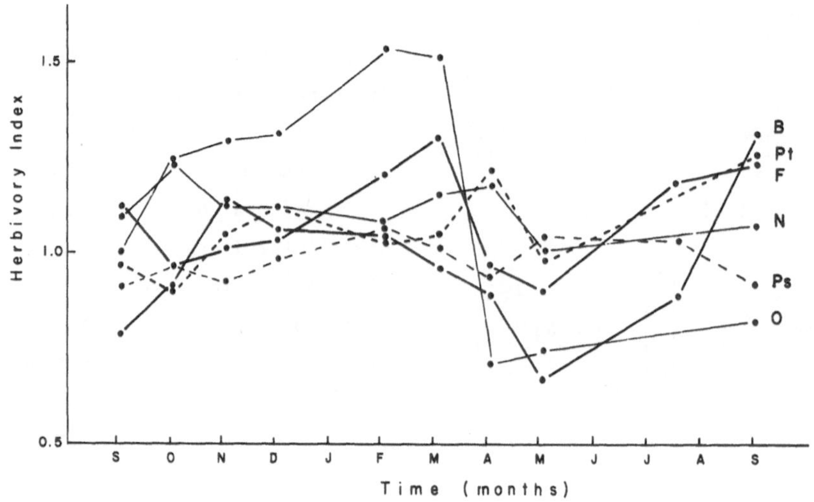

FIGURE 4. The index of herbivory through time for six plant species observed in permanent quadrats at Los Tuxtlas, Mexico. N=*Nectandra ambigens*; Ps=*Psychotria chiapensis*; Pt=*Pterocarpus hayesii*; F=*Faramea occidentalis*; O=*Omphalea oleifera*; B=*Brosimum alicastrum* (From R. Dirzo, in preparation).

concerned, we have found great variation both within and between species. Reliable estimations are also difficult to obtain because insect predators develop within the seed and the extent of damage is very much dependent upon the length of time that the seed takes to germinate in relation to the development of the predator. In our surveys, we rarely find germinating seeds when damage is high (i.e., in the order of 50% or more of the seed live weight removed), but we do see germinating seeds when damage is shifted towards the lower end (i.e., 10% or less). Some implications of this 'partial predation' are discussed below.

3. THE IMPACT OF INSECT HERBIVORY ON PLANT PERFORMANCE.

The relevant matter here is to define what happens to a plant's fitness when an animal removes a certain amount of plant material; however, this is a very difficult question to answer: a 50% leaf area removal may mean inevitable death to a seedling growing at high seedling density in the shaded conditions of the forest understory, while the same unit area removed from a seedling growing at low density in a forest gap may be an almost insignificant fraction of a quickly replaceable leaf. The impact of folivory will vary depending on intrinsic variables such as plant phenostage, the relative contribution of photosynthate by leaves of different ages, and the amounts of foliage eaten; it will also vary depending on extrinsic variables like competing neighbors, light and soil conditions, etc. So, with that many interacting variables, how does one measure the impact of folivory? This is an area in which it is mandatory to apply some sort of experimental manipulation. In this section I would like to show how ecologists can attempt to evaluate the impact of herbivory on seeds, seedlings and mature plants.

3.1. Seeds.

The seed is one of the stages of the plant's life cycle in which damage can be lethal. However, even at this vulnerable stage, not all of the attacked seeds will necessarily die provided that the embryo is not eaten. This situation can be experimentally simulated (e.g., Janzen 1976) and Fig. 5 shows some results of an experiment (see Dirzo 1984) in which holes of different sizes were drilled into the seeds of Omphalea oleifera to simulate seed weight

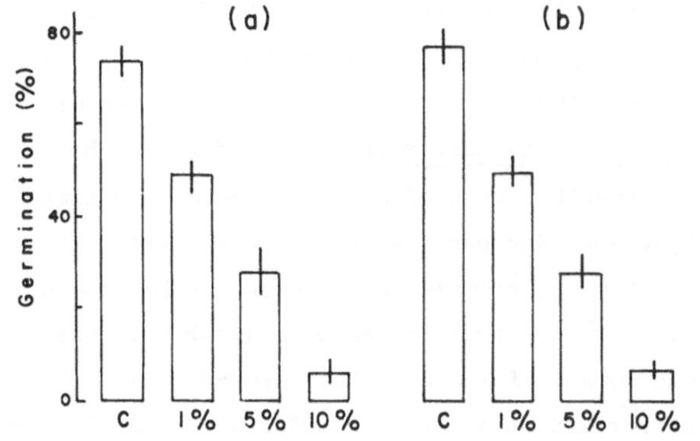

FIGURE 5. The germination ($\bar{X}$ ± S.E.) of seeds of Omphalea oleifera subjected to three levels of artificial seed predation (1, 5, and 10%; C=Control). Values are given for germination at 29 (a) and 40 (b) days after sowing (from Dirzo 1984).

removal of 1, 5, and 10%. Seeds were germinated under controlled conditions and the results show that germination potential was increasingly reduced with the intensity of damage. This suggests that, other things being equal, an advantage would be conferred to those genotypes that allocate resources to reduce damage to a minimum. As an extension of the experiments, the seedlings that emerged from the different seed treatments were grown in competition under controlled conditions. The growth of the seedlings in the mixture is shown in Fig. 6.

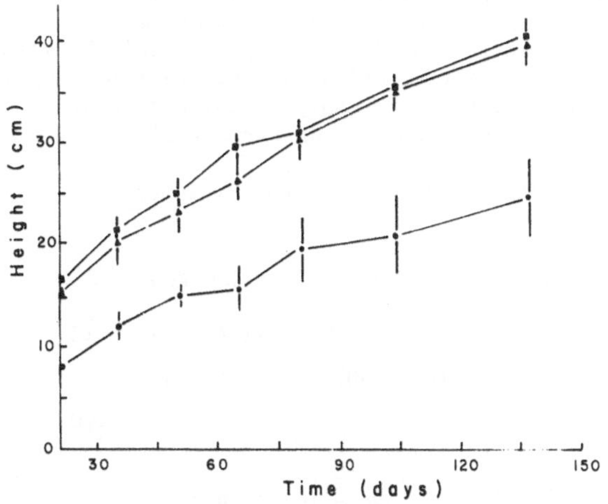

FIGURE 6. The growth of seedlings of Omphalea oleifera that emerged from seeds subjected to three levels of artificial seed predation: (■), control; (▲), 1% seed weight removed; (●), 5% seed weight removed (from Dirzo 1984).

Seedlings that emerged from the control and 1% damaged seeds, showed no differences in height while those that emerged from the 5% damaged seeds had a very reduced growth. It is easy to see the ecophysiological relevance of these differences in plant performance particularly in the conditions of the forest understory.

3.2 Seedlings.

Some data on the performance of seedlings in the field have been obtained from the continuous measurement of growth, survival and herbivory on individual plants and leaves of six species in permanent quadrats (R. Dirzo, in preparation). From these observations a great deal of variation in plant growth and survival both within and between species has been detected and I have been able to show that some of this variation can be correlated to herbivory. For example, from the analysis of individual leaves, comparisons have been made of the fate of damaged and undamaged leaves from one recording date to a subsequent one. Table 3

TABLE 3. A contingency analysis (a) for the fate (survival or death) of damaged and undamaged leaves of Pterocarpus hayesii between two subsequent recording dates; (b) a summary of results of the same analysis for the other species under study (from Dirzo 1984).

(a) Contingency analysis for P. hayesii

Leaves	Damaged	Undamaged	Σ
Surviving	12 (18.5)[a]	21 (14.5)	33
Dead	16 (9.50)	1 (7.5)	17
Σ	28	22	40

$X^2 = 10.20$; $P < 0.01$

(b) Summary for all species

Species	X^2	P
Nectandra ambigens	6.62	0.025
Omphalea oleifera	2.12	n.s.[b]
Pterocarpus hayesii	10.20	0.01
Faramea occidentalis	3.73	0.05
Psychotria chiapensis	1.32	n.s.

[a]Expected numbers in parenthesis

[b]n.s. = non significant

shows that the risks of a leaf dying between two recording dates (i.e., a two-month interval) are increased if, initially, it had been eaten by insects. The same tendency has been observed for all the species under study although statistical significance may be sometimes reached only after longer time intervals. However, although a significant relationship between herbivory and plant performance can be shown, we still have the problem of defining how much of the variation in plant performance is accounted for by herbivory or other interacting variables. Thus, on the basis of field observations in the permanent quadrats, a defoliation experiment was designed (B. Zagorin and R. Dirzo, in preparation). The experiment included the following variables: (i) two light conditions, sun and shade, to stimulate the light regimes of gap and forest understory; (ii) two contrasting densities, mimicking the mean and maximum seedling densities found in the field and (iii) three levels of simulated herbivory; one to mimic the mean observed in the permanent quadrats plus two extremes, a minimum and a maximum, to simulate the action of an herbivore which is more and less deterred than on average. The response of plants was analyzed in terms of modules (leaves) and whole individuals (see Dirzo 1984). Some results at the leaf population level are shown in Fig. 7. The survivorship of four cohorts of leaves of Psychotria chiapensis is shown and it can be seen that the risk of death increases and life expectancy decreases with the defoliation increase. Curiously, the effects of increased herbivory appear to be more noticeable in the last cohort of leaves. The correct interpretation of these results will undoubtedly require the participation of plant physiologists.

As a next step, one would like to assess how these subtle responses translate into the performance of the whole plant. Fig. 8 summarizes results in which all the experimental variables are incorporated and their impact measured in terms of plant survival. It can be seen that in the absence of light stress, plants at low density experienced no mortality at all; plants at high density experienced some mortality and this was correlated to the increase in defoliation. In the shade, there was no effect of defoliation if the plants were at low density; however, under conditions of high density, increasing defoliation dramatically reduced plant survival. These results indicate that the real impact of herbivory may only be revealed when it is

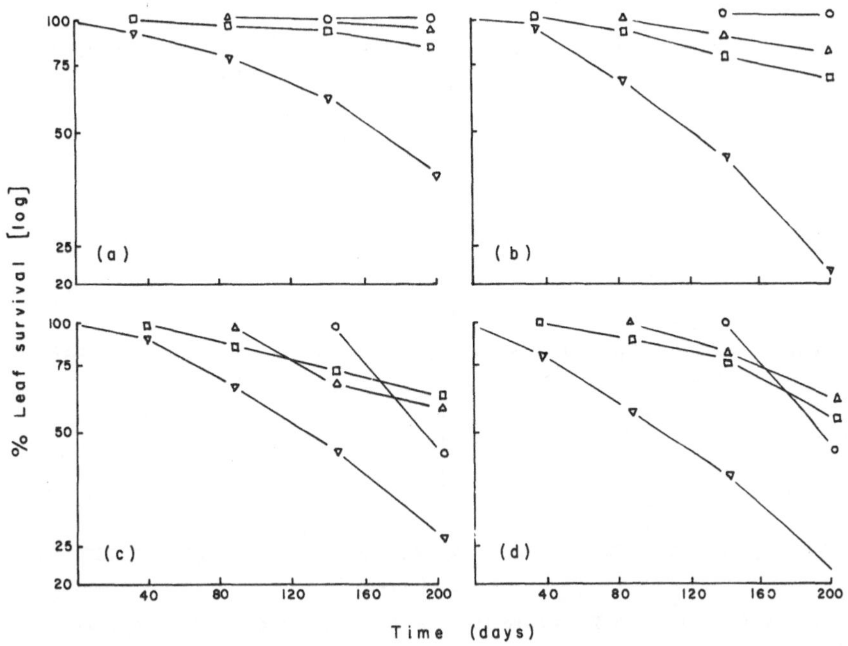

FIGURE 7. The survivorship of four cohorts of leaves of Psychotria chiapensis under four regimes of artificial defoliation: (a), control; (b), 5% defoliation; (c), 25% defoliation; (d), 75% defoliation. (▽), first cohort; (□), second cohort; (△), third cohort; (o), fourth cohort (from Dirzo 1984).

observed in conjunction with competition and other relevant stresses from the physical environment. For this particular system then, to assess the ecophysiological response to herbivory, we would need to put it together with the ecophysiological effects of competition and shading. Had the plants not been grown at high density (mimicking the maximum field densities), one would have been led to conclude that even 75% defoliation had no effect on plant performance.

3.3 Mature plants.

Most of the literature concerned with the effects of insect defoliation on the fitness of mature, wild tropical trees is limited to casual observation. In Los Tuxtlas, death of mature plants due to defoliation appears to be a rare event and when it does happen, it is usually under very special circumstances, for example, after continuous massive defoliation of plants growing in the immediate vicinity of the nests of leaf-cutting ants. However, I suspect that the most common effect of intense defoliation is reduction of growth or reproductive capacity. That intense defoliation levels (approaching 90%) are not so uncommon has been suggested by Janzen (1981) and Rockwood (1973); however, quantitative evaluations of the impact of such defoliations are poorly documented. The only relevant experiment appears to be that of Rockwood (1973). Mature plants of six species were hand-defoliated completely twice a year, thus his experiments mimicked those episodes of defoliation typical of leaf-cutting ants as described above. The effects, measured in terms of fruit production, are summarized in Table 4. There were marked differences between the

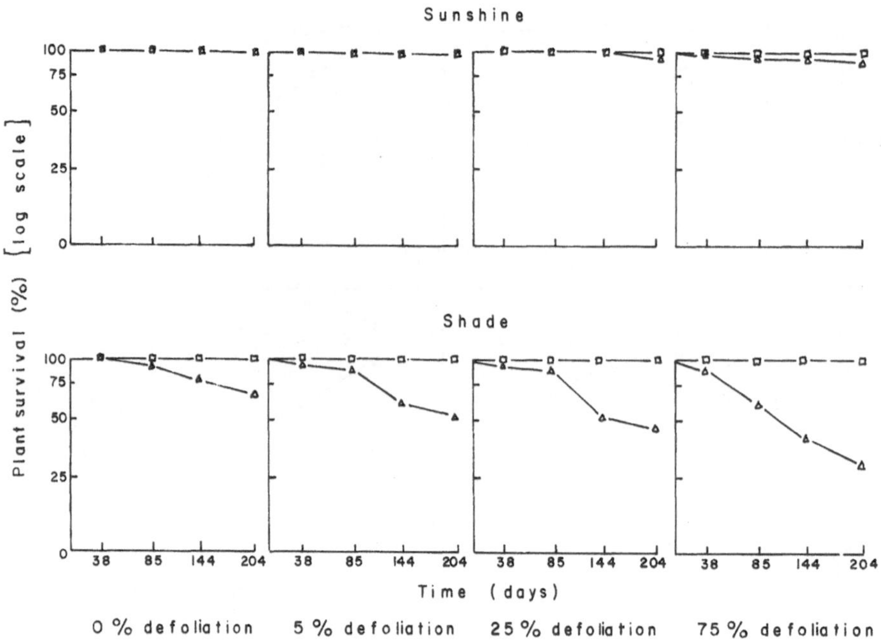

FIGURE 8. The survivorship of seedlings of Omphalea oleifera grown under two light conditions (sunshine and shade), two plant densities, low (□) and high (Δ) and four defoliation treatments (0, 5, 25 and 75% leaf area removed) (from Dirzo 1984).

TABLE 4. The mean fruit number produced per tree after intensive hand defoliation (after Rockwood 1973).

Species	N	Mean no. of fruits		Significance
		Control	Experimental	
Acacia farnesiana	10	155.4	0.7	P<0.02
Bauhinia ungulata	10	24.9	1.6	P<0.003
Cochlospermum vitifolium	10	3.8	1.8	P<0.18
Gliricidia sepium	10	419.5	96	P<0.02
Spondias purpurea	9	1182.0	0	P<0.17
Crescentia alata	8	36.0	4.6	P<0.01

control and defoliated plants except in Cochlospermum vitifolium and Spondias purpurea; in the latter only two controls and no experimental plants did set fruits. No data or even casual observations are reported for the impact of defoliation levels close to the modal values (i.e., less than 10%) of several tropical systems.

The cases I have documented, illustrate situations in which herbivory or seed predation have a detrimental effect (although I have suggested that frequently herbivory per se can have a trivial, or no effect unless it acts in conjunction with other factors). I will not attempt to discuss in detail those cases in which, in contrast, it has been claimed that herbivory has a beneficial effect on the performance (mainly productivity) of plants. Quite clearly, this is an area of great importance in the study of the interaction and I suspect that physiologists are likely to make the major contributions. Nevertheless, I still have to be convinced that these responses truly increase the plant's fitness in the field. There have been some isolated field studies that present reasonably convincing evidence that herbivory increases plant performance (e.g., MacNaughton 1979). However, as a population biologist, I still see a major drawback for these cases: if animals are devices for the plant to increase its performance, it becomes wholly dependent on an external agent to perform a given physiological task (e.g., rupture of apical dominance, alteration of sink/source ratios, etc.) that could otherwise be handled by the plant itself. Janzen (1979) has discussed this situation and quite appropriately states that under these conditions the plant lacks a proper control of its performance and it is extremely difficult to see how the external agent can operate the physiological machinery of the plant with the same precision and plasticity as could be done by the plant itself. I reiterate, however, the need for more field and experimental work on this area.

4. SOME ECOPHYSIOLOGICAL FEATURES OF TROPICAL PLANTS AND THEIR POSSIBLE RELATION TO HERBIVORY.

Bazzaz and Pickett (1980) list several ecophysiological characteristics of tropical plants; in this section some of these will be analyzed in order to explore some of their possible relationships with herbivory.

4.1 Seeding patterns and seed characteristics.

As Janzen (1969b) has noted, there are two extreme contrasting seeding patterns in tropical plants. On the one hand there is synchronous seeding in which extremely large seed crops are produced within a rather restricted period of time. On the other hand there are species in which a small seed crop is produced rather intermittently throughout the year. These two extreme behaviors undoubtedly demand contrasting ecophysiological adjustments. The patterns show some correlation with seed predation: Janzen (1969b) has shown that Central American legumes can be assigned to two groups; the first being characterized by large crops of seeds, of which the vast majority are destroyed by seed beetles. The second group produces smaller crops of seed that are able to escape beetle attack because of their heavy load of toxic secondary metabolites. At least for the case of legumes, seed size is also involved; the first group of plants characteristically produces small seeds while the second produces larger seeds. In the first group predator satiation appears to be the plant's strategy to overcome the beetles selective pressure while in the second chemical defense seems to be the strategy. Thus, the physiological adjustments for reproductive behavior and seed size must be seen in the context of costs and advantages associated to predation escape. These and other aspects of seed biology (e.g., dormancy) in relation to predation are discussed in more detail by Vázquez-Yánes in this volume. However, one other feature of seed biology that apparently has not been considered, is that of the possible relation between mode of germination and herbivory. Tropical plants may have an epigeal, hypogeal or Durian type of germination (Ng 1978). Presumably, in epigeal germination, seedlings have their cotyledons exposed to herbivores while the other two types keep their cotyledons protected. I suspect that amongst other things, the defensive chemistry of these types of seedlings is different. Some preliminary observations (R. Dirzo, unpublished), suggest that the cotyledons of hypogeal seedlings are relatively more acceptable to herbivores (grasshoppers) than those of epigeal seedlings. The implications of these interactions warrant more detailed study both by physiologists and ecologists.

4.2 Photosynthesis and Respiration.

There is essentially no information available for these two crucial physiological aspects of tropical plants as related to herbivory; for example, we do not know whether costs associated to the production and storing of secondary metabolites (if indeed these costs can be measured appropriately) are negatively correlated with the plant's photosynthetic capacity. At the moment we can only infer some correlative patterns like the one shown in Table 5. There are marked differences in photosynthesis and respiration between plants of different successional status, and both processes show a negative relationship with measured degrees of herbivory. How could we explain these apparent correlations? What are the physiological adjustments involved? Again, this is an area in which the joint expertise of ecologists and physiologists is required.

4.3 Growth.

As an adaptation to gap invasion, there is a very marked contrast between the growth of early successional and mature-forest species (see Bazzaz, this volume). This difference is undoubtedly related to the much more rapid accumulation of nutrients in the pioneers; these species apparently allocate a great proportion of their resources to growth and, consequently, a quick attainment of the reproductive size with little allocation of resources to other ends (e.g., defense). The opposite is suggested for mature forest species which grow slowly and devote comparatively more resources to defense (see Cates, Orians 1975; Feeny 1976; Rhoades, Cates 1976). These differences correlate well with the fact that early successional species are more acceptable to herbivores under experimental conditions (Hartshorn 1978) and show greater values of herbivory in the field (cf. Table 5).

Besides these between-species differences,

TABLE 5. Photosynthesis (Ps), Respiration (Rd), Herbivory Index (HI) and Grazing Rates (GR) of plants of different successional status.

	Early successional trees	Canopy trees	Understory plants
[1]Ps (mg CO_2.dm^{-2}.h^{-1})	14.1	6.9	2.9
[1]Rd (mg CO_2.$dm^{-2}hr^{-1}$)	2.0	1.0	0.3
[2]HI (% area eaten)	10.9 ± 1.8	7.3 ± 1.7	-
[3]GR (% eaten-day^{-1})	0.47 ± 0.16	0.04 ± 0.02	-

[1]From Bazzaz and Pickett (1980)

[2]From R. Dirzo et al. (unpublished)

[3]From Coley (1980)

intraspecific variations in growth occur in relation to environmental heterogeneity, the most obvious of which is sun vs. shade. This heterogeneity is particularly important for seedlings. The respiration of seedlings growing in the shade is quite high and may exceed photosynthesis (Bazzaz, Pickett 1980); most photosynthates are used for maintenance and little or no growth occurs; these seedlings remain alive ('dormant') under the canopy 'waiting' for the aperture for a gap to occur. Herbivory is likely to have a much greater impact on seedlings that are growing in the shade, in comparison to those growing in gaps (or sunflecks) of the forest. This situation has been investigated by means of defoliation experiments (referred to previously) on seedlings growing under shaded and full-sun conditions. The combined effects of defoliation and shading were measured in terms of cost. Cost was defined as 1-Ex/Cx, where Ex is the plant response under the experimental treatment (i.e., defoliation plus shading) and Cx is the response in the control situation (i.e., no-defoliation plus full-sun). Some of the results obtained with two plant species (B. Zagorin and R. Dirzo, in preparation) are shown in Table 6. For Psychotria chiapensis, there was a

TABLE 6. The joint costs of herbivory and shading on the growth (leaf population rate of increase and height) of seedlings of two plant species (see text for details)

Species	Leaf population rate of increase	Growth (height)
Psychotria chiapensis	0.64	0.997
Omphalea oleifera	0.22	0.070

very high cost in the rate of increase of the population of leaves, and also there was a high cost in plant growth. For Omphalea oleifera, there was a significant cost in the rate of increase while the cost in growth was trivial (P>0.05). These results indicate the joint effects of herbivory and shade on growth, and also indicate interspecific differences in shade tolerance.

4.4 Leaf toughness.

Bazzaz and Pickett (1980) also mention a number of contrasting leaf characteristics for pioneer and late successional species. Among these, they indicate that leaves of pioneers have softer tissue than late successionals. Measurements of leaf texture (toughness) for sixty-five species at Los Tuxtlas, give evidence of this difference (Fig. 9a). Tougher leaves

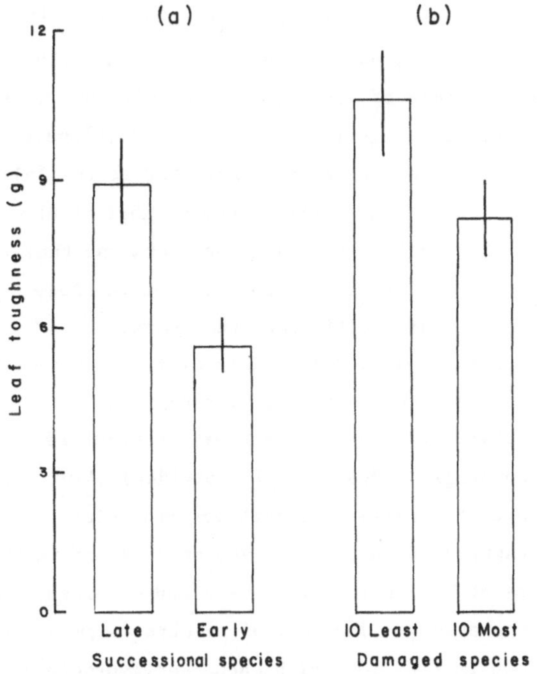

FIGURE 9. The mean leaf toughness (± S.E.) in relation to successional status (a), and the degree of damage (b) for plants from Los Tuxtlas, Mexico (from R. Dirzo et al., in preparation).

may be correlated to low water content (Cherrett

1968) or high tannin content (Feeny 1970). Whether it is due to these or other reasons, there is a tendency for tougher leaves to be less damaged than softer leaves (Fig. 9b).

5. OTHER SPECIAL FEATURES OF TROPICAL SYSTEMS AND HERBIVORY.

In this section I would like to draw attention to the occurrence of some characteristic features of tropical systems for which I suspect herbivory may be quite relevant.

5.1 Spatial heterogeneity.

Spatial heterogeneity is a major, though perhaps not exclusive, feature of tropical systems, the most obvious being that generated by the falling of trees or branches. This heterogeneity has been discussed by several contributors of this volume (e.g., Bazzaz), however, there are even finer levels of heterogeneity within this mosaic of gaps and forest. Some workers distinguish three contrasting zones associated to the fallen tree (see Bazzaz, this volume): that of the uplifted root, that along the trunk and that of the fallen crown. The environment in these zones is quite different, therefore, a colonizing plant may perform differently in each of these conditions, and different ecophysiological responses may be required accordingly. The fate of individual plants and leaves in relation to herbivory is being investigated in the root and crown zones in two gaps at Los Tuxtlas (J. Nuñez and R. Dirzo, in preparation). The sort of results emerging is shown in Fig. 10, which shows survivorship of two pioneer species in the two zones of the gap. There is a more or less similar behavior of the two species but the major contrast occurs between the sites; the crown zone implies a much greater risk of mortality for the two species. Associated with these differences in biotic characteristics as shown in Table 7. There is a rather poor floristic similarity between the two zones; diversity is greater and herbivory lower in the zone of roots and plant survival is considerably higher. These data indicate that besides the differences in physical factors within a gap, associated biotic differences may interact affecting the probability of survival of the colonizing plants. Ecophysiological work here undoubtedly would be rewarding.

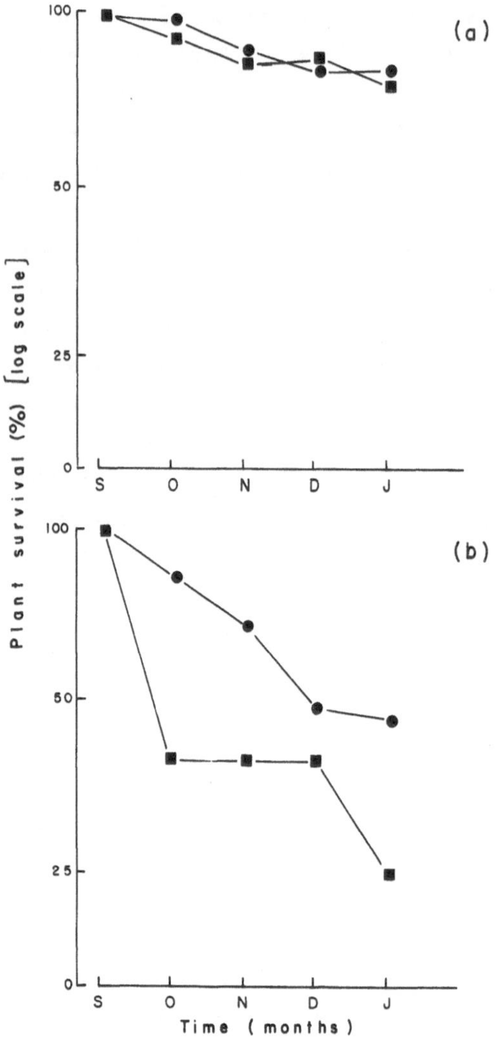

FIGURE 10. The survivorship of colonizing seedlings of Cecropia obtusifolia (●) and Heliocarpus appendiculatus (■) in the zone of roots (a) and crown (b) of gaps in the forest of Los Tuxtlas, Mexico (from J. Nuñez and R. Dirzo, in preparation).

TABLE 7. Biotic within-gap heterogeneity and characteristics (herbivory and survival of Cecropia obtusifolia seedlings.

Site	Between-Site-Floristic Similarity[1]	Diversity[2] ($\bar{H}$)	Herbivory Index	Plant survival to 6 mo. (%)
1: roots	19.4 - 34.5	0.96 ± 0.24	0.99 ± 0.35	83.6
2: crown		0.62 ± 0.10	2.50 ± 0.51	49.3

[1] Sorensen's Similarity Coefficient

[2] Shannon-Wiener's Index

5.2 Dioecy.

Another tropical feature I would like to address is that of dioecy. Recent work has revealed a great prevalence of dioecy in tropical systems (e.g., Bawa 1980). The differentiation of labor should translate into a number of ecophysiological adjustments of male and female plants to their environment. I presume that many of these adjustments come about as a consequence of the sometimes very discrepant allocation of resources to reproduction; female plants generally allocating a lot more. If this were so, it would be expected that females would reduce allocation to other ends such as defense. This situation is being investigated in Los Tuxtlas with the dioecious palm Chamaedorea tepejilote (K. Oyama and R. Dirzo, unpublished). This palm shows a characteristic damage on its leaves and we have discovered a chrysomelid beetle which is its specific herbivore. Damage can be sometimes very intense and may have a profound effect, especially at the young stages of the plant. We carried out some feeding trials to investigate the possible differences in acceptability of male and female plants from both sexes under experimental conditions. The results of these trials (Table 8) show that there is a significant preference for the tissue of female plants; in addition, a preliminary analysis of secondary metabolites in the foliage has revealed some inter-sexual differences, particularly with phenolic compounds. Other ecophysiological differences between sexes are currently being investigated.

TABLE 8. Selective feeding by a chrysomelid beetle on the leaves of male and female plants of the palm Chamaedorea tepejilote. Data are given for feeding trials of 24 (a) and 36 hours (b). Means of six replicates.

	Leaf area eaten (cm^2)			
	Trial (a)		Trial (b)	
	Female	Male	Female	Male
$\bar{X}$	4.63	2.72	10.58	7.16
S.D.	1.46	1.15	3.90	1.75
t-test	$P<0.025$		$P<0.05$	

6. CONCLUSIONS

To conclude, I hope I have shown a sample of aspects in which studies of herbivory (particularly by insects) on tropical plants, may enormously benefit from the participation of ecophysiologists. I very much hope that in a near future the interaction of workers on herbivory and ecophysiology becomes as tight as the interaction between herbivores and plants themselves.

ACKNOWLEDGMENTS

I am very grateful to Beatriz Zagorin, Ken Oyama, Juan Nuñez, Miguel A. Armella and Araceli Vargas-Mena for allowing me to use some of their unpublished data. My special thanks are due to Victor Jaramillo for critically reading this paper and offering useful suggestions. This paper is dedicated to the memory of Beatriz Zagorin.

REFERENCES

Bawa KS (1980) Evolution of dioecy in flowering plants, Ann. Rev. Ecol. Syst. 11, 15-39.

Bazzaz FA and Pickett STA (1980) Physiological ecology of tropical plants succession: a comparative review, Ann. Rev. Ecol. Syst. 11, 287-310.

Cates RG and Orians GH (1975) Successional status and the palatability of plants to generalized herbivores, Ecology 56, 410-418.

Cherrett JM (1968) A simple penetrometer for measuring leaf toughness in insect feeding studies, J. Econ. Ent. 61, 1736-1739.

Coley PD (1980) Effects of leaf age and plant life-history patterns on herbivory. Nature 284, 545-546.

Derr JA (1980) Coevolution of the life, history of a tropical seed-feeding insect and its food plants, Ecology 61, 881-892.

Detling JK, Winn DT, Proctor-Gregg C and Painter EL (1980) Effects of simulated grazing by below-ground herbivores on growth, CO_2 exchange and carbon allocation patterns of Bouteloua gracilis. J. Appl. Ecol. 17, 771-778.

Dirzo R (1984) Herbivory: a phytocentric overview. In Dirzo R and Sarukhan J, eds. Perspectives in plant population ecology (in press). Sinauer Associates, Sunderland, Mass.

Feeny PP (1970) Seasonal changes in oak leaf tannins and nutrients as a cause of spring feeding by winter moth caterpillars, Ecology 51, 565-581.

Feeny PP (1976) Plant apparency and chemical defense. In Wallace JW and Mansell RL, eds. Recent advances in phytochemistry, Vol. 10: Interactions between plants and insects, pp. 1-40. Plenum Press, New York.

Gilbert LE (1977) Development of theory in the analysis of insect-plant interactions. In Horn DH, Mitchel RD and Stiles GR, eds. Analysis of ecological systems, pp. 117-154, Ohio State University Press, Ohio.

Gilbert LE and Raven PH, eds. (1975) Coevolution of animals and plants, University of Texas Press, Austin.

Hartshorn GS (1978) Tree falls and tropical forest dynamics. In Tomlinson PB and Zimmermann MH, eds. Tropical trees as living systems, pp. 617-665. Cambridge University Press, London.

Janzen DH (1966) Coevolution of mutualism between ants and acacias in Central America, Evolution 20, 249-275.

Janzen DH (1969a) Allelopathy by myrmecophytes: the ant Azteca as an allelopathic agent of Cecropia, Ecology 50, 147-153.

Janzen DH (1969b) Seed-eaters versus seed size, number, toxicity and dispersal, Evolution 23, 1-27.

Janzen DH (1974) Epiphytic myrmecophytes in Sarawak: mutualism through the feedings of plants by ants, Biotropica 6, 237-259.

Janzen DH (1976) Reduction of Mucuna andreana (Leguminosae) seedling fitness by artificial seed damage, Ecology 57, 826-828.

Janzen DH (1979) New horizons in the biology of plant defenses. In Rosenthal GA and Janzen DH, eds. Herbivores, their interaction with secondary compounds, pp. 331-350. Academic Press, New York.

Janzen DH (1981) Patterns of herbivory in a tropical deciduous forest, Biotropica 13, 271-282.

Johnston IM (1981) Consumption of leaves by herbivores in mixed mangrove stands; Biotropica 13, 252-259.

McNaughton SJ (1979) Grazing as an optimization process: grass-ungulate relationships in the Serengeti, Amer. Nat. 113, 691-703.

Ng FSP (1978) Strategies of establishment in Malayan forest trees. In Tomlinson PB and Zimmermann MH, eds. Tropical trees as living systems, pp. 129-162, Cambridge University Press, London.

Raven PH (1977) Perspectives in tropical botany: concluding remarks, Ann. Missouri Bot. Gard. 64, 746-748.

Ridsdill Smith TJ (1977) Effects of root feeding by scarabeid larvae on growth of perennial rye-grass plants, J. Appl. Ecol. 14, 73-80.

Rhoades DF and Cates RG (1976) A general theory of plant anti-herbivore chemistry. In Wallace JW and Mansell RL, eds. Recent advances in phytochemistry, vol. 10: Interactions between plants and insects, pp. 168-213, Plenum Press, New York.

Rockwood L (1973) The effect of defoliation on seed production of six Costa Rican tree species, Ecology 54, 1363-1369.

ASSESSING THE EFFECTS OF HERBIVORY

P. A. MORROW (University of Minnesota, U.S.A.)

1. INTRODUCTION

Plant biologists long dismissed the importance of herbivorous insects in plant communities on the assumption that the amount of tissue they consume in non-outbreak years is inconsequential. In the last 15 years, the role of insects in driving the evolution of plant chemical and physical defenses has become appreciated. Other roles that insects may play are less well established: some feel that the impact of herbivores is small enough to ignore while others consider their affects on physiology, population and community dynamics as too important to dismiss.

The basis for assuming that insects have little direct affect on plant growth is the common observation that less than 10% of leaf area is normally removed by insects. Two kinds of questions arise from this observation: (1) Does this visual estimate accurately indicate the leaf area loss to insects? Do such losses have an insignificant affect on the growth and reproduction of isolated plants, on plants in different stages of development, on plants in competitive environments? Is the amount of leaf material recycled by herbivores insignificant compared to other pathways of energy and nutrient flow? The importance of a given amount of damage may be trivial at one level but critical at another. (2) To what extent are observed levels of damage a consequence of features of individual plants, e.g., life history traits and defenses, and emergent properties of communities such as species diversity and dispersion of conspecifics? Below I suggest the kinds of data needed to begin answering these questions. For convenience, the discussion is divided into sections on the assessment of damage to individual plants, responses of individuals, and population and community consequences. There is considerable overlap between these topics and the divisions are fairly arbitrary. The kinds of data needed to understand how community structure affects levels of damage do not fall into a single category and they are discussed at various places.

2. ASSESSMENT OF DAMAGE TO INDIVIDUAL PLANTS

We need many more estimates of leaf damage to plants of different ages and species in different vegetation types, on various substrates and on different continents. The amount of leaf area missing is generally estimated from a sample of about 100 leaves taken from each of 5 to 10 individuals of a species toward the end of the growing season. The amount of leaf area missing can be determined from measures of actual and potential leaf area using a leaf area meter. However, this is a time consuming method and for many leaf types visual estimates of damage are very similar and much faster (Fox, Morrow 1983).

Missing area is estimated by sorting leaves into 7 classes: 0%, <1%, <10%, <25%, <50%, <75%, and <100% of leaf area missing. Damage is calculated as the mean percentage of area missing on each plant, assuming the area lost from every leaf in a category is the midpoint of that category, e.g., 5.5% of the 1-10% class (Fox, Morrow 1983). These categories provide better resolution of lightly damaged leaves and thus are less prone to overestimating damage than classes with equal damage intervals. Damage estimates can also be made on leaves in litter samples (Proctor et al. 1983b).

The proportions of heavily (>25%) or lightly (<10%) damaged leaves on several species of

Eucalyptus are correlated with the mean leaf area missing from a tree. If such relationships exist for other species, damage among individuals of those species can be rapidly assessed because scoring only the extreme cases is faster than placing leaves into seven categories. While less accurate than sorting leaves into more categories, the level of accuracy may be sufficient for many studies given the variability of leaf damage among species and habitats (Fox, Morrow 1983).

Whether on living or shed samples of leaves, discrete samples underestimate total damage because they ignore leaves that have been entirely consumed, do not measure loss of photosynthate to insects piercing phloem and mesophyll and ignore damage to, or loss of, buds and stems. However, we have so few estimates of insect damage in different plant communities that the ease and widespread use of this kind of assessment makes these data valuable in spite of their obvious problems. Certainly they will help us to better evaluate the widely held assumption that herbivorous insects do much more damage in lowland tropical rainforests than in temperate plant communities, a generalization not currently supported by the available data (summarized in Proctor et al. 1983b, Lowman 1982, Fox, Morrow 1983).

Better estimates of loss of leaf tissue may be obtained by repeated measures of damage on the same leaves over one or more growing seasons. In Australian rainforests, long term measures on six tree species at three sites gave estimates of leaf area loss from 1.2 to as much as 5 times higher than losses estimated by discrete sampling (Lowman 1982). At Los Tuxlas the two methods gave similar estimates for the one species assessed (Dirzo, personal communication). In addition to whatever other units are used to express leaf damage in long term studies, e.g., leaf area lost per day (Coley 1983a,b), it would be valuable to report both cumulative leaf area loss by a species over a growing season and damage levels on discrete samples so that comparisons can be made with other studies. Long term measures of leaf area loss do not tell us how much the productivity of a plant is reduced by insect feeding but they do indicate whether or not insects do more damage than one assumes from casual observation.

Insect frass and bits of leaves and buds excised but not eaten by herbivorous insects fall into litter traps but even in dry temperate vegetation these fractions are very difficult to separate. Thus frass fall probably cannot be used to estimate insect feeding rates (J. Landsberg, personal communication).

In temperate forests insecticides have been used to assess the impact of herbivory on productivity (Cantlon 1969, Morrow, LaMarche 1978). Use of contact insecticides is impracticable in tropical lowland forests because of the heavy rainfall. However, systematic insecticides may prove to be viable tools.

Herbivorous insects, especially early larval and nymphal stages, generally eat young leaves which are higher in nitrogen and water and are more easily cut and chewed than mature leaves (Rockwood 1974, Scriber 1977, White 1978, Lowman, Box 1983, but see Stanton 1975). Thus herbivores must locate not only a host plant but must find one during the period that tissue of appropriate age is available.

We need data on schedules of leaf production, especially the timing, duration and variability of leaf initiation such as those collected by

Frankie, et al. (1974), Lowman (1982) and Lieberman (1982) to help evaluate the effects of phenological patterns on insect abundance and damage. Frankie et al. (1974) recognized three categories of leaf phenology in their Costa Rican forests: (1) continuous growth-species which continuously produce a small number of new leaves and have no concentrated period of leaf drop, (2) discontinuous production of new leaves alternating with production of many, few or no new leaves and no concentrated leaf drop or leafless period, (3) marked flushing and leaf drop.

These categories indicate the temporal availability of different classes of tissue to insects. The categories would be even more useful if, in addition, researchers indicate whether leaf production is determinate, i.e., leaves are produced from a predetermined number of leaf primordia in the bud (flushing behavior of categories 2 and 3) or indeterminate, i.e., leaves continuously initiated as the shoot elongates. Indeterminate growth could be continuous, seasonal or opportunistic (categories 2 and 1). The length of time during which young leaves are continuously available is likely to be longer on species with indeterminate than with determinate shoots.

Many biologists feel that biotic interactions are more specific and/or of greater overall importance in the functioning of tropical wet than of temperate vegetation. Ecosystem biologists could contribute more to our evaluation of this assertion if data on plant biomass, litterfall and nutrient contents were reported in categories that are closer to those recognized by the fauna. For instance, flowers, fleshy (animal dispersed) and dry (wind dispersed) seeds or fruits support different segments of the fauna so their biomass values should not be combined in a "flowers and fruits" category. Similarly, young and old leaves may be very different resources for herbivores so, where possible, both categories should be reported rather than combined as "leaves". Specific densities of leaves are valuable data since they give an indication of leaf toughness and nitrogen content that are important to herbivorous insects.

Other kinds of data obtained by ecosystem ecologists are of value to population and community ecologists. Reports that give species and taxonomic diversity of a community (Proctor et al. 1983a, Lieberman 1982) provide a rough indication of the diversity of secondary chemical classes in the vegetation which serve to both help and hinder herbivores locate their host species (e.g., Ehrlich, Raven 1965). Comparisons of communities having different levels of species and taxonomic diversity, evenness, spatial patterning and characteristic damage levels may help us to evaluate the importance of emergent properties of communities on the herbivore trophic level, revealing patterns that justify more careful examination.

It has been suggested that the frass, honeydew and feeding scraps falling from the canopy of a plant benefit that plant by increasing the rate at which it can recycle its minerals (Harris 1974, Petelle 1981, Owen 1980). It seems unlikely that individual plants in a competitive environment could benefit from this kind of recycling since their roots may not exclusively or even largely dominate the space beneath their crowns (D. Janos, personal communication) where these quickly recycling materials are deposited. In contrast, internal retranslocation is an efficient and controllable recycling process. However, insects may increase the rate of nutrient recycling in the system as a whole and this may benefit the general "health" of the vegetation (Mattson, Addy 1975) if not the individuals being eaten.

3. RESPONSE OF INDIVIDUALS TO HERBIVORE ATTACK

Studies of the influence of leaf shape, size, orientation and pubescence on energy balance and gas exchange have demonstrated a sensitive coupling of leaf to environment. Leaf characteristics also may have evolved in response to insect attack. The extreme intra- and interspecific variability in leaf shape of Passiflora, a genus of ca. 350 species of tropical vines, may be an evolutionary response to Heliconius butterflies, a group of monophagous, visually searching herbivores (Gilbert 1975).

Insects have typical ways of feeding. For instance, beetles and larvae of many moth and butterfly species chew leaves from the edge, removing large portions of the lamina and leaving smooth edges. Many adult beetles may make deep, narrow incisions while flea beetles make small pits where they scrape away the lower cuticle and mesophyll. Other insects skeletonize leaves or begin feeding from the leaf center, making several large holes or a large number of small holes which may be either smooth or jagged. Do these different patterns of leaf removal have any consequences for the efficiency of xylem and phloem transport within the damaged leaf? Are different venation patterns more or less affected by different types of damage? Are damaged leaves more susceptible to water stress? Is leaf morphology of a plant species correlated with the ways in which its major herbivore species feed?

Colonies of phloem feeding insects create a sink for carbon and nitrogen to which a plant may respond with increased rates of photosynthesis (Way, Cammell 1970). Are rates of photosynthesis also affected when leaf area is removed by chewing insects and do the patterns of chewing have different effects on these rates? Lowman (1983) found that seedlings of a rainforest tree recover from defoliation faster if tissue is removed from each leaf than when the same area is removed by clipping whole leaves.

The observation that leaves of defoliated plants may increase rates of photosynthesis has led to two suggestions: (1) leaves usually fix carbon at rates well below those of which they are capable (Harper 1977, p. 392). This view suggests that plants have greater in situ capacity for carbon fixation than they generally require. (2) Resources from other parts of a damaged plant are reallocated to leaves which then generate the additional photosynthetic capacity. This suggestion assumes that plants maintain only the photosynthetic machinery required to meet existing carbon demands. Experiments designed to test these assumptions would be valuable.

As stated earlier, insects often eat young expanding tissue. Is the cost of a young, nitrogen rich leaf that is still importing photosynthate greater than a bite of a fully functional leaf that has less nitrogen per unit mass?

In response to grazing, canopies of crop plants and grasses may regenerate by changing the allocation of resources between roots and shoots. How does this affect an individual's long term productivity as opposed to the productivity of the vegetation as a whole?

Plants produce a diverse array of chemical compounds that repel or poison non-adapted herbivores. Speculation as to their cost is widespread. Plants producing high concentrations of indigestible carbon based compounds such as lignin, tannins and phenolics are considered to have "expensive" defenses because of the concentrations involved (up to 60% of leaf dry mass). Compounds such as the nitrogen containing alkaloids are present in much smaller concentrations (often less than 2% of leaf dry mass) and are considered to be relatively cheap (Cates, Rhoades 1977). However, concentrations provide no indication of the energetic cost of synthesis, degradation, transport or storage of static and recycling defensive chemicals or the scarcity of resources allocated to these chemicals (Fox 1981). Estimating the cost of these compounds to the plant will in general be a difficult laboratory exercise or a deduction from knowledge of the biosynthetic pathways of the chemicals. Determining cost should not be a major priority of work in the tropics at this time.

Two issues were raised in the workshop which are tantalizing but for which we have almost no data. (1) Roots may have higher concentrations of defensive chemicals than shoots of the same plant. This suggests that grazing of roots may be heavy. Dirzo (this symposium) summarizes the few data available. (2) Within-plant variability in leaf (and other) characteristics may make a plant more difficult for an insect species to exploit. Recent work by Whitham and Slobodchikoff (1981) suggests that some of the variability within an individual plant may result from somatic mutations. We should be aware of this possibility and become more observant rather than dismiss the hypothesis out of hand.

4. CONSEQUENCES OF HERBIVORY AT POPULATION AND COMMUNITY LEVELS

The rapid growth of plants when canopy gaps occur, the density of seedlings on the floor of tropical forests and the order of magnitude thinning of individuals as canopy closure occurs suggest that competition is an important force in these communities. Even minor levels of herbivore damage may influence patterns of growth (Louda 1983), survival and dominance when competition becomes intense. To assess the importance of herbivorous insects on population and community structure we need to know how damage is distributed between competing individuals of the same and different species and between seedlings, saplings and canopy trees. Whether damage is distributed evenly between individuals and leaves or is concentrated on a small proportion of them will have different competitive consequences.

Competition experiments should be done between plants artificially defoliated at different levels to learn how different amounts of leaf loss affect competitive abilities. The experiments should be done at different planting densities, between the same and different species and under a range of light and nutrient levels found in the community before we say that a leaf area loss of 5 or 10% is insignificant.

The effects of herbivores on plant populations can be assessed by following changes in populations of marked seedlings in monospecific and mixed species groups, in both shade and gaps, and calculating the probability of survival given different histories of damage by herbivores (see Dirzo, this symposium). These studies need to be long term since insect populations may fluctuate markedly between years just as they do in temperate zones (Wolda 1978).

As the density of individuals decreases from seedling through sapling and canopy tree stages, there are probably important changes in the genetic structure of populations. Some genotypes are selectively removed by insects. However, in some cases the thinning may be completely haphazard, due as much to the chance of a seedling being damaged by one of the branch falls common in tropical rainforests. We know little about changes in the genetic structure of populations as they age and no easy solutions to the problem were offered by workshop participants.

Herbivore damage to leaves affects the reproductive potential of plants. Heavily to completely defoliated trees set few or no seeds (Rockwell 1973) and trees protected from insects produced seed at least two years earlier than unprotected trees (Morrow, unpublished data). Removal of apparently minor amounts of endosperm from seeds resulted in major effects on germination and survival (Dirzo, this symposium).

The technology required to assess the effects of herbivorous insects on population and community structure are straightforward, labor intensive and long term. The skill lies in posing tractable hypotheses, in experimental design and statistical analysis.

5. CONCLUDING REMARKS

Herbivorous insect/plant interactions have been studied mostly from the perspective of the animals. Although an insignificant proportion of plant species are immune to insect attack, plant ecologists have rarely dealt with the possibility that insects may commonly influence the course of community development or mold the physiological responses of individual plants and leaves. We need data to support our slighting of this trophic level before we can dismiss it as unimportant, especially as we construct models of ecosystem function or cost optimization by individual plants.

If we must set priorities, I suggest we concentrate our efforts in two areas. We need (1) estimates of discrete leaf area damage from a wide variety of plant species and tropical vegetation types in new and old world tropics so that we may ascertain whether broad patterns of damage exist, and (2) long term studies of herbivore impact at population and community levels such as those described in section 4. While questions of physiological response to insect attack are important, the initial studies can be done in temperate systems, where the appropriate instrumentation is concentrated.

A number of ecologists are currently studying insect herbivores in the tropics. Their studies together with the above data will form the foundations on which to build a second generation of research priorities.

ACKNOWLEDGEMENTS
I thank the workshop participants for a stimulating discussion, Laurel Fox, Yan Linhart and Kathleen Shea for comments on the manuscript, and Jean Cavanagh for typing the manuscript. Financial support from a National Science Foundation grant, DEB7904953, and a fellowship from the John Simon Guggenheim Foundation are gratefully acknowledged.

LITERATURE CITED

Cantlon JE (1969) The stability of natural populations and their sensitivity to stress. Diversity and stability in ecological systems. Brookhaven Symposia in Biology 22, 197-205.

Cates RG and Rhoades DF (1977) Patterns in the production of antiherbivore chemical defenses in plant communities, Biochemical Systematics and Ecology 5, 185-193.

Coley PD (1983a) Intraspecific variation in herbivory on two tropical tree species, Ecology 64, 426-433.

Coley PD (1983b) Herbivory and defensive characteristics of tree species in a lowland tropical forest, Ecological Monographs 53, 209-233.

Dirzo R This symposium.

Ehrlich PR and Raven PH (1965) Butterflies and plants: a study in coevolution, Evolution 18, 586-608.

Fox, LR (1981) Defense and dynamics in plant-herbivore systems, American Zoologist 21, 853-864.

Fox LR and Morrow PA (1983) Estimates of damage by herbivorous insects on Eucalyptus trees, Australian Journal of Ecology, in press.

Frankie GW, Baker HG and Opler PA (1964) Comparative phenological studies of trees in tropical wet and dry forests of the lowlands of Costa Rica, Journal of Ecology 62, 881-919.

Gilbert LE (1975) Ecological consequences of a coevolved mutualism between butterflies and plants. In Gilbert LE and Raven PH, eds. Coevolution of animals and plants, pp. 210-240. Austin, Texas, Univ. Texas Press.

Harper JL (1977) Population biology of plants. Academic Press, London. 892 pp.

Harris P (1974) A possible explanation of plant yield increases following insect damage, Agro-Ecosystems 1, 219-225.

Lieberman D (1982) Seasonality and phenology in a dry tropical forest in Ghana, Journal of Ecology 62, 791-806.

Louda SM 1983 Chrysomelid herbivory in the performance of a native crucifer, Bulletin of the Ecological Society of America 64, 98.

Lowman MD (1982) Leaf growth dynamics and herbivory in Australian rain forest canopies, Ph.D. thesis, University of Sydney.

Lowman MD (1983) Effects of different rates and methods of leaf area removal on rain forest seedlings of coachwood (Ceratopetalum apetalum), Australian Journal of Botany 30; 477-483.

Lowman MD and Box JD (1983) Variation in leaf toughness and phenolic content among five species of Australian rain forest trees, Australian Journal of Ecology 8, 17-25.

Mattson WJ and Addy ND (1975) Phytophagous insects as regulators of forest primary productivity, Science 190, 515-522.

Morrow, PA and LaMarche VC (1978) Tree ring evidence for chronic insect suppression of productivity in subalpine Eucalyptus, Science 210, 1244-1245.

Owen, DF (1980) How plants may benefit from the animals that eat them, Oikos 35, 230-235.

Petelle M (1981) More mutualisms between consumers and plants, Oikos 38, 125-127.

Proctor J, Anderson JM, Chai P and Vallack HW (1983a) Ecological studies in four contrasting lowland rainforests in Gunung Mulu National Park, Sarawak. I. Forest environment, structure and floristics, Journal of Ecology 71, 237-260.

Proctor J, Anderson JM, Fogden SCL and Vallack HW (1983b) Ecological studies in four contrasting lowland rainforests in Gunung Mulu National Park, Sarawak. II. Litterfall, litter standing crop and preliminary observations on herbivory, Journal of Ecology 71, 261-283.

Rockwood LL (1973) The effects of defoliation on seed production of six Costa Rican tree species, Ecology 54, 1363-1369.

Rockwood LL (1974) Seasonal changes in the susceptibility of Crescentai alata leaves to the flea beetle Oedionychus sp., Ecology 55, 142-148.

Scriber JM (1977) Limiting effects of low leaf-water content on the nitrogen utilization, energy budget, and larval growth of Hyalophora cecropia (Lepidoptera: Saturniidae), Oecologia (Berl.) 28, 269-287.

Stanton N (1975) Herbivore pressure on two types of tropical forests, Biotropica 7, 8-11.

Way MJ and Cammell M (1970) Aggregation behaviour in relation to food utilization by aphids. In Animal populations in relation to their food resources. 10th Symposium of the British Ecological Society. pp. 229-247. Oxford, Blackwell.

White TCR (1978) The importance of relative shortage of food in animal ecology, Oecologia (Berl. 33, 71-86.

Whitham TG and Slobodchikoff CN (1981) Evolution by individuals, plant-herbivore limitation and competition, Ecology 59, 287-292.

Wolda H (1978) Fluctuations in abundance of tropical insects, American Naturalist 112, 1017-1045.

DYNAMICS OF WET TROPICAL FORESTS AND THEIR SPECIES STRATEGIES

F. A. BAZZAZ
Department of Plant Biology, University of Illinois
Urbana, IL 61801 U.S.A.

1. INTRODUCTION

The physiological ecology of tropical vegetation in general and that of the rainforest in particular is poorly understood. Several factors have contributed to this situation, among which are the forests' low accessibility, limited availability of accurate and easily portable field instrumentation for ecophysiological research, and the concentration of most physiological ecologists in the temperate zone. The opportunities for innovative research in tropical forests are thus excellent, for these forests are an essentially unexplored ecosystem with very high diversities of species, genotypes, strategies, and biological interactions.

Mooney et al. (1980) have reviewed the status of the physiological ecology of tropical forests and identified several major areas in which research should be done. They have drawn attention to the fact that these forests are severely threatened by human exploitation for timber production, agriculture and industry. Among the areas that require much more research, the authors emphasized environmental triggers for reproduction, regenerative capacity of different growth forms, seed dormancy, carbon, water and nutrient balances, mechanisms of seedling protection and mycorrhizal and rhizobial associations.

This paper is intended to provide a background against which the details of the physiological ecology of plants in the wet tropical forest may be viewed. The paper relies heavily on, and in places updates a review of the physiological ecology of tropical succession by Bazzaz , Pickett (1980). The paper overlaps in several places with material in the preceding chapters of this book and therefore also provides a summary of the physiological ecology of plants in wet tropical forests. Here I propose several testable hypotheses and predictions about the physiological ecology of species in gaps and mature forests. I discuss forest dynamics as a process of gap creation and filling, and point out the importance of the process to the evolution of species strategies with emphasis on ecophysiological behaviours and some of their correlated demographic features. The basic premise is that the evolution of many of the observed species adaptations is highly influenced by the processes of forest dynamics as they dictate much of the physical and biological interactions in wet tropical forests.

2. THE DYNAMIC NATURE OF TROPICAL FORESTS

Ecologists have long recognized the dynamic nature of tropical forests. They viewed the forest as a patchwork of gaps in various stages of succession (Aubre'ville 1938, Richards 1952). The process of gap creation and filling dominates the dynamics of the forest and plays a major role in the evolution of life-history features of the plant populations and in the distribution of individuals in the forest (Schulz 1960, Ashton 1969, Paijmans 1970, Hubbell 1979).

Attempts have been made to identify differing strategies among species for performance in different gap types. These strategies are generally thought to range from sun to shade strategies and include a whole spectrum of intermediate ones. Some authors have suggested that different species of tropical forests specialize on different gap types and this specialization is a major contributor to the high species diversity of these forests (Ricklefs 1977, Strong 1977, Connell 1978, Denslow 1980). Many, if not all species of the tropical forest benefit from or require gaps at least at some stage of their life. Only permanent understory species seem to be exempt from this requirement (Whitmore 1975, 1978). Forest structure, spatial pattern (Knight 1975, Hartshorn 1978), and life history features of germination, growth, time of flowering, etc. are influenced by gap characteristics (Schulz 1960, Whitmore 1975, Bazzaz, Pickett 1980, and references therein). Furthermore, the activities of seed dispersers and predators, herbivores, pathogens and pollinators are influenced by gap characteristics. Thus the process of gap creation and filling has received much attention from tropical ecologists (reviews of Pickett 1983, Brokaw 1984a). A body of literature is now accumulating regarding gap size, shape, frequency of occurrence, and agents of their creation in several tropical forests. These characteristics, together with environmental heterogeneity in the gap and the intensity of disturbance, are biologically important aspects of gaps.

Gaps have been defined as breaches in the canopy but there has not been an agreement on how deep the opening need be. Brokaw (1982a) suggests that the hole should extend through all levels of the forest to a height of 2 meters above the ground. A standard definition is needed in order to accurately estimate the turnover rates in various tropical forests.

There are several causes of gap formation in tropical forests. Small gaps are created by the fall of tree branches. Larger gaps may be caused by tree falls due to old age, disease, and wind. The death of monocarpic individuals after reproduction creates single tree gaps as well (Foster 1977). Fire, which can create gaps, usually occurs after the accumulation of debris created by hurricane force wind (Stocker, Mott 1981). Earthquakes cause landslides and destroy large tracts of forest, especially in mountainous areas (Veblen, Ashton 1978, Garwood et al. 1979). Lightning may create single tree gaps as well (Brunig 1964).

In general, gap creation by tree fall is seasonal. Peak tree fall occurs in the rainy season (Oldeman 1972) because of loose anchorage and heavy rainload on the trees (Brokaw 1982b). High winds during the rainy season also aid in the creation of tree falls (Sarukhán 1978). Small tree falls are more common than large ones and older forests tend to have larger gaps (Brokaw 1982b). Furthermore, turnover rate, which is the time between consecutive gap formation in the same location, is also greater in older forests. Tropical forests turn over faster than do temperate forests and neotropical forests seem to turn over faster than those in Malesia (Table 1). Apparently, however, turnover rates differ within tropical forests and various patches turnover at different rates. In Los Tuxtlas, Veracruz, Mexico the turnover rate inside the forest is $\sim$ 100 years (Torquebiau 1981) but close to the edge of the forest it is only about 33 years (M. Martinez, personal communication). Medium to large gaps are common enough in tropical forests to maintain relatively large populations of secondary forest species and pioneers (Knight 1975, Hartshorn 1978).

Various gap-filling guilds contribute differently in different gap sizes. Very small gaps are usually filled by increased growth of branches of adjacent trees and by reiteration. Also, saplings may contribute to small gap filling (Kramer 1933, Whitmore 1978). As gap size increases the gaps are usually filled by advance regeneration (suppressed seedlings that are present on the forest floor before canopy opening) and by resprouting, which is apparently common in tropical trees (Stocker 1981, Uhl et al. 1981). In larger gaps the seed bank contributes significantly to gap filling, as many secondary

TABLE 1. Mean gap size and turnover rates in some tropical forests

Location	Mean gap size m^2	Turnover rate (Y)	Reference
La Selva, Costa Rica	89 $\pm$ 88	80 - 130	Hartshorn 1978
BCI, Panama	85	114	Brokaw 1982a, b
Malesia	400	250	Poor 1968
Amazonia	--	100	Uhl, Montgomery 1980
Los Tuxtlas, Mexico	91	100	Torquebiau 1981

forest species have long-lived seeds (Bazzaz, Pickett 1980, Vazquez-Yanes in this volume, and references therein). In very large gaps the new colonizers, whose seeds arrive after the creation of the gap, are the dominant gap-filling guild (Fig. 1). Generally, increasing severity of the disturbance that creates the gap in effect moves the relative contribution of various guilds to the right in Fig. 1 because of the destruction of advance regeneration, reduction of resprouting probabilities, and loss of seed bank. In the case of large clearing and severe disturbances the sites may be invaded by several herbaceous, weedy grasses and forbs and the process of gap filling is delayed. Nearly pure stands of some pioneer trees, e.g. Cecropia, Heliocarpus, Piper, Trema, Ochroma, may form even-aged stands.

The shapes of gaps vary considerably as well. The death and collapse of a single tree creates a nearly circular gap. Tree falls create linear gaps and these may get larger and change in shape due to fall of some trees at the initial gap edge as they become more exposed to high winds. The direction of tree falls determines the orientation of the resulting gap.

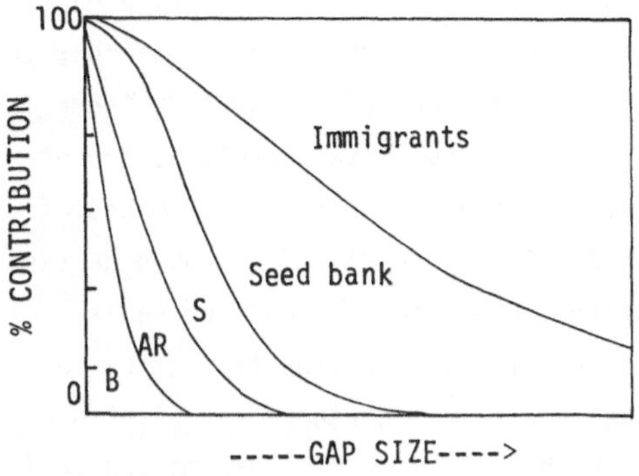

FIGURE 1. Relationship between gap size and the relative contribution of various guilds to gap filling. Increased severity of disturbance during gap creation moves the time axis to the right. B = branches, AR = advance regeneration, S = sprouts.

3. GAP ENVIRONMENT

Gap size, shape, and orientation determine the gap environment, especially with regard to light and its distribution within the gap. Obviously, larger gaps receive more light than do small gaps and elongated gaps receive more light when they are oriented east-west than when oriented north-south. Gaps that receive high light will also experience high evaporation rates, low air humidity, greater variation in daily temperature and more wind. In very large gaps with severe disturbance there is usually much bare soil (Uhl et al. 1982). In these situations, soil temperature during the day may become much higher than air temperature (Schulz 1960, Longman, Jenik 1974).

More often than not, a gap has a heterogeneous environment. The level of this heterogeneity is determined by gap size, shape, orientation, severity of disturbance and the mode of gap creation. Thus environmental gradients are extremely complex; and simple, discrete concentric circles of environments and species abundances are highly unlikely. The "Chablis" of Oldeman (1978) has the crown gap (epicenter), the tip up, and a narrow area between them where the tree trunk lies. The substrate within this complex is heterogeneous and may be made up of exposed mineral soil, downed wood, leaf litter, etc. (Hartshorn 1978). The level and dynamics of nutrient release will vary considerably within these gaps (Bazzaz 1983).

Although plants in the center of a gap usually receive more light than plants near the edge, individuals located at equal distance from the center but at opposite directions do not necessarily receive the same total amount of daily irradiance (Fig. 2). This could be caused by asymmetry in the daily amount of solar radiation received, resulting from afternoon cloudiness and rain. Furthermore, plants located in the west side of a gap may be in the sun in the morning when their leaves are turgid and therefore are able to photosynthesize maximally, while those located in the east side of a gap may be in full sun in the afternoon when the relative humidity of the air has declined and their water potentials are more negative, perhaps causing some reduction in their photosynthetic rates. The direction of the wind and the structure of turbulence are very complex in gaps, and plants located at various parts of the gap will experience continually changing air temperature, humidity and CO_2 concentration.

The ecophysiological activities of the plants, e.g. carbon, water, and nutrient acquisition and deployment patterns, will be much influenced by the heterogeneity of the gap environment.

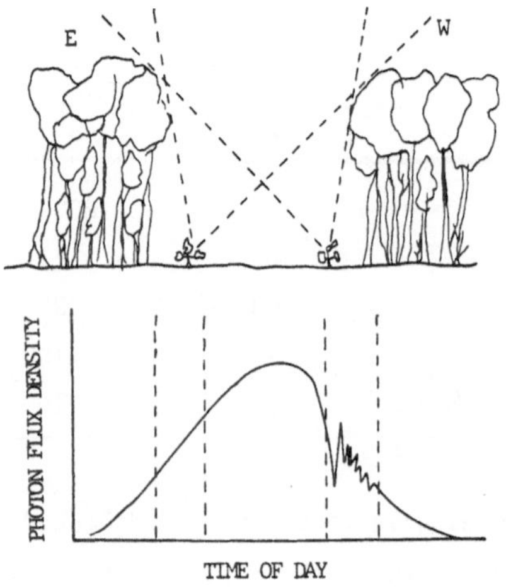

Figure 2. Unequal daily irradiance received by two plants at equidistance from the gap center but at opposite locations on an east-west transect. Afternoon cloudiness and rain reduces the amount of total irradiance received by the individual at the left.

4. PHYSIOLOGICAL ADAPTATIONS OF PLANTS TO GAPS

Tropical ecologists have recognized that different species have specific gap requirements (reviews in Bazzaz, Pickett 1980, Pickett 1983, Brokaw 1984b). Whitmore (1975) recognized four groups of plants with regard to gap requirements. They are: 1) species that establish and grow beneath closed canopies, 2) species that establish and grow beneath closed canopies but benefit from gaps, 3) species that establish under closed canopies but require gaps to mature and reproduce, and 4) species that establish, grow and reproduce only in gaps. But Whitmore and others recognize that these are points on a continuum of gap preferences and that each species may be unique in its preference (Pickett 1983).

Some specific physiological adaptations of species to gaps have been identified (Table 2) and there is good experimental evidence for them from various tropical systems, but much research is still required before these characteristics are established as truly general. Furthermore, there seem to be many similarities in the behaviour of gap species and pioneers in tropical and temperate forests (Bazzaz 1979, Bazzaz, Pickett 1980).

TABLE 2. Physiological characteristics of gap species and pioneers

1. Long seed and seedling dormancy
2. Germination is enhanced by light, decreased F_r/R ratios, temperature fluctuations and nutrients
3. Mostly epigeal germination; photosynthetic cotyledons
4. High rates of photosynthesis, respiration, transpiration, high conductances, high N-content
5. Continuous production of leaves; fast leaf turnover rates; leaves arranged in flat crowns, and are not multilayered
6. Rapid growth; low density wood; large leaves
7. Highly branched, intensive deep root system; low dependence on mycorrhizae; mostly NO_3 users
8. Early and long flowering time
9. Rapid response to changes in resource levels
10. High acclimation potential
11. High susceptibility to herbivores and pathogens

Seeds of pioneers and some gap species have long dormancy and usually germinate synchronously during the early part of the rainy season (Garwood 1983). Lianas seem to behave similarly. Non-pioneers in Malesia (Ng 1978) and shade-tolerant plants on BCI, Panama (Garwood 1983) have staggered germination. Canopy removal by gap formation creates conditions suitable for the germination of early successional species in tropical forests. Increased light, higher R/F_r ratio, increased temperature fluctuation, released nutrients, availability of germination sites, and reduced competition enhance germination of these species. (For a comprehensive review of seed germination of tropical trees see Vazquez-Yanes, in this volume.)

The process of gap occupation is usually episodic. Cohorts of new seedlings from the seed bank, new dispersals, and advance regeneration from the seedling bank all grow quickly and occupy the lower depth of the gap. Release from canopy light limitation, reduced competition from established individuals, and nutrient pulses (e.g. Smythe 1970) enhance the growth rates of the seedlings. There is usually a rapid development of a dense high leaf-area index which later is reduced by extension growth (Ewel 1971). In many instances this rapid growth causes early competitive interaction among the juveniles. Some will overtop others, deprive them of resources and even eliminate them altogether. Continued rapid growth with high photosynthetic rates, energetically inexpensive soft and light wood in the branches, and usually long petioles generates a thin, nearly mono-layered canopy of relatively large leaves. Thus the canopy permits a fair amount of light penetration to the forest floor, and further recruitment of new individuals, usually of the more shade-tolerant, late successional species. The susceptibility of a gap to invasion is therefore very high early, declines sharply in a short time and increases again later (Fig. 3).

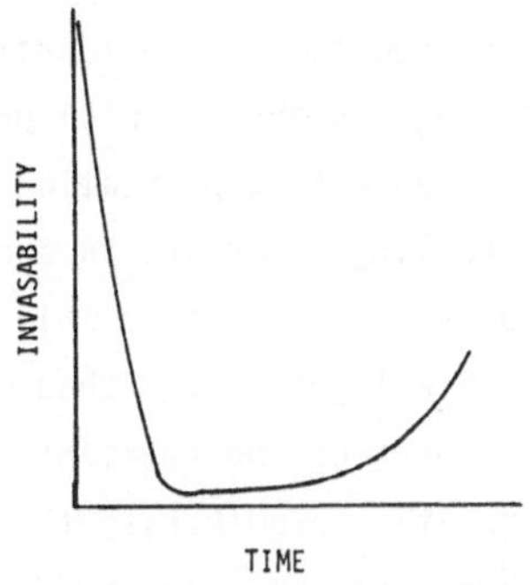

FIGURE 3. Changes in the availability of a gap to invasion by plants.

The rapid changes in the light environment of a given location in the intact forest by gap creation and during gap filling profoundly influence photosynthetic and water exchange rates of advance regeneration and other understory plants. It is not known whether these plants respond to this large change by acclimation of current leaves to the new light environment (e.g. Fetcher et al. 1983, Oberbauer 1983) or whether they discard their shade leaves and replace them with sun-adapted ones, e.g. as in some temperate forest herbs (Bazzaz, Carlson 1982). The position of the leaf on the photosynthesis/age curve (Fig. 4) may determine the fate of these individual leaves when the canopy opens and, in case of acclimation, the leaf must at least make some adjustment in dark respiration, light compensation and saturation points, and the attending structural and biochemical attributes, e.g. chlorophyll concentration, Chl a/b ratio and ribulose bisphosphate carboxylase activity (Boardman 1977). If a leaf has a high life expectancy (e_x) and high V_x (the carbon acquisition value of an individual leaf of age x) (Fig. 4) it would be advantageous for the individual plant to retain it, but if its expected contribution is low, it is damaged by herbivores or it has many epiphytes it may be dropped. Of course this will have to be balanced with the costs of defending the leaf against herbivores and pathogens whose impact may increase or decrease after gap creation and the costs of building new ones. The location of an individual seedling in the gap will undoubtedly influence its response to gap creation. It is expected that seedlings that are in areas of high nutrient pulses will be more able to use the newly available light energy than would seedlings that experience no nutrient pulse. The increased relative growth rate that ought to occur with resource release may be differentially experienced by various plant parts. Initial allocation to roots (higher root relative growth

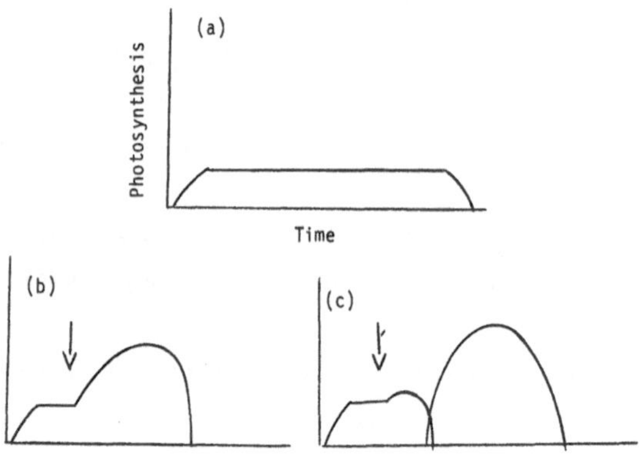

FIGURE 4. Relationship between photosynthetic rate, leaf age, and the light environment near the forest floor. (a) A shade leaf of a seedling in the understory. (b) A leaf exposed to a canopy gap early in life increases its photosynthetic rate to near that of a sun-leaf and reduces its life expectancy. (c) The shade leaf is dropped after gap creation and a new sun-adapted leaf is produced. Time of gap creation is indicated by arrows.

rate) may be more advantageous in some situations while allocation to new leaves may be more so in other situations. The shift in allometric relationships with age observed for some gap species, e.g. Astrocaryum mexicanum (Pinero et al. 1982), should be investigated. Clearly the demographic and physiological ecology of advance regeneration is a fertile area of research which would significantly enhance our understanding of tropical forest dynamics.

In large gaps the physical environment, the identity of the occupants, and the mode of their colonization vary as discussed earlier (Fig. 1). Within these gaps population structure and other aspects of community organization differ at different distances from the gap edge, which may be the major source of new colonists. Near the edge there is usually high seedling density, high conspecific competition, higher density-dependent mortality, and lower genotypic diversity than away from the edge (Bazzaz 1983). Also because of high density near the edge, small advances in germination time may be crucial to the seedling's physiology, its resource acquisition capacity, and its survival.

Clumped individuals of some species may have similar ages and sizes because of episodic invasion. In some instances they may die within a short period of time because of disease and/or old age. The die-back phenomenon of Metrosideros polymorpha in Hawaii is one example (Mueller-Dombois 1980). The germination, growth, and reproduction of individuals of the species and seedlings that are suppressed under the intact forest is enhanced (Burton, Mueller-Dombois 1983). Experimental removal of different percentages of the canopy showed that germination of M. polymorpha seeds was highest under 100% canopy removal, average height increment was highest at ∿ 55% removal. Extension growth per unit of irradiance was constant between 15 and 60% irradiance and declined drastically at higher levels.

Gap creation changes the biological interaction among species of differing trophic levels as well. Pathogen populations and identity must be different in the gaps than in the intact forest. In BCI, Panama, seedling mortality of some species is caused mainly by pathogens in the intact forest but not in gaps (Augspurger 1983a, b). Dispersal, seed predation and pollination must also differ in gaps and in intact forest because there are, for example, bird species that specialize on gaps (Schemske, Brokaw 1981).

The costs of defense against pathogens, predators and herbivores may be high in tropical forests. The formation of seedling banks (Whitmore 1975, Ng 1978) instead of seed banks by mature forest species provides a mode of escape from seed predators and pathogens. Apparently, the probability of survival for these individuals must be higher as seedlings than as seeds, but this needs investigation.

It is generally assumed that the level of herbivory is higher in tropical than in temperate forests. In several tropical forests insects seem to be the most important herbivores (Leigh, Smythe 1978, Coley 1983). Leaves of early successional and gap species are more susceptible to herbivory than are leaves of mature forest species (Hartshorn 1978, Coley 1983). However, young leaves of both groups are more susceptible to herbivore damage than are mature leaves (Coley 1980, 1983) although the former have higher phenol concentrations. Herbivores may remove different amounts of the area of a leaf. This damage may occur by removal of large pieces, as is done by cutting ants and Lepidopteran larvae, or by creating many small holes in the leaf lamina. The consequences of these different modes of herbivory to photosynthetic rate, water-use efficiency, longevity, and susceptibility to pathogen damage are not known (as is discussed elsewhere in this volume). The time at which a leaf is eaten is also important to its function, especially its contribution to the plants' carbon economy. A leaf devoured early in its life will clearly contribute less than one damaged later in life. By knowing the leaf's age-specific photosynthetic rate and the time at which damage is done, it would be possible to calculate the impact of herbivores on the carbon economy of the plant using conventional demographic formulations (Bazzaz 1984) (Fig. 4). The costs of defense against herbivores include fiber and legnin synthesis, formation of pubescence, and chemicals e.g. phenolics and tannins. It would be a rewarding but a difficult task to estimate these costs, especially because these characteristics have other roles beside their defensive function (e.g. Mooney, Gulmon 1982, Mooney in this volume). Relating herbivory to gap filling, which involves a shift from continuous leaf production of many pioneers to seasonal production of mature forest species, shifts in leaf turnover rates, and other phenological features of the plants would enhance our understanding of plant-herbivore interactions in tropical forests.

The physiological ecology of reproduction in tropical forest is poorly understood. The strong interactions between the plants and their pollinators and dispersers must involve some precise patterns of carbon and nutrient allocation to reproductive parts. Many plants in seasonal forests, and some in wet forests, flower at a certain time of the year and some produce large, showy, short-lived flowers, many of which produce large amounts of nectar. The cost of construction and pollination of such flowers is high because nectar contains, in addition to sugar, some costly compounds, e.g. proteins, amino acids, lipids, organic acids, alkaloids, glycosides, phenolics, etc. (Baker, Baker 1975). Furthermore, because these flowers are non-green they must depend on the rest of the plants for their required carbon, unlike some temperate trees (Bazzaz et al. 1979).

The emergents of tropical forests present a challenge to physiological ecologists. Their specific adaptations to the changing environment during their life must be better understood. These species, which may attain great sizes, more often than not require gaps for establishment (Richards 1952, Jones 1956, Schulz 1960, Burgess 1972). Their seedlings may be present in large numbers beneath the forest canopy. They grow upward when there is increased light above them, penetrate the canopy through gaps and emerge above it in an open, relatively windy environment. These species apparently have different leaf morphologies at different times of their life. For example, Shorea curtisii (Dipterocarpaceae) seedlings in the understory have large, thin leaves with good stomatal control, but emergent individuals have small, sclerophyllous waxy leaves (Ashton 1978) that apparently have no stomatal control (Kenworthy 1971). Many of these giant trees are considered early successional; they grow fast and have soft wood, e.g. Terminalia amazonia, Ceiba pentandra and several other Bombacaceae, while others, e.g. Lecythis, Aspidosperma and Couratari have characteristics of late successional trees.

The physiological ecology of other life forms in tropical forest still needs much research. Lianas, epiphytes and mycorrhizae play a significant role in forest productivity and nutrient capture and cycling (Benzing; Janos, in this volume). Epiphylls, which are common on leaves of many rainforest species, must influence CO_2 and water exchange characteristics of their hosts and may be important in their nitrogen economy, as many epiphylls undoubtedly are nitrogen fixers.

It is now possible to make some general predictions about the ecophysiological characteristics of gap and pioneer species of tropical rainforests (Table 2). However, much research is required to rigorously test them before they can be accepted as valid generalizations.

As a summary I propose the following list of areas for future ecophysiological research in wet tropical forests.

1. The physical environment of intact forests and gaps in various stages of gap filling needs much quantification. Light, temperature, soil moisture, air humidity, CO_2 concentrations and levels of nutrient availability must be studied horizontally, vertically and temporally, including measurements within and above the canopy where appropriate.
2. Environmental heterogeneity, especially in forest gaps of different sizes, shapes, orientations and severity of disturbance must be investigated with special emphasis on the patterns and levels of resource release. Various species will respond

differently to this heterogeneity. Thus the identity, spatial and temporal distribution, competition and many other interactions will be influenced by the ecophysiological responses of the species involved.

3. The ecophysiological responses to field and controlled environmental conditions of several species representing various gap filling strategies and life forms (e.g. lianas, emergents and epiphytes) should be determined. The effects of resources and controllers, e.g. light, temperature, soil moisture, air humidity, CO_2 and nutrients, on rates of photosynthesis, respiration, transpiration, growth and reproduction must be investigated. The deployment of carbon, nitrogen, and phosphorus to various plant parts and the way it changes under different environmental conditions is practically unknown. The role of mycorrhizae, pathogens, and herbivores in resource acquisition, use, and redistribution needs much more research.
4. The response of some key species to the changing physical and biological environment as seedlings, saplings, and mature individuals in the canopy or above it as emergents must be determined. The degree of acclimation to rapid changes in the levels of environmental resources and the ecophysiological bases of that acclimation should be investigated and related to the predictability of the forest environment.
5. Extreme environmental conditions such as drought and unusually cold temperatures occasionally experienced by many tropical forests have significant impact on these forests. Therefore, the ecophysiological response surfaces of species should reflect these conditions.
6. The biophysical and biochemical bases of ecophysiological responses should be considered whenever possible.
7. The strength of biological interactions in tropical forests dictates that the ecophysiological bases for reproduction, dispersal, herbivore defense, pathogen resistance, etc. must be investigated. The multiple strategies that tropical plants may adopt for resolving the often opposing demands on the carbon and nutrient pools of the individuals would make this approach highly profitable.
8. The above mentioned studies would be most profitable if they adopt, when appropriate, the concept of cost-benefit of morphological and physiological features of responses, especially to competition, predation, pollination, and herbivory.

REFERENCES

Ashton PS (1969) Speciation among tropical forest trees: some deductions in the light of recent evidence, Biol. J. Linn. Soc. 1, 155-196.

Ashton PS (1978) Crown characteristics of tropical trees. In Tomlinson PB and Zimmerman MH, eds. Tropical trees as living systems, pp. 591-615. Cambridge, Cambridge Univ. Press.

Aubre'ville A (1938) La Foret Coloniale; Les Forets d'Afrique Equatoriale, Boise & For. Trop. 2, 24-35.

Augspurger CK (1983a, in press) Seed dispersal by the tropical tree Platypodium elegans, and the escape of its seedlings from fungal pathogens, J. Ecol.

Augspurger CK (1983b, in press) Pathogen mortality of tropical tree seedlings: experimental studies of the effects of dispersal distance, seedling density, and light conditions, Oecologia.

Baker HG and Baker I (1975) Nectar-constitution and pollinator-plant coevolution. In Gilbert LE, Raven PH, eds. Coevolution of animals and plants, pp. 100-140. Austin, Texas University of Texas Press.

Bazzaz FA (1979) The physiological ecology of plant succession, Ann. Rev. Ecol. Syst., 10, 351-371.

Bazzaz FA (1983) Characteristics of populations in relation to disturbance in natural and man-modified ecosystems. In Mooney HA and Godron M, eds. Disturbance and ecosystems--components and response, pp. 259-275. Heidelberg, Springer-Verlag.

Bazzaz FA (1984) Demographic consequences of plant physiological traits: some case studies. In Sarukhan J and Dirzo R, eds. Perspectives in plant population ecology. Colorado, Sinaur Publishers.

Bazzaz FA and Carlson RW (1982) Photosynthetic acclimation to variability in the light environment of early and late successional plants, Oecologia 54, 313-316.

Bazzaz FA and Pickett STA (1980) The physiological ecology of tropical succession: a comparative review, Ann. Rev. Ecol. Syst. 11, 287-310.

Bazzaz FA, Carlson RW and Harper JL (1979) Contribution to reproductive effort by photosynthesis of flowers and fruit, Nature 279, 554-555.

Boardman NK (1977) Comparative photosynthesis of sun and shade plants, Ann. Rev. Plant Physiol. 28, 355-377.

Brokaw NVL (1982a) The definition of treefall gap and its effect on measures of forest dynamics, Biotropica 11, 158-160.

Brokaw NVL (1982b) Treefalls: frequency, timing, and consequences. In Leigh EG, Jr, Rand AS and Windsor DM, eds. The ecology of a tropical forest, pp. 101-108. Washington, DC, Smithsonian Institute Press.

Brokaw NVL (1984a, in press) Treefalls, regrowth, and community structure in tropical forests. In Pickett STA and White PS, eds. Natural disturbance: an evolutionary perspective. New York, Academic Press.

Brokaw NVL (1984b, in press) Gap-phase regeneration in a tropical forest, Ecology.

Brunig EF (1964) A study of damage attributed to lightning in two areas of Shorea albida forest in Sarawak, Empire Forestry Review 43, 134-144.

Burgess PF (1972) Studies on the regeneration of the hill forest of the Malay Peninsula: the phenology of dipterocarps, Malay. For. 35, 103-123.

Burton PJ and Mueller-Dombois D (1983, in press) Response of Metrosideros polymorpha seedlings to experimental canopy opening, Oecologia.

Coley PD (1980) Effects of leaf age and plant life history patterns on herbivory, Nature 284, 545-546.

Coley PD (1983) Herbivory and defensive characteristics of tree species in a lowland tropical forest, Ecol. Monogr. 53, 209-233.

Connell JH (1978) Diversity in tropical rain forests and coral reefs, Science 199, 1302-1310.

Denslow JS (1980) Gap partitioning among tropical rainforest trees, Biotropica 12 (suppl.), 47-55.

Ewel JJ (1971) Biomass changes in early tropical succession, Turrialba 21, 110-112.

Fetcher N, Strain BR and Oberbauer SF (1983, in press) Effects of light regime on the growth, leaf morphology, and water relations of seedlings of two species of tropical trees, Oecologia.

Foster RB (1977) Tachigalia versicolor is a suicidal neotropical tree, Nature 268, 624-626.

Garwood NC (1983) Seed germination in a seasonal tropical forest in Panama: a community study, Ecol. Monogr. 53, 159-181.

Garwood NC, Janos DP and Brokaw N (1979) Earthquake caused landslides: a major disturbance to tropical forests, Science 205, 997-999.

Hartshorn GS (1978) Tree falls and tropical forest dynamics. In Tomlinson PB and Zimmerman MH, eds. Tropical trees as living systems, pp. 617-638. Cambridge, Cambridge Univ. Press.

Hubbell S (1979) Tree dispersion, abundance and diversity in a tropical dry forest, Science 203, 1299-1309.

Jones EW (1956) Ecological studies on the rain forest of Southern Nigeria. IV. The plateau forest of the Okomu Forest Reserve, J. Ecol. 44, 83-117.

Kenworthy JB (1971) Water and nutrient cycling in a tropical rain forest. In Wilkinson HR, ed. The water relations of Malaysian forest, pp. 49-65. Aberdeen, Inst. South-East Asian Biol.

Knight DH (1975) A phytosociological analysis of species-rich tropical forest on Barro Colorado Island, Panama, Ecol. Monogr. 45, 259-284.

Kramer F (1933) De natuurlijke verjonging in het Goenoeng-Gedehcomplex, Tectona 26, 156-185.

Leigh EH and Smythe N (1978) Leaf production, leaf consumption and the regulation of folivory on Barro Colorado Island. In Montgomery GG, ed. The ecology of arboreal folivores, pp. 33-50. Washington, DC, Smithsonian Institute Press.

Longman KA and Jenik T (1974) Tropical forest and its environment. London, Longman. 196 pp.

Mooney HA and Gulmon SL (1982) Constraints on leaf structure and function in reference to herbivory, BioScience 32, 198-206.

Mooney HA, Bjorkman O, Hall AE, Medina E, and Tomlinson PB (1980) The study of the physiological ecology of tropical plants-- current status and needs, BioScience 30, 22-26.

Mueller-Dombois D (1980) The Ohio dieback phenomenon in the Hawaiian Rain Forest. In Cairns J, Jr., ed. The recovery process in damaged ecosystems. Ann Arbor, Mich., Ann Arbor Science.

Ng FSP (1978) Strategies of establishment in Malayan forest trees. In Tomlinson PB and Zimmerman MH, eds. Tropical trees as living systems, pp. 129-162. Cambridge, Cambridge Univ. Press.

Oberbauer SF (1983) The ecophysiology of Pentaclethra macroloba, a canopy tree species in the rainforests of Costa Rica, Ph.D. Thesis, Duke University.

Oldeman RAA (1972) L'architecture de la vegetation ripicole forestiere des fleuves et criques guyanais, Adansonia (New Series) 12, 253-265.

Oldeman RAA (1978) Architecture and energy exchange. In Tomlinson PB and Zimmerman MH, eds. Tropical trees as living systems, pp. 535-560. Cambridge, Cambridge Univ. Press.

Paijmans K (1970) An analysis of four tropical rainforest sites in New Guinea, J. Ecol. 58, 77-101.

Pickett STA (1983, in press) Differential adaptation of tropical tree species to canopy gaps and its role in community dynamics, Trop. Ecol.

Piñero D, Sarukhán J and Alberdi P (1982) The costs of reproduction in a tropical palm, Astrocaryum mexicanum, J. Ecol. 70, 473-481.

Poore MED (1968) Studies on Malaysian rain forest. 1. The forest on triassic sediments in Jengka Forest Reserve, J. Ecol. 56, 143-196.

Richards PW (1952) The tropical rain forest: an ecological study. London, Cambridge Univ. Press.

Ricklefs RE (1977) Environmental heterogeneity and plant species diversity: a hypothesis, Amer. Natur. 111, 376-381.

Sarukhán J (1978) Studies on the demography of tropical trees. In Tomlinson PB and Zimmerman MH, eds. Tropical trees as living systems, pp. 163-184. Cambridge, Cambridge Univ. Press.

Schemske DW and Brokaw NVL (1981) Treefalls and the distribution of understory birds in a tropical forest, Ecology 62, 938-945.

Schulz JP (1960) Ecological studies on rainforest in Northern Surinam. Amsterdam, North Holland. 267 pp.

Smythe N (1970) Relationships between fruiting seasons and seed dispersal methods in a neotropical forest, Amer. Natur. 104, 25-35.

Stocker GC (1981) Regeneration of a North Queensland rainforest following felling and burning, Biotropica 13, 86-92.

Stocker GC and Mott JJ (1981) Fire in the tropical forests and woodlands of Northern Australia. In Gill AM, Groves RH and Noble IR, eds. Fire and the Australian biota. Canberra, Australia, Aust. Acad. Sci.

Strong DR (1977) Epiphyte loads, treefalls, and perennial forest disruption: a mechanism for maintaining higher tree species richness in the tropics without animals, J. Biogeogr. 4, 215-218.

Torquebiau E (1981) Analyse architecturale de la foret du Los Tuxtlas (Veracruz), Mexique, Ph.D. Thesis, Universite des Sciences et Techniques du Languedoc, Academie de Montpellier.

Uhl C and Montgomery PG (1980) Composition, structure, and regeneration of a Tierra Firme Forest in the Amazon Basin of Venezuela, (in press).

Uhl C, Clark H and Clark K (1982) Ecosystem recovery in Amazon Caatinga Forest after cutting, cutting and burning, and bulldozer clearing treatments, Oikos 38, 313-320.

Uhl C, Clark K, Clark H and Murphy P (1981) Early plant succession after forest cutting and burning in the Upper Rio Negro region of the Amazon Basin, Agro-Ecosystems 7, 63-83.

Veblen TT and Ashton DH (1978) Catastrophic influences on the vegetation of the Valdivian Andes, Chile, Vegetatio 36, 149-167.

Whitmore TC (1975) Tropical rain forests of the Far East. Oxford, Clarendon Press. 278 pp.

Whitmore TC (1978) The forest ecosystems of Malaysia, Singapore and Brunei: description, functioning and research needs. In UNESCO, Tropical forest ecosystems, pp. 641-653. Paris, UNESCO.

Subject index

D

N

O

P

Q

R

S

T